Advanced Structured Materials

Volume 65

Series editors

More information about this series at http://www.springer.com/series/8611

Lucas F.M. da Silva
Editor

Materials Design and Applications

Springer

Editor
Lucas F.M. da Silva
Department of Mechanical Engineering
Faculty of Engineering of the University
of Porto
Porto
Portugal

ISSN 1869-8433 ISSN 1869-8441 (electronic)
Advanced Structured Materials
ISBN 978-3-319-50783-5 ISBN 978-3-319-50784-2 (eBook)
DOI 10.1007/978-3-319-50784-2

Library of Congress Control Number: 2016960268

Printed on acid-free paper

This Springer imprint is published by Springer Nature
The registered company is Springer International Publishing AG
The registered company address is: Gewerbestrasse 11, 6330 Cham, Switzerland

Preface

This volume of *Advanced Structured Materials* contains selected papers presented at the 1st International Conference on Materials Design and Applications 2016 (MDA 2016), held in Porto, Portugal, during June 30–July 1, 2016. The goal of the conference was to provide a unique opportunity to exchange information, present the latest results, as well as discuss issues relevant to materials design and applications. The focus is on fundamental research and application areas in the field of the design and application of engineering materials, predominantly within the context of mechanical engineering applications such as automobile, railway, marine, aerospace, biomedical, pressure vessel technology, and turbine technology. This includes a wide range of materials engineering and technology, including metals, e.g., lightweight metallic materials, polymers, composites, and ceramics. Advanced applications include manufacturing in the new materials, testing methods, and multi-scale experimental and computational aspects (e.g., microscale and nanoscale techniques). Approximately, 150 papers were presented by researchers from nearly 30 countries.

In order to disseminate the work presented in MDA 2016, selected papers were prepared which resulted in the present volume dedicated to 'Materials Design and Applications.' A wide range of topics are covered resulting in 31 excellent papers dealing with metals, ceramics, composites, design, power generation, additive manufacturing, machining, and joining. This book is state of the art of materials design and applications and also serves as a reference volume for researchers and graduate students working with advanced materials.

The organizer and editor wish to thank all the authors for their participation and cooperation, which made this volume possible. Finally, I would like to thank the team of Springer-Verlag, especially Dr. Christoph Baumann, for the excellent cooperation during the preparation of this volume.

Porto, Portugal
September 2016

Lucas F.M. da Silva

Contents

Part I Metals

Selected Properties of P/M Ti-6Al-2Sn-4Zr-6Mo Alloy After Hot Deformation ... 3
M. Wojtaszek, T. Śleboda and G. Korpała

Positron Annihilation Study on Nanocrystalline Copper Thin Films Doped with Nitrogen ... 15
P.M. Gordo, M.F. Ferreira Marques and M.T. Vieira

The Influence of Microstructure on the Mechanical Behaviour of Dual Phase Steels ... 25
R. Amaral, A.D. Santos, J.A. Sousa and A.B. Lopes

Part II Ceramics

Application of Cutting Edges with High Durability Made of Nanocrystalline Cemented Carbides ... 39
Maciej Jan Kupczyk, Piotr Siwak and Jedrzej Komolka

Design of Nanocrystalline Cemented Carbides with High Hardness ... 49
Maciej Jan Kupczyk

Probing Oxygen Vacancies in $BaTiO_3$ Powders and Single Crystals by Micro-Raman Scattering ... 65
Hiroaki Fukushima, Hirotaka Oka, Hiroki Moriwake, Hiroshi Uchida, Hiroshi Funakubo and Ken Nishida

Part III Composites

Investigation of Film Formation and Electrical Properties of PS Latex/MWCNT Nanocomposites ... 79
Saziye Ugur, Okan Yildiz and Selim Kara

Cellulose Nanowhiskers Obtained from Waste Recycling of Paper Industry 101
Alana G. Souza, Daniel B. Rocha and Derval S. Rosa

Coffee Powder Reused as a Composite Material 113
Vasco Canavarro, Jorge L. Alves and Bárbara Rangel

Comparison of Mechanical Properties of Polyester Composites Reinforced with Autochthonous Natural Fibres: Flax and Hemp 125
J. Rocha, J.E. Ribeiro and L. Queijo

Advanced Epoxy-Based Anticorrosion Coatings Containing Graphite Oxide 135
Monica Periolatto, Elisabetta Di Francia, Marco Sangermano, Sabrina Grassini and Pasquale Russo Spena

Part IV Design

A Numerical Study of Fenestral Otosclerosis 147
B. Areias, M.P.L. Parente, F. Gentil and R.M. Natal Jorge

Development and Validation of a Numerical Model for the Optimization of a Brace for Lower Limb 157
G. Bellavita, M. Cocconcelli, D. Castagnetti and R. Rubini

Part V Power Generation

Electrical and Geometrical Optimization for a 2DoF Non-linear Energy Harvester 173
E. Boco, R. Frizzell and J. Punch

Nonlinear Analysis of a Two-Degree-of-Freedom Energy Harvester 187
Valeria Nico, Ronan Frizzell and Jeff Punch

Experimental Investigations of MR Fluids in Air and Water Used for Brakes and Clutches 197
Alin Ilie Bosioc, Trandafir Emanuel Beja, Sebastian Muntean, Istvan Borbáth and Ladislau Vékás

Hydrodynamic Investigations in a Swirl Generator Using a Magneto-Rheological Brake 209
Sebastian Muntean, Alin Ilie Bosioc, Raul Alexandru Szakal, Ladislau Vékás and Romeo Florin Susan-Resiga

Part VI Additive Manufacturing

Direct Digital Manufacturing: A Challenge to the Artistic Glass Production ... 221
R. Felismina, M. Silva, A. Mateus and C. Malça

Post-process Influence of Infiltration on Binder Jetting Technology ... 233
Edwin Ocaña Garzón, Jorge Lino Alves and Rui J. Neto

Development of Plaster Mixtures Formulations for Additive Manufacturing ... 257
D.E. Caetano, J.L. Alves, R.L. Neto and T.P. Duarte

Part VII Machining

FE Modal and Harmonic Analysis of Micro Drill with Ultrasonic Horn ... 281
Muddu Allaparthi, Mohammed Rajik Khan and Syam Narayana Addepalli

Optimization of Machining Parameters to Minimize Surface Roughness in the Turning of Carbon-Filled and Glass Fiber-Filled Polytetrafluoroethylene ... 295
Muhammet Emre Sanci, Serdar Halis and Yavuz Kaplan

Investigation and Application of Fe–Co–Cu Based Diamond Cutting Tools with Different Bronze Content Used in Marble Production ... 307
Berrak Bulut, Onur Tazegul, Murat Baydogan and Eyup Sabri Kayali

Investigation of Surface Roughness and Tool Wear in End Milling of Al7075-SiC Co-continuous Composite ... 315
A.S. Prasanth and R. Ramesh

Machinability of an Aluminium Cast Alloy Using PCD Tools for Turning ... 329
R.B. Soares, A.M.P. de Jesus, R.J.L. Neto, P.A.R. Rosa, M. Machado and A. Reis

Optimization of Geometric Quality in a 5 Axis Machining of Curved Surfaces in a EN-AW-7075 Alloy by Taguchi Method ... 347
K. Castro, L.J. Segura, S.D. Castellanos and J. Lino Alves

Part VIII Joining

The Production-Related Influence of Iron Oxides on Steel Surfaces on the Adhesion of Fusion-Bonded Hybrid Structures ... 363
T. Reincke, S. Kreling and K. Dilger

Comparison of Stepped, Curved, and S-Type Lap Joints Under Tensile Loading 377
Şerif Çitil

Bonding Strength of Hot-Formed Steel with an AlSi Coating and Approaches to Improve It by Laser Surface Engineering 389
Alexander Wieczorek, Matthias Graul and Klaus Dilger

Micro Cork Particles as Adhesive Reinforcement Material for Brittle Resins 399
A.Q. Barbosa, L.F.M. da Silva, A. Öchsner, E.A.S. Marques and J. Abenojar

Magnetic Pulse Welding of Dissimilar Materials: Aluminum-Copper 419
I.V. Oliveira, A.J. Cavaleiro, G.A. Taber and A. Reis

Part I
Metals

Selected Properties of P/M Ti-6Al-2Sn-4Zr-6Mo Alloy After Hot Deformation

M. Wojtaszek, T. Śleboda and G. Korpała

1 Introduction

Titanium alloys are widely used for manufacturing responsible structural components, mainly for space technology, aviation and the automotive industry [1–3]. This results from favourable properties of these materials, including low specific gravity, high strength, crack resistance, fatigue strength and corrosion resistance. In case of aviation applications, the phase stability up to a temperature of 600 °C, and resistance to varying loads in high or cyclically varying speed conditions, is also important [4]. The factor limiting the application of products made of titanium and its alloys, in industry branches other than transport, is the high cost, resulting from high reactivity of titanium with oxygen and nitrogen, the fact that it does not occur in pure form [5] and problems with machining [6]. Consequently, products other than structural components are considered exclusive, being produced in limited quantities and finally, they are expensive.

The Ti-6Al-2Sn-4Zr-6Mo (Ti-6-2-4-6) alloy belongs to the group of two-phase $\alpha + \beta$ alloys. Its density amounts to 4.65 g/cm^3, and its chemical composition was designed considering the requirements for parts rotating at high speed in elevated temperature conditions, mainly for engine turbine subassemblies. As a result, the

M. Wojtaszek (✉) · T. Śleboda
Faculty of Metals Engineering and Industrial Computer Science,
AGH University of Science and Technology, Av. Mickiewicza 30,
30-059 Cracow, PL, Poland
e-mail: mwojtasz@metal.agh.edu.pl

T. Śleboda
e-mail: sleboda@agh.edu.pl

G. Korpała
Institut für Metallformung, Technische Universität Bergakademie Freiberg,
Bernhard-Von-Cotta-Str. 4, 09599 Freiberg, DE, Germany
e-mail: grzegorz.korpala@imf.tu-freiberg.de

L.F.M. da Silva (ed.), *Materials Design and Applications*,
Advanced Structured Materials 65, DOI 10.1007/978-3-319-50784-2_1

products made of this alloy are predisposed to work in the conditions of long-lasting loads at temperatures up to 400 °C as well as short-duration loads at temperatures up to 540 °C [7].

The most commonly used technology of industrial scale production of the Ti-6-2-4-6 alloy products is die forging realised as hot forming or isothermal forming. Isothermal forging is usually realised in a single operation, while hot forging, where the temperature of tools is elevated, but significantly lower than that of a stock, in one or several operations. The application of plastic working allows to eliminate the machining, its effect being not only giving the required shape, but also ensuring high product properties [8]. However, in order to achieve this, it is necessary to know the proper process parameters, especially the temperature of material and tools, amount of deformation and strain rate, but also to use the technology, which allows to precisely control the above mentioned parameters [9].

Currently, the feedstock for forging of responsible structural components of the Ti-6-2-4-6 alloy is usually a casting wrought to the form of a billet or a bar. However, an alternative may be the use of P/M semi-finished products. Various forming methods can be applied to prepare such a stock, e.g. pressing and then hot sintering in a protective atmosphere or in a vacuum [10], hot pressing of powders, including isostatic pressing [1, 11], hydrogen sintering [12] and others. Plastic working of such a stock can be realised without any modifications of existing forging processing lines [9, 13]. It is possible to obtain products showing a fine-grained microstructure and favourable properties, but also to reduce costs of production. However, in this case, the type of stock is of particular importance. This may be, alternatively, the alloy powders obtained by means of atomisation or the mixtures of elemental powders. Usually, the microstructure and properties of products based on alloy powders are stable, but in this case, the production cost is high, especially when the chemical composition of the alloy is complex. Manufacturing a mixture of elemental powders is significantly easier and less costly, but there is a risk of the occurrence of chemical composition inhomogeneities, which may lead to the deterioration of product properties. In order to avoid this, the uniform mixing of ingredients has to be guaranteed and the proper conditions of forging of compacts should be selected, which would favour the homogenisation of chemical composition within the volume of a product. The problem of obtaining a high quality of products based on elemental powders was discussed in a number of works [9, 13, 14].

This paper is focused on the study of the evaluation of favourable parameters of hot forging of the Ti-6-2-4-6 alloy compacts. For this purpose, a series of modelling investigations was realised, based on upsetting tests carried out in isothermal conditions and with strictly controlled parameters. Based on the test results, the flow curves were obtained under the conditions of strain rates and temperatures assumed in the tests. The metallographic examinations of specimens after upsetting were also performed, and the relationships between thermomechanical test conditions and state of the microstructure were determined. The obtained flow curves were used as a database for the description of the behaviour of the Ti-6-2-4-6 alloy compacts, depending on the upsetting conditions.

Part of the paper is dedicated to the thermomechanical simulation of forming of a forging of a selected shape, with application of the QForm commercial software, based on the finite element method (FEM). Numerical modelling of the die forging process allowed to select a favourable feedstock geometry and to indicate thermomechanical parameters advantageous for manufacturing a forging of an assumed shape, from the investigated material.

The results obtained based on upsetting tests, microstructure examinations as well as from modelling, were verified by comparing them with the experimental data, which is also reported. The forging tests realised in a commercial processing line were successful. It testifies that the information collected in this paper may be useful for the selection of suitable parameters of plastic working of the Ti-6-2-4-6 alloy compacts, especially by means of hot forging.

2 Experimental Work

2.1 Material for Research and Investigations Procedure

As the initial material to be investigated, a mixture of titanium powder with other elemental powders (i.e. aluminium, tin, zirconium and molybdenum) was used. The powders were mixed in proportions adequate for obtaining the chemical composition required for the Ti-6-2-4-6 alloy, which is given in Table 1. Figure 1 presents a photograph of the mixture of elemental powders. The manufacturing of compacts was realised with the application of a hot pressing test stand, which is a part of the

Table 1 Chemical composition of Ti6Al4V alloy (AMS 498 1B)

Al	Fe	H	Mo	N	O	Sn	Zr	Ti
5.5–6.5	0.15	0.0125	5.5–6.5	0.04	0.15	1.75–2.25	3.5–4.5	bal

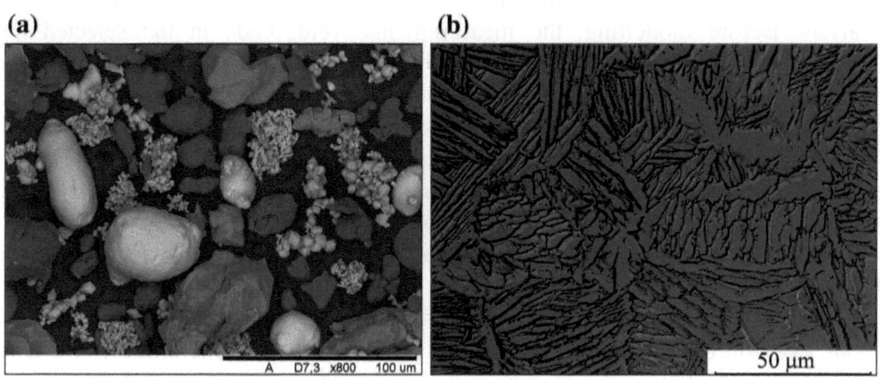

Fig. 1 Mixture of elemental powders (**a**) and microstructure of the compact after hot pressing (**b**)

equipment of AGH-UST in Krakow, Poland. The mixture was placed in a device chamber and pressed, in an argon protective atmosphere, for 3 h, at a temperature of 1200 °C and under a pressure of 25 MPa.

The selected properties of compacts were evaluated. The relative density of compacts, determined by means of the Archimedes method, amounted to 99.1%. The average hardness of this material was equal to 360.4 ± 4.7 HV2, and the compression strength: 1006 ± 96 MPa.

The upsetting tests of compacts were carried out at Institut für Metallformung, TU Bergakademie Freiberg, Germany, using the Bähr MDS 830 test stand for simulation of plastic working processes. The device is capable of deforming the material in isothermal conditions as well as with controlled cooling rate. The maximum upsetting load amounts to 250 kN, the attainable strain rates range from 0.01 to 100 s^{-1} and the tests are possible to carry out at temperatures of up to 1500 °C. The upsetting tests were realised in isothermal conditions. Cylindrical specimens of a 10 mm diameter and a 12 mm height, prepared with the application of spark machining, were used in the tests. The specimens were induction-heated to the assumed test temperature, applying a 2.5 Ks^{-1} heating rate, then held for 10 s at that temperature, and finally deformed with the assumed strain rate. During the tests, the following temperatures were assumed: 900, 950, 1000 and 1100 °C, and strain rates of 0.1, 1, 10 and 80 s^{-1} were applied. In order to freeze the microstructure after upsetting, the specimens were cooled, applying a 100 °C/s cooling rate.

The observations of the microstructure of specimens after upsetting tests were carried out by means of optical microscopy (LEICA DM 4000 M microscope). The examinations were performed on cross-sections of upset specimens, in the plane passing through the axis of symmetry. Two-stage etching was applied (1st stage: 6% HF + 96% H_2O, 2nd stage: 2% HF + 2% HNO_3 + 96% H_2O).

The flow curves constructed based on plastometric tests were used as a database, which was employed in the next stage for the modelling of material behaviour during hot die forging. The numerical analysis was performed with the application of the Form 2D/3D software, based on the finite element method (FEM). A mass-produced forging in the form of a ring with a flange was selected for the analysis. Before modelling, the measurements were made in the selected hot forging processing line (ATI ZKM Forging, Poland), in order to determine the abilities and limitations occurring in real industrial conditions. In consequence, among other things, the temperature of the lower tool was assumed to be 300 °C (maximum attainable temperature while heating with gas burners), and 100 °C was assumed for the upper tool. For modelling the device used in the process, the kinematic characteristic adequate for a crank press of 1000 t maximum load was assumed. The moving speed of the upper tool (press ram) was assumed to be 1.0 ms^{-1} at the moment of deformation.

Hot forging tests were realised at the ATI ZKM Forging company in Stalowa Wola, Poland. The stock was heated and held in an induction furnace, to a temperature of 1000 °C for 20 min, then transferred to the die impression, held to reach a temperature of 950 °C and forged in one operation. After forging, the specimens were cooled to ambient temperature in forced air circulation conditions. In order to compare the results obtained from FEM numerical simulation with the experimental ones, the last forging operation, which is flash trimming, was not realised.

2.2 *Results and Discussion*

2.2.1 Results of Plastometric Tests

Based on the data obtained from upsetting tests performed with the application of the MDS 830 simulator, the flow curves of the Ti-6-2-4-6 alloy were constructed, depending on the temperature and the strain rate assumed in the tests. The example variations of true stress as a function of strain are shown in Fig. 2. It was observed that, in case of deformation realised at lower temperatures, after reaching the maximum stress, further deformation proceeded at constantly dropping stress values. In case of deformation at a low strain rate (Fig. 2a), this effect is the most evident at a temperature of 900 °C, while becoming less intensive with increasing temperature. At a higher strain rate (Fig. 2b), the distinct drop of stress with proceeding deformation was also observed at a higher temperature (950 °C). The character of the curve obtained in these conditions, for strain ranging from 0.1 to 0.4, is qualitatively similar to that obtained at 900 °C, for the same strain rate.

It should be noticed that the behaviour of the material subjected to plastic deformation is influenced not only by the chemical composition of stock, but also by the method of its manufacturing. Therefore, the flow curves available in databases, which were obtained for the most common case of cast and wrought material, cannot be used for a proper description of forming of hot-pressed powder mixtures. For this reason, plastometric tests realised for this kind of stock allowed to develop a database necessary for a correct numerical modelling of hot forming of such material.

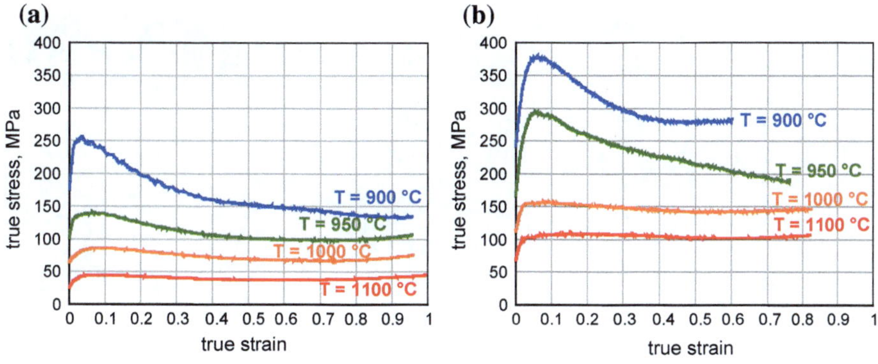

Fig. 2 Example true stress—true strain curves obtained from upsetting tests of Ti-6-2-4-6 alloy compacts, with application of MDS 830 simulator, at the strain rate of: 0.1 s^{-1} (**a**) and 10 s^{-1} (**b**)

2.2.2 Metallographic Examination

Figure 3 presents the images of microstructures of specimens after upsetting in some selected variants of temperature and strain rate conditions. The results of the metallographic examination showed the significant influence of upsetting parameters on the state of the microstructure of the Ti-6-2-4-6 alloy compacts. The

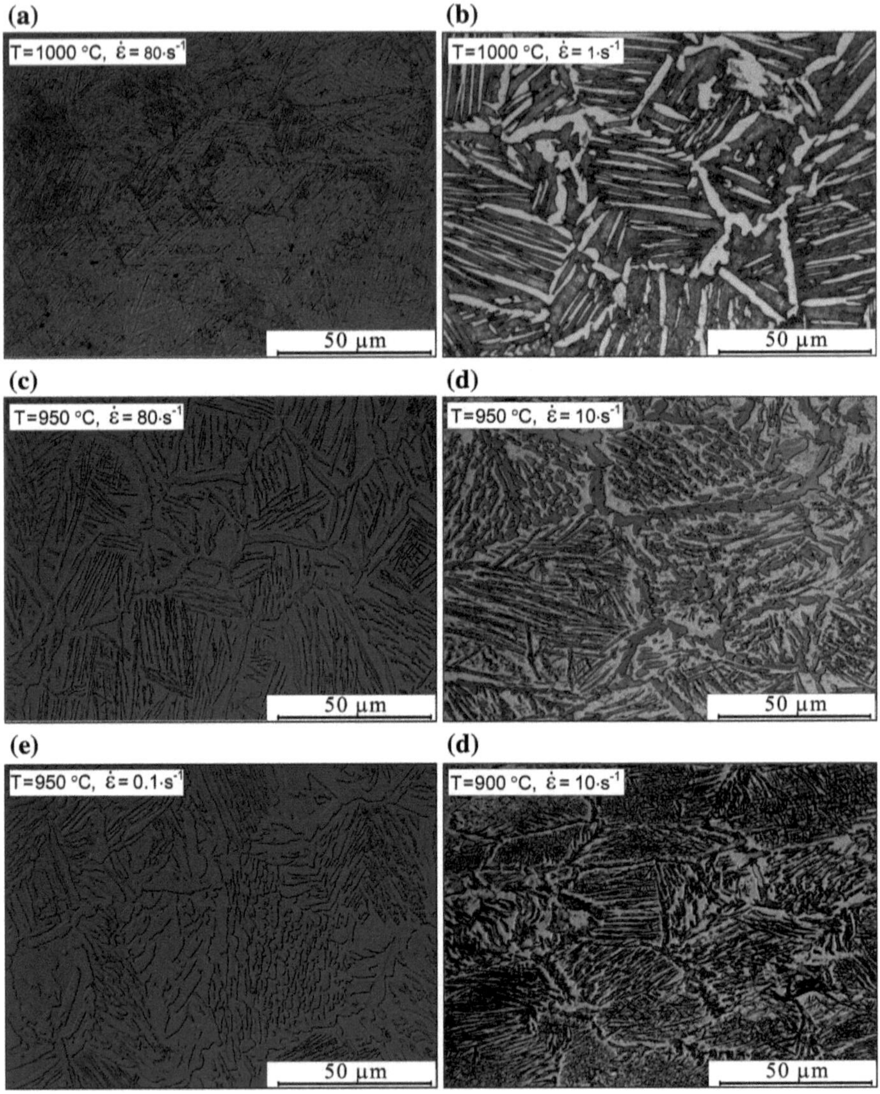

Fig. 3 Effect of strain rate and temperature applied in plastometric tests on the microstructure of Ti-6-2-4-6 alloy specimens obtained using powder metallurgy method. Cross-sections, etched, central region

microstructure is of a lamellar or a necklace character, showing different length and being more or less massive, depending on the temperature and the strain rate applied. It is different from the microstructure of a compact (Fig. 1b), which is composed of large, much more massive lamellae of the α phase, in the β phase matrix. As a result of upsetting test realised at a temperature higher than the transformation temperature (β transus temperature) and at a high strain rate (Fig. 3a), a microstructure was obtained, which is composed of large-sized primary grains of the β phase with colonies of very fine and elongated α + β lamellae visible inside. Decreasing strain rate with the temperature held constant resulted in the formation of more massive lamellae and smaller primary grains of the β phase. This is probably caused by the increased time of exposure of a specimen to heat, resulting from a lower strain rate, thus facilitating the dynamic recrystallisation to take place. Decreasing the test temperature with the strain rate held constant (Fig. 3a, c) resulted in the formation of more massive α + β lamellae of locally reduced size. As a result, beside α + β lamellae, also the microstructure of a necklace character was observed. In this case, a tendency is evident to form more massive lamellae as well as α phase precipitates composing the grain boundaries of the primary β phase, with the size of these grains remaining unchanged.

2.2.3 FEM Numerical Modelling of Hot Die Forging Process

A series of the finite element method (FEM) numerical simulations was designed and performed, considering different variants of stock geometry and forging process conditions, within the range possible to be realised in industrial conditions. Based on the analysis of the results, the stock dimensions and forging conditions, favourable for the alloy being analysed and the selected shape of a forging, were evaluated. Figures 4 and 5 present example results of modelling of forging of the Ti-6-2-4-6 alloy compact, which were selected as the most favourable ones. The cylindrical stock was assumed, with a diameter of 26 mm and a height of 50 mm, which was formed at an initial temperature of 950 °C. Figure 4a shows the initial

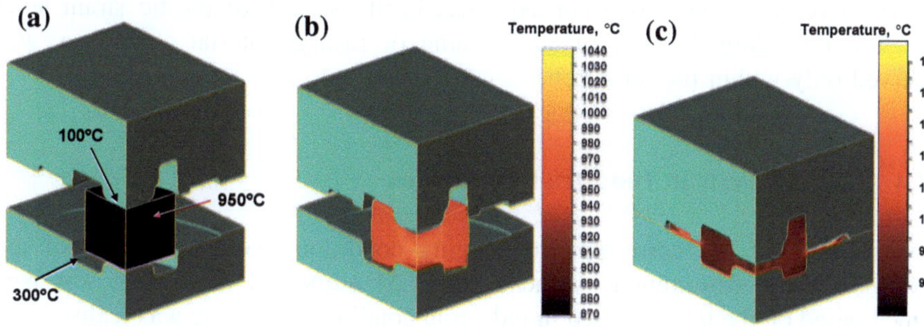

Fig. 4 Stages of die forging of Ti-6-2-4-6 alloy compact at 950 °C, obtained from FEM numerical modelling, with corresponding forging temperature distribution: initial state (**a**), intermediate stage (**b**) and final stage (**c**)

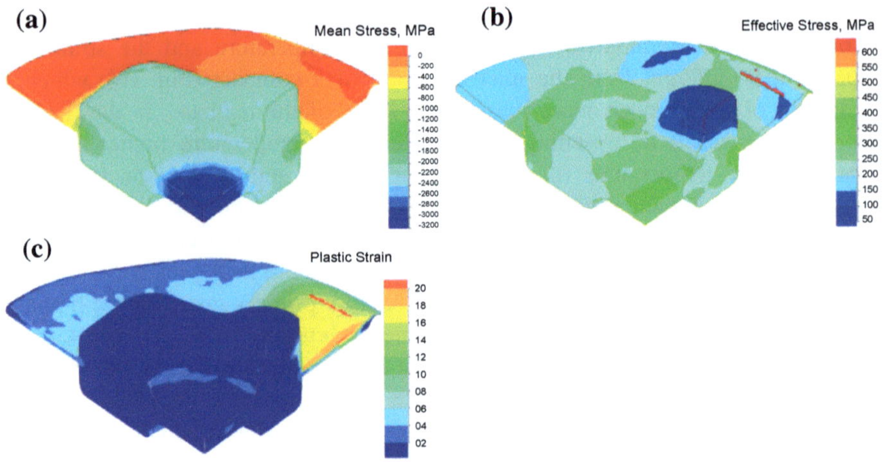

Fig. 5 Distributions of mean stress (**a**), effective stress (**b**) and plastic strain (**c**) obtained from FEM numerical modelling of die forging of Ti-6-2-4-6 alloy compact at 950 °C

state, i.e. the position of a stock in the working zone between tools, at the moment when forging begins, as well as the temperatures of stock and tools assumed in the discussed modelling variant. Figures 4b, c present the intermediate and final forging stages, respectively, along with corresponding temperature distributions within the volume of a forging. The zones of decreased temperature are evident, resulting from the contact with a cooler tool and from giving up heat to the environment. The regions of locally increased temperature can also be observed, which is the effect of conversion of the work of plastic deformation into heat, due to high process velocity. The intensive temperature rise observed in the final stage (Fig. 4c) is limited to the frontal part of a flash, which is caused by a high velocity of the upper tool and, as a consequence, a high speed of metal flow in that region.

Figure 5 presents the selected results of FEM numerical simulation of forging at 950 °C, obtained in the final stage, i.e. in the last step of calculations. Based on the obtained results of modelling, it can be found that high concentration of mean stress and effective stress and, as a consequence, local increase of plastic strain level (Fig. 5c), indicating the risk of lap forming or losing material coherence, was observed only within the flash zone.

2.2.4 Forging Tests in Industrial Conditions

A verification of the correctness of forging conditions for the product of the Ti-6-2-4-6 alloy, determined by means of FEM numerical modelling, was performed based on the tests realised in industrial conditions. Forging was realised in a commercial processing line located at the ATI ZKM Forging company in Stalowa Wola. The model presenting the shape of a forging selected for the tests, along with photographs of the stock and hot-forged product, is presented in Fig. 6. Figure 7 and shows the microstructure of Ti-6-2-4-6 forging obtained from the PM compact. The die forging process realised in the assumed conditions resulted in obtaining the product of a correct shape. The quality assessment of the state of a forging did not

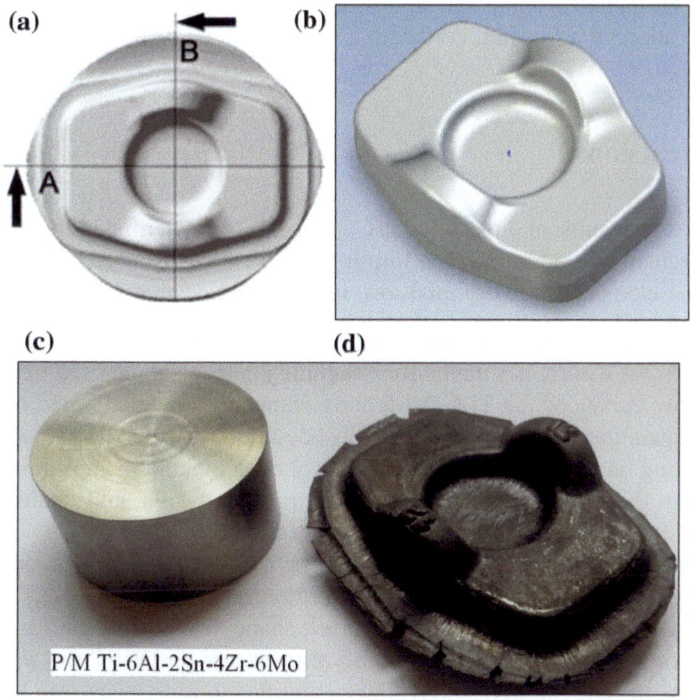

Fig. 6 The model of flange forging (**a**, **b**) with the photographs of a forging stock (**c**) and a forging just after forging at 950 °C (**d**)

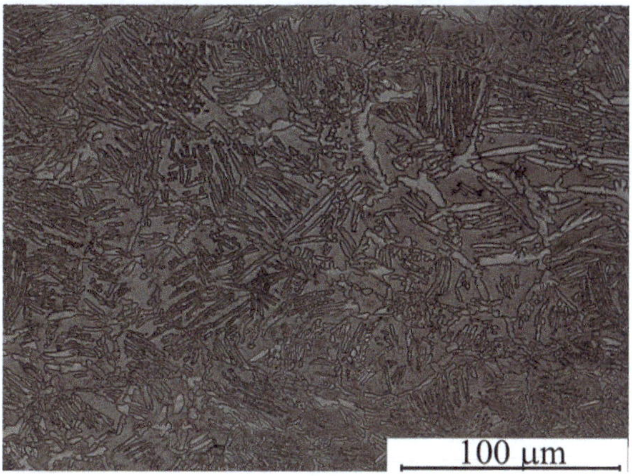

Fig. 7 Microstructure of a PM forging. Cross-sections, etched, central part of a forging

show any surface discontinuities or other visible defects, which would result from improperly selected forming process parameters. Cracks were observed only in the flash region, which confirmed the results of the FEM numerical simulation. This is not a technological problem, since the last stage of the manufacturing process is the removal of a flash, which is a waste material. No cracks and pores were also observed on metallographic specimens, thus confirming the conclusions based on the observation of a forging, concerning its quality.

3 Conclusions

Based on the results of investigations aiming at the determination of favourable thermomechanical conditions for forming the Ti-6-2-4-6 alloy, obtained by hot pressing of elemental powder mixtures, it can be found that:

1. Hot pressing of the powder mixture of a chemical composition corresponding to the Ti-6-2-4-6 alloy is suitable for obtaining compacts to be used in hot forging.
2. The flow curves of the Ti-6-2-4-6 alloy compacts, obtained based on upsetting tests, quantitatively and qualitatively describe the behaviour of the alloy depending on the assumed combinations of the temperature and strain rates, thus creating the database necessary for a proper design and realisation of a numerical simulation of the forming process.
3. FEM modelling of the die forging process allowed to select favourable feed-stock geometry and to evaluate the thermomechanical parameters suitable for manufacturing of a forging of an assumed shape, from the investigated material. Based on the results of modelling, at a temperature of 950 °C that was assumed to be the most favourable, the risk was identified to form a lap or to lose material coherence only within the flash zone, which is a waste material.
4. The correctness of the selection of forging parameters was verified in industrial conditions. The forging tests realised in a commercial processing line were successful and forgings without any visible defects were obtained, which was confirmed in a metallographic examination.

The results of investigations show that the proposed method of combining powder metallurgy and plastic working processes can be applied for the production of high-quality structural components of the Ti-6Al-2Sn-4Zr-6Mo alloy. A reduction of product costs is also possible, due to the application of a mixture of elemental powders of an equivalent chemical composition instead of the expensive alloy powder.

Acknowledgements Financial support of the Polish Ministry of Science and Higher Education is gratefully acknowledged (AGH-UST statutory research project no. 11.11.110.292).

References

1. Chunxiang, C., BaoMin, H., Lichen, Z., Shuangjin, L.: Titanium alloy production technology, market prospects and industry development. Mater. Des. **32**, 1684–1691 (2011)
2. Ivasyshyn, O.M., Aleksandrov, A.V.: Status of the titanium production, research and applications in the CIS. Mater. Sci. **44**, 311–327 (2008)
3. Sha, W., Malinov, S.: Titanium Alloys: Modelling of Microstructure. Woodhead Publishing Limited, Cambridge, UK, Properties and Applications (2009)
4. Wojtaszek, M., Śleboda, T., Czulak, A., Weber, G., Hufenbach, W.: Quasi-static and dynamic tensile properties of Ti-6Al-4V alloy. Arch. Metall. Mater. **58**, 1261–1265 (2013)
5. Lutjering, G., Williams, J.C., Gysler, A.: Microstructure and mechanical properties of titanium alloys. Microstruct. Prop. Mater. **2**, 1–74 (2000)
6. Heinl, P., Müller, L., Körner, C., Singer, R.F., Müller, F.A.: Cellular Ti-6Al-4V structures with interconnected macro porosity for bone implants fabricated by selective electron beam melting. Acta Biomater. **4**, 1536–1544 (2008)
7. Boyer, R., Welsch, G., Collings, E.W. (eds.): Materials Properties Handbook: Titanium Alloys. ASM International, USA (1994)
8. Chakrabarti, A.K., Burn, M., Fournier, D., Kuhlman, G.: Microstructure and mechanical property optimization through thermomechanical processing in Ti-6-4 and Ti-6-2-4-6 Alloys. In: Lacombe, P., Tricot, R., Beranger, G. (eds.) 6th World Conference on Titanium, Les Editions de Physiques, Paris, pp. 1339–1344 (1988)
9. Śleboda, T., Wojtaszek, M.: Design and verification of thermomechanical parameters of P/M Ti6Al4V alloy forging. J. Alloy. Compd. **615**, 546–550 (2014)
10. Delo, D.P., Piehler, H.R.: Early stage consolidation mechanisms during hot isostatic pressing of Ti-6Al-4V powder compacts. Acta Mater. **47**, 2841–2852 (1999)
11. Haasea, C., Lapovoka, R., Pang Nga, H., Estrina, Y.: Production of Ti–6Al–4V billet through compaction of blended elemental powders by equal-channel angular pressing. Mater. Sci. Eng. A, **550**, 263–272 (2012)
12. Paramore, J.D., Fang, Z.Z., Sun, P., Koopman, M., Ravi Handran, K.S., Dunstan, M.: A powder metallurgy method for manufacturing Ti-6Al-4V with wrought-like microstructures and mechanical properties via hydrogen sintering and phase transformation (HSPT). Scripta Mater. **107**, 103–106 (2015)
13. Wojtaszek, M., Śleboda, T.: Thermomechanical processing of P/M Ti-6Al-4V alloy. In: METAL 2013: 22nd International Conference on Metallurgy and Materials, pp. 364–369. Ostrava, TANGER (2013)
14. Ivasishin, O.M., Bondareva, K.A., Bondarchuk, V.I., Gerasimchuk, O.N., Savvakin, D.G., Gryaznovb, B.A.: Fatigue resistance of powder metallurgy Ti–6Al–4V alloy. Strength Mater. **36**, 225–230 (2004)

Positron Annihilation Study on Nanocrystalline Copper Thin Films Doped with Nitrogen

P.M. Gordo, M.F. Ferreira Marques and M.T. Vieira

1 Introduction

In the last decade a strong development in the nanomaterials field has been observed due to the unique and specific mechanical and physical properties of these nanomaterials comparatively to their coarser grain size materials [1–3]. On the other hand, thin metallic films play also an important role in many areas of the electronic industry for different applications, such as protective coatings [4], sensors [5] and contacts [6, 7]. Particularly, nanocrystalline copper is a promising material for automotive and electronic industries.

Among different fabrication processes, sputtering is well known for producing nanocrystalline thin films. However, maintaining the nanocrystalline character during processes or applications is still not an easy task due to the tendency towards grain growth exhibited by nanomaterials (copper presents a low tendency for nanocrystallity due to the high adatoms mobility). On the other hand, the addition of solutes with a strong affinity for grain boundary segregation can inhibit grain growth, particularly during the manufacturing process, but it is still not well known the effects of impurities and dopants on the mechanical and physical properties of the nanomaterial. The presence of doping elements during the formation of sputtered metallic films could have an important role on the grain size decrease,

P.M. Gordo (✉) · M.F. Ferreira Marques
CFisUC, Department of Physics, University of Coimbra,
3004-516 Coimbra, Portugal
e-mail: pgordo@ci.uc.pt

M.F. Ferreira Marques
e-mail: fatima@coimbra.lip.pt

M.T. Vieira
CEMUC®, Department of Mechanical Engineering, University of Coimbra,
Coimbra, Portugal
e-mail: teresa.vieira@dem.uc.pt

L.F.M. da Silva (ed.), *Materials Design and Applications*,
Advanced Structured Materials 65, DOI 10.1007/978-3-319-50784-2_2

particularly if they present low solubility in the metallic matrix, as is the case of N and C on Cu [8], and low tendency for the formation of compounds (Gibbs free energy > 0).

Few years ago, nanocrystalline copper thin films were successfully produced by magnetron sputtering through controlled additions of nitrogen in order to inhibit the grain growth [9]. The most likely mechanism for the production of nanosized grains was attributed to segregation of nitrogen to the copper grain boundaries anchoring their movements and avoiding grain growth. However, it was impossible to determine the localization of nitrogen, even after a detailed Transmission Electron Microscopy (TEM) study. In fact, Energy Dispersive Spectroscopy (EDS) and Electron Energy Loss Spectroscopy (EELS) analysis did not reveal any vestiges of nitrogen inside the grains or in the grain boundaries [9].

Positron annihilation spectroscopy is a well established tool for the study of electronic and defect properties of solids [10–12]. The crux of this technique is that in solid materials positrons are efficiently trapped at defects such as vacancies or voids, which makes them a very sensitive probe for studying these defects. After trapping the positron will annihilate with an electron from the immediate surrounding of the defect, thus producing two 511 keV gamma-photons. The defect-related information is obtained from properties such as the lifetime of the positrons, the Doppler broadening of the photo-peak, or the angular-correlation between the two photons. Furthermore, the development in slow positron beam methods allows the extension of traditional techniques to investigation of thin films, layered structures and surfaces [12, 13]. Slow positrons are produced in an ultrahigh vacuum environment by the slowing down of positrons from a 5 mCi β^+ emitting ^{22}Na source in a 4 μm thick polycrystalline tungsten moderator foil and subsequent reemission of the positrons from the surface of the moderator. The positron are magnetically guided and near the samples are accelerated by an electric field to kinetics energies ranging between 50 eV and 25 keV entering into the sample. Varying the incident energy results in a depth-resolved study of materials properties in the first 1.5 μm surface region of the sample. (For full description of the Coimbra variable energy positron beam, see Ref. [14].)

When positrons are implanted in a sample, the stopping profile is generaly assumed to be Makhovian with a mean implantation depth approximately proportional to $E^{1.6}$, where E represents the positron energy [15]. After slowing down and some diffusion the positrons implanted in the sample either annihilate in a defect-free region or become trapped in a defect and annihilate there. The energy of the annihilation radiation of the positron may deviate slightly from the value of 511 keV as a result of the nonzero value of the momentum of the electron at the moment of annihilation (Doppler broadening). The electron momentum distribution in defects is in general different from the one in defect-free material, resulting in a different Doppler broadening of the 511 keV annihilation peak. The momentum distribution is characterized by the line shape parameters $S(W)$ [12] defined as the relative number of annihilation events in the centroid (wings) of the 511 keV line. Annihilations with low (high) momentum electrons fall to the energy window of

S(*W*). Therefore, mainly valence electrons contribute to *S* whereas only core electrons are represented in *W*. The annihilation parameters at the defect can be used as fingerprints of the open volume of the defects. The larger the open volume, the lower the core annihilation parameter and the higher the valence annihilation parameter. In general the *S* value is also proportional to the defect concentration in the sample (although the proportionality is not necessarily linear), i.e., a higher *S* corresponds to a higher defect concentration. A positron in a defect sees a different electron density than in a defect-free material, owing to the lower concentration of core electrons. Generally, the most important effect of appreciable positron trapping is a narrowing of the momentum curves and, therefore, a narrowing of the 511 keV peak width, which is equivalent to an increase of the *S* parameter [12]. The positron diffusion length is also dependent on the defect density. A higher defect concentration leads to a shorter diffusion length.

S- and *W*-parameters measurements as a function of incident positron energy give an indication of the depth-resolved defect regions in a sample [13].

Since nanocrystalline materials due to the small grain size contain a large free volume fraction associated with grain interfaces, vacancies and nanopores [14–17], which constitute trapping centers for positrons, can be study with positrons.

In the present study we employ positron annihilation Doppler broadening measurements to charaterize the dependency of nitrogen content on the open volume defects present in nanocrystalline copper thin films and try to identify the possible nitrogen location in these nanomaterials. The experimental details are presented in Sect. 2. In Sect. 3 we present the results concerning the identification of the positron trapping centers in the films. The conclusions are summarized in Sect. 4.

2 Experimental

Copper and nitrogen doped copper thin films were growth onto glass substrate by dc magnetron sputtering using a pure copper target. The nitrogen doped copper films were produced in reactive mode using 1:60, 1:30 and 1:2 nitrogen/argon partial pressure ratios (P_{N2}/P_{Ar}). The thickness of films was measured to be between 2–3 μm. The deposition was carried out at 0.3 Pa total pressure and applying to the copper target a constant power density of 3.33×10^4 Wm^{-2}. During the deposition, and to avoid the grain growth, the temperature of the substrate was maintained lower than 373 K by promoting the heat flow through the substrate's holder. The sputtering chamber was evacuated to 2×10^{-4} Pa before admitting pure argon and nitrogen gases.

All the films studied in this work are identical to those used in Ref. [9].

Carbon, hydrogen, nitrogen and sulphur contents of the N doped films were determined by elemental analysis using a Fisons EA1108 mass spectrometer.

In situ positron annihilation Doppler broadening measurements as a function of energy were carried out at room temperature with a magnetically guided slow positron beam. The energy range of the implanted positron was between 0.1 and 25 keV. This energy range corresponds to a mean penetration depth, {z = 45 [*E* (keV)]$^{1.6}$ Å}, of up to 0.8 μm in copper (assuming that the copper film density is similar to bulk copper density, 8.92 g.cm^{-3}). Positrons are stopped well before the substrate since all the films are more than 2 μm thick. For each positron energy, the 511 keV annihilation line was measured using an intrinsic Ge detector with an energy resolution of 1.2 keV at 497 keV gamma energy (^{103}Ru). The measurements were performed in the Coimbra Variable Energy Positron beam. This facility provides positrons with a typical flux of 10^2 s^{-1}cm^{-2}. The background pressure in this vacuum system is about 10^{-7} Torr. The structural properties of the films are investigated by implantation of monoenergetic positrons. After thermalization the positron diffuse through the material. In defective materials most of them will be trapped at defects. Ultimately, they annihilate with an electron, thus producing two gamma-photons of about 511 keV. The information on the defects is obtained from the Doppler broadening of the photo-peak, which is related to the momentum of the annihilated particles.

The photo-peak is characterized using two parameters, the shape parameter (*S*) and the wing parameter (*W*), defined in Fig. 1. The shape parameter is associated with annihilation with low momentum (valence) electrons [12]. The wing parameter corresponds to annihilation with high momentum (core) electrons [12]. Thus the information about the structure of defects of the thin films is provided by the energy-dependent data, *S*(*E*) and *W*(*E*).

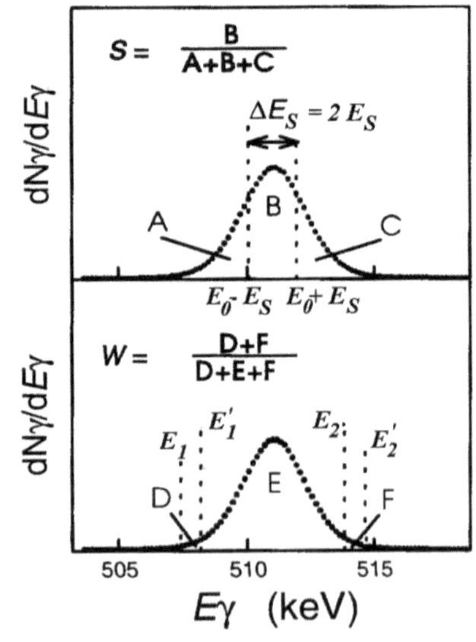

Fig. 1 Definition of shape parameter *S* and wing parameter *W* of the photo-peak. The areas A, B, C, D, E, F indicate sections defined with the aid of a fixed integration window (ΔE_S, E_1 and E_2). The figure shows a typical experimental spectrum

3 Results and Discussion

The introduction of nitrogen in the sputtering chamber was successfully used to produce nanocrystalline copper thin films. The presence of nitrogen, even in small amount, results in a reduction in the grain size and, simultaneously, in a homogeneous grain size distribution as observed by Transmission Electron Microscopy [9].

In Table 1 is presented the nitrogen content and the grain size of the films as a function of the nitrogen/argon partial pressure ratio during the film deposition. A continuous grain size decrease down to 30 nm is observed as the nitrogen content increases up to 3.5 at.%. For the highest nitrogen content a slight grain size increase is observed, which might be attributed to the formation of the Cu_3N nitride phase. In fact, as previously reported in reference [9], for the highest P_{N2}/P_{Ar} ratio studied a shoulder corresponding to the (111) plane of the Cu_3N phase is identified, besides the fcc Cu phase. It must be reported here that the nitrogen content presented in Table 1 is an estimated average value, since the observed nitrogen content value at the top surface of the film is higher than that observed at the bottom where the film began to grow. This gradient depends on the nitrogen content and it is larger for the films richer in nitrogen.

In Fig. 2 the Doppler *S* parameter and *W* parameter values—as a function of positron implantation energy—are shown for pure and for different nitrogen content Cu thin films. As can, immediately, be observed the films with nanocrystalline structure (3.5 and 7.0% of nitrogen content, respectively) reveal a quite different *S*(E) and *W*(E) curves than the films with microstructure (pure copper and 1.5% nitrogen content). Particularly, far from the surface of the film (positron energy higher than 15 keV) the *S*-parameter value, Fig. 2a, is significantly higher for the nanocrystalline films comparatively to the value of the same parameter for the films with larger grain size.

The effect of increasing of the *S*—parameter value in the region of the film is related to the presence of the large free volume fraction of these nanocrystalline films comparatively to the microcrystalline ones, namely the presence of vacancy type defects as will be discussed later. Also, Fig. 2a shows that the film with lower nitrogen content has a similar *S*(E) curve as the pure copper film meaning that the positron annihilates with pure Cu film characteristics. Analogous behaviour is observed in *W*(E) curve.

Table 1 Nitrogen contend and grain size of pure and N doped Cu thin films deposited onto glass substrate under different N_2/Ar partial pressure ratio

P_{N2}/P_{Ar}	Nitrogen average content (%)	Grain size (nm)
0	–	480
1/60	1.5	140
1/30	3.5	30
1/2	7.0	50

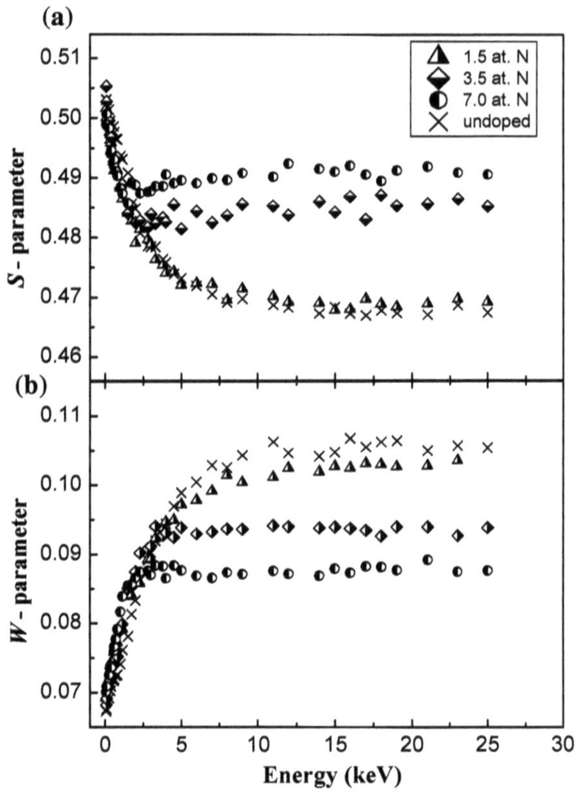

Fig. 2 Doppler *S* and *W* parameters versus incident positron energy for pure and nitrogen doped copper thin films with 1.5, 3.5 and 7.0% average nitrogen content **a** *S*—parameter, **b** *W*—parameter

Using the VEPFIT code [18] to fit the *S*(E) curves showed in Fig. 2a, the characteristic Doppler parameter, S_{film}, of each film were obtained and, also, the positron diffusion length. Table 2 shows these parameters for all the films. The analyses were performed considering one or two layers to characterize the films: the films with microstructure (pure and with 1.5% N) were well defined considering only one layer and for the films with higher content of nitrogen a second layer (top layer, immediately below the surface) was needed to consider. So, for the pure and low doped films the positron annihilate with 2 different characteristic *S*—parameter values: surface and film; and for the high doped films a third characteristic *S*—parameter value was needed, related to the top layer immediately below the surface. The normalization of the *S*—parameter values of the films, S_{film}, to the reference pure copper (undoped) film value gives 1.001, 1.038 and 1.050 for the films with 1.5, 3.5 and 7.0% nitrogen content, respectively. The increasing observed in the *S* parameter for the films with higher nitrogen content is a typical evidence of the presence of open volume type defects. The comparatively small values of the diffusion length of the positron is an indication that positrons annihilate in a saturation trapping regimen: all positron annihilate after to be captured in some type of open volume defect.

Table 2 Positron related parameters derived from VEPFIT analysis for pure and N doped Cu thin films

Film	S_{lay}	L_{lay} (nm)	d_{lay} (nm)	S_{film}	L_{film} (nm)
Undoped	–	–	–	0.4679 (2)	23 (1)
1.5% at. N				0.4691 (9)	21 (2)
3.5% at. N	0.4388 (4)	12 (3)	17 (3)	0.4856 (3)	51 (17)
7.0% at. N	0.4523 (5)	10 (3)	15 (2)	0.4911 (3)	23 (9)

The determined parameters are the shape-parameter *S* for the top layer, S_{lay}, and film, S_{film}, and thickness of top layer, d_{lay}, and diffusion length of positrons in the top layer, L_{lay}, and in the film, L_{film}. The surface *S*—parameter is not presented. The errors are also shown between parentheses

For nanocrystalline metallic materials, it has been reported in the literature [16, 17, 19] the observation, mainly, of two positron annihilation lifetimes associated with vacancy sized defects at the interfaces between grains ($\tau_2 \sim$ 150–200 ps) and vacancy clusters or nanovoids ($\tau_3 \sim$ 300–400 ps) at the intersection of interfaces (triple-lines). Sometimes, a third components is referred ($\tau_4 \sim$ 1–2 ns) as due to the large pores or missing grains [20].

The trapping of positrons at the interfaces between the grains it is expected to act as a shallow trapping center [12, 21] and the *S* parameter observed from the annihilation of positron trapped in these centers does not differ much from the free nitrogen copper film, which can explain the similar S_{film} values observed for pure Cu and 1.5% N doped Cu film. Also, the positron diffusion length are quite similar (23 ± 2 and 21 ± 2 nm, respectively) in both films. In fact this value for the positron diffusion length is much smaller than the one observed in defect-free metals (around 100 nm, [17, 19]) and since the grain size for these films are 480 and 140 nm respectively, the results suggest that some king of positron trapping center (vacancy type defect) must be present in these films. On the contrary, a high S_{film} values is observed for both 3.5 and 7.0% N doped films. These results reveal the presence of vacancy defects, like vacancy clusters or nanovoids at the intersection of interfaces (triple-lines) as observed in positron lifetime spectroscopy ($\tau_3 \sim$ 300–400 ps) experiments for nanocrystalline Cu [17, 19]. In fact, the existence of these nanovoids was, clearly, observed in the our film with 7.0% nitrogen content by High Resolution Transmission Electron Microscopy (HRTEM) as reported in reference [9]. The fitting of the *S*(E) curves requires, at least for the nanocrystalline Cu thin films (3.5 and 7.0% of N content), a minimum of two layers to describe the films, which suggested a non-uniform film structure in depth. It can be seen in Fig. 2, for incident positron energy at around 2.5 keV, a minimum in the *S*(E) curve for these films. The fitted thickness of the first top-layer is 15–17 nm and the positron diffusion length in this layer is around 10–12 nm. This result suggests a strong trapping of positron in this region with different positron annihilation characteristics. The application of the *S*-*W* plot analysis allows clarifying this interpretation.

In Fig. 3 we have plotted the *S*(*E*) and *W*(*E*) data of the different copper thin film as a trajectory in the *S*—*W* plane, using the implantation energy as a running parameter. The arrows denote the direction of increasing implantation energy.

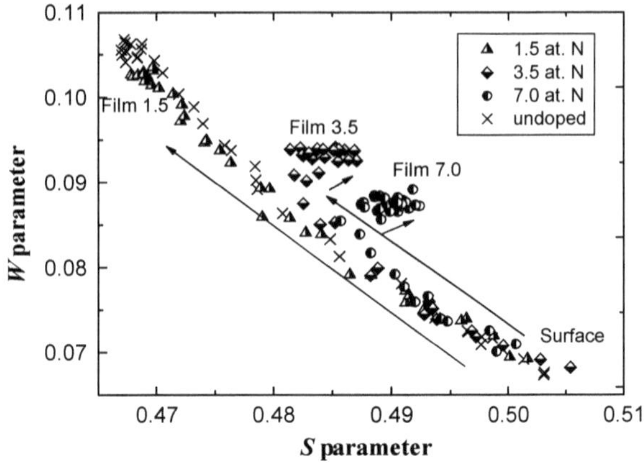

Fig. 3 *S-W* trajectories for positrons implanted in pure and nitrogen doped copper thin films with 1.5, 3.5 and 7.0% average nitrogen content. *Arrows* indicate the direction of increasing implantation energy

The different trapping layers are characterized by (*S*, *W*) coordinates instead of a single *S*-value. For example, for the high-energy implantation all positrons annihilate with the *S* and *W* values characteristics of the film. In the *S*–*W* plane this is seen from the clustering of the experimental data.

The convenience of interpretation of the *S-W* plot stems from the linearity property of *S* and *W* which is a consequence of the particular choice of their definition (Fig. 1) [22]. The value of this property for the interpretation of the *S-W* trajectory can be seen as follows. Suppose that we have a system where the positrons annihilate in two different trapping layers only, which will be designed *A* and *B*. These layers are situated at different depth and are characterized by different (*S*, *W*) coordinates, (S_A, W_A) and (S_B, W_B). The implanted positrons thus will be distributed over the *A* and *B* layer. This distribution can be varied by changing the implantation energy. Because of the linearity property of the shape and wing parameters, such a variation corresponds to a straight line trajectory in the *S-W* plane running from the coordinates (S_A, W_A) towards (S_B, W_B),

$$S_M = f_A S_A + (1 - f_A) S_B \tag{1}$$

$$W_M = f_A W_A + (1 - f_A) W_B \tag{2}$$

where f_A denotes the fraction of positrons trapped in layer *A* and S_M and W_M are the measured values. The presence of an additional, trapping layer with another depth distribution can now be immediately established from the trajectory: it should be curved unless the characteristic (*S*, *W*) coordinates of this trapping layer is accidentally located on the line (S_A, W_A) – (S_B, W_B).

With the aid of the above interpretations of straight and curved trajectories, we can easily explain our experimental data shown in Fig. 3. The *S-W* trajectories for the pure copper and the 1.5% nitrogen content films show a linear behavior and this can be interpreted as the positron annihilates either with the characteristic values of the surface or of the characteristic values of the film:. the two annihilation places are surface and film. On the other hand, for the 3.5 and 7.0% nitrogen content copper films the trajectory deviates from the straight line and shows a curved trajectory revealing the existence of a third layer. This result is consistent with the analyses referred above related to the fit of the *S*(*E*) curve that has revealed the existence of a top-layer with a thickness of the about 15–17 nm. At very low energy the positron annihilate in the surface of the film or in that top layer states and for higher energies the positrons annihilate with the characteristic values of the film (clustering of points).

In our opinion the reduction in the *S* parameter in the sub-surface region comparatively to the S_{film} value must be related to the partial occupancy of the free volume (vacancy clusters or nanovoids) with nitrogen. In fact, this behaviour agrees with the observed gradient in nitrogen content reported before and it seems to indicate that during the film deposition some of the nitrogen is released from the bottom to the top of the film surfaces as was also reported in [23, 24]. This hypothesis can be, eventually, confirmed by positrons measuring the *S*(E) and *W*(E) curves in the bottom surface of the films, which will be one of the next steps of the investigation.

4 Conclusion

The presence of nitrogen during the formation of Cu films by magnetron sputtering had an important role decreasing the grain size and in the production of nanocrystalline thin films. The positron annihilation measurements have identified the interfaces between the grains and the nanovoids at the intersection of interfaces as the main positron trapping centers and, consequently, the main lattice defects in these films. A non-uniform film structure was also observed and it was related to the gradient of the nitrogen content in depth. The reduction observed in the *S* parameter near the top surface of the nitrogen doped films suggests the partial occupancy of the open volume defects with nitrogen.

References

1. Lu, L., Shen, Y., Chen, X., Qian, L., Lu, K.: Ultrahigh strength and high electrical conductivity in copper. Science **304**, 422–426 (2004)
2. Kumar, K.S., Swygenhoven, H.Van, Suresh, S.: Mechanical behavior of nanocrystalline metals and alloys. Acta Mater. **51**, 5743–5774 (2003)
3. Su, J.Q., Nelson, T.W., Sterling, C.J.: A new route to bulk nanocrystalline materials. J. Mater. Res. **18**, 1757–1760 (2003)
4. Rohella, R.S., Swain, B.C., Murty, J.S.: Cathodic protection system: Protecting under-water steel piles of an iron ore berth. Anti-corrosion Meth. Mater. **38**, 4–7 (1991)

5. Dimitrov, D.A., Zahariev, A.L., Georgiev, J.K., Kolev, G.A., Petrinski, J.N., Ivanov, T.: Thin film platinum resistance thermometers: calibration and mathematical description of T(R) function. Cryogenics **34**, 487–489 (1994)
6. Tokura, H., Window, B., Neely, D., Swan, M.: Microstructure and mechanical properties of sputtered platinum films. Thin Solid Films **253**, 344–348 (1994)
7. Peto, G., Anderson, T.: Preparation of a Pt-GaAs Schottky contact by ion plating. Solid State Electron. **34**, 591–592 (1991)
8. Hansen, M., Anderko, K. (eds.): Constitution of Binary Alloys. McGraw-Hill, New York (1958)
9. Calinas, R., Vieira, M.T., Ferreira, P.J.: The effect of nitrogen on the formation of nanocrystalline copper thin films. J. Nanosci. Nanotechnol. **8**, 1–6 (2008)
10. Brandt, W., Dupasquier, A.: Positron Solid State Physics. North Holland, Amsterdam (1983)
11. Dupasquier, A., Mills Jr., A.P.: Positron Spectroscopy of Solids. IOS, Amsterdam (1995)
12. Krause-Rehberg, R., Leipner, H.S.: Positron Annihilation on Semiconductors—Defect Studies. Springer, Berlin (1999)
13. Ishii, A.: Positrons at Metallic Surfaces. Trans Tech, Aedermannsdorf (1992)
14. de Lima, A.P., Lopes Gil, C., Gordo, P.M., Duarte Naia, M.: Initial results with a variable energy positron system at coimbra. In: OECD/NEA—Workshop Proceedings on Ion and Slow Positron Beam Utilisation, pp. 121–128 (1998)
15. Asoka-Kumar, P., Lynn, K.G.: Implantation profile of low-energy positrons in solids. Appl. Phys. Lett. **57**, 1634–1635 (1990)
16. Schaefer, H.-E., Würschum, R., Birringer, R., Gleiter, H.: Structure of nanometer-sized polycrystalline iron investigated by positron lifetime spectroscopy. Phys. Rev. B **38**, 9545–9554 (1988)
17. Eldrup, M., Sanders, P.G., Weertman, J.R.: Positron annihilation study of the influence of grain size and purity on the annealing behaviour of nano-crystalline copper. Mat. Sci. Forum **255–257**, 436–438 (1997)
18. van Veen, A., Schut, H., de Vries, J., Hakvoort, R.A., Ijpma, M.R.: Positron Beams for Solids and Surfaces, pp. 171–198. AIP, New York (1990)
19. Mizuno, M., Kihara, T., Araki, H., Shirai, Y., Onishi, T.: Identification of lattice defects in Cu thin films by positron annihilation spectroscopy. Phys. Stat. Sol (c) **4**, 3550–3553 (2007)
20. Würschum, R., Scheytt, M., Schaefer, H.-E.: Nanocrystalline Metals and Semiconductors Studied by Positron Lifetime Spectroscopy. Phys. Stat. Sol. (a) **102**, 119–126 (1987)
21. Kuriplach, J., Melikhova, O., Hou, M., Petegem, S.Van, Zhurkin, E., Šob, M.: Positron annihilation in vacancies at grain boundaries in metals. Appl. Surf. Sci. **255**, 128–131 (2008)
22. Asoka-Kumar, P., Lynn, K.G., Welch, D.O.: Characterization of defects in Si and $SiO_2 - Si$ using positrons. J. Appl. Phys. **76**, 4935–4982 (1994)
23. Ji, A., Li, C., Du, Y., Ma, L., Song, R., Huang, R., Cao, Z.: Formation of a rosette pattern in copper nitride thin films via nanocrystals gliding. Nanotechnology **16**, 2092–2095 (2005)
24. Nosaka, T., Yoshitake, M., Okamoto, A., Ogawa, S., Nakayama, Y.: Copper nitride thin films prepared by reactive radio-frequency magnetron sputtering. Thin Solid Films **348**, 8–13 (1999)

The Influence of Microstructure on the Mechanical Behaviour of Dual Phase Steels

R. Amaral, A.D. Santos, J.A. Sousa and A.B. Lopes

1 Introduction

Transportation industry and particularly the automotive industry is facing in the coming future a number of challenges, including more stringent legislation on efficiency, fuel consumption, emissions and safety. These requests include the need to reduce the weight of automobiles and improvement of aspects related to safety and security. These issues are related and involve the selection and use of appropriate materials with higher strength properties, so that thinner components will have similar or better characteristics to previous thicker parts with additional weight. Other important material properties include high energy absorption and good formability behaviour, in order to obtain the component with the required geometry.

Among high-strength materials, dual phase (DP) steels are very important engineering materials due to their exceptional combination of high strength and

R. Amaral (✉) · A.D. Santos
INEGI, Institute of Science and Innovation in Mechanical and Industrial Engineering, R. Dr. Roberto Frias, 400, 4200-465 Porto, Portugal
e-mail: ramaral@inegi.up.pt

A.D. Santos
e-mail: abel@fe.up.pt

A.D. Santos
FEUP, Faculty of Engineering, University of Porto, R. Dr. Roberto Frias, 4200-465 Porto, Portugal

J.A. Sousa · A.B. Lopes
CICECO Aveiro Institute of Materials, Department of Material and Ceramic Engineering, University of Aveiro, Campus Universitário de Santiago, 3810-193 Aveiro, Portugal
e-mail: joseandrealvessousa@gmail.com

A.B. Lopes
e-mail: augusto@ua.pt

L.F.M. da Silva (ed.), *Materials Design and Applications*, Advanced Structured Materials 65, DOI 10.1007/978-3-319-50784-2_3 good formability thus allowing the production of both thin and strong components for many applications. The microstructure of these advanced high strength steels (AHSS), which are currently used in automotive industry, is predominantly composed by a soft ferritic matrix, that ensures the good formability, combined with hard martensite particles acting as a strengthening phase [1, 2].

On the other hand, the use of numerical methods is today a basic requirement when analyzing manufacturing processes, such as sheet metal forming. The corresponding advantages include the possibilities to improve processing and the quality of final product, as well as permitting an evident increase in efficiency both for time and cost reductions. However, the accuracy of numerical analysis of the plastic forming operations depends on a correct characterization of material properties.

During metal forming operations the material is usually loaded under different strain paths, which may change along the process and most probably will affect its formability limits by promoting damage or necking and thus leading to rejection of the product.

Mechanical behaviour of material is defined in each instant by the evolution of dislocation structures and crystallographic textures, which in turn will depend on different parameters like crystal structure, presence of solutes or second phases, loading path, pre-strain values, amplitude of the strain path change, etc. [3, 4].

Although its extensive use, the behaviour of dual phase steels has still many questions to be solved, particularly the effect of martensitic hard particles on the evolution of dislocation structures and crystallographic textures.

This chapter presents a study on the mechanical and microstructural behaviour of dual phase steels sheets with different amounts of martensite. The mechanical behaviour was characterized by tensile tests along different orientations in respect to rolling direction of the sheets.

Also, the mechanical behaviour exhibited by material during these tests was analyzed in terms of dislocation microstructure and crystallographic texture evolutions using scanning electron microscopy (SEM), electron backscattered diffraction (EBSD), nanoindentation tests and calculations with viscoplastic self-consistent (VPSC) model proposed by Lebensohn and Tome [5].

2 Microstructural Characterization

The materials used in this work were three DP steels sheets with 0.8 mm thickness and different amounts of martensite, namely DP500, DP600 and DP780.

The SEM observations performed on the initial materials (Fig. 1) showed that the microstructure is defined by the presence of hard martensite particles (white areas) dispersed in the soft ferritic matrix.

The average size of ferritic grains and volume percentage of martensite for each dual phase steel, obtained from microstructural observations are presented in Table 1.

Nanoindentation tests performed in different points (Fig. 2) showed that materials present very different hardness values along the analyzed areas, corresponding

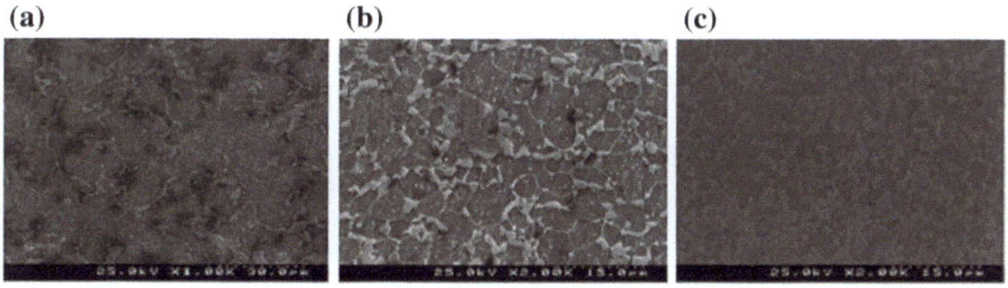

Fig. 1 SEM images of initial materials **a** DP500, **b** DP600 e **c** DP780

Table 1 Average size of ferritic grains (D) and volume percentage of martensite (V_M) for the initial materials

Material	D (μm)	V_M (%)
DP500	22	13
DP600	10	18
DP780	6	25

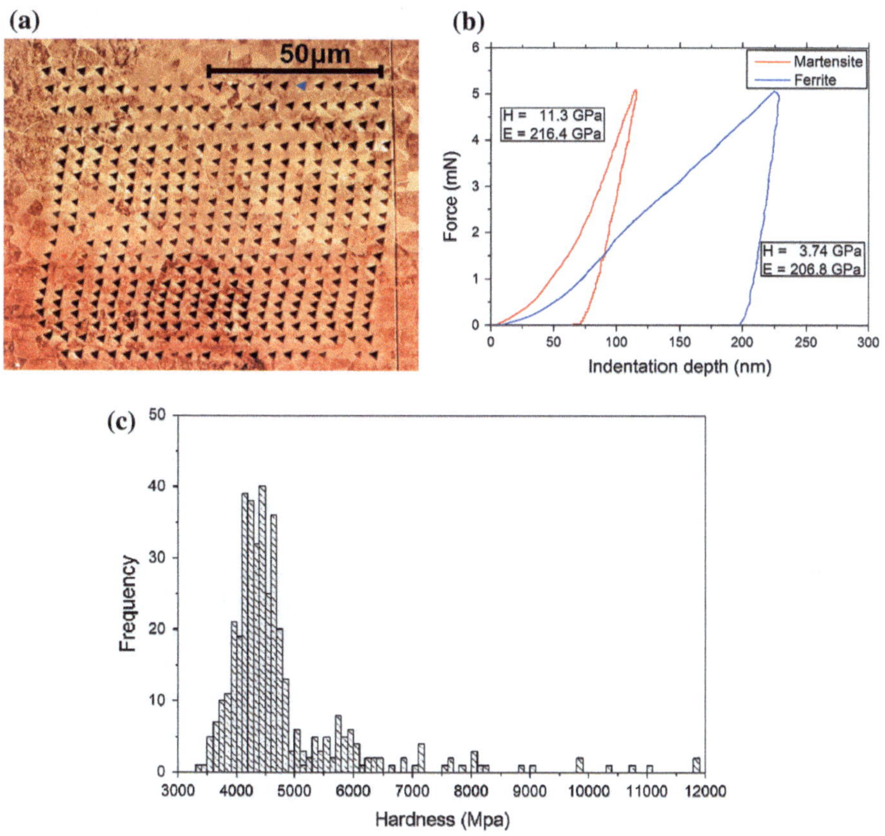

Fig. 2 Results of nanoindentation tests in different points of the sample (DP500); **a** image obtained by atomic-force microscopy (AFM) for the area under analysis; **b** curves force-penetration for the higher and lower hardness points; **c** frequency histogram for hardness values

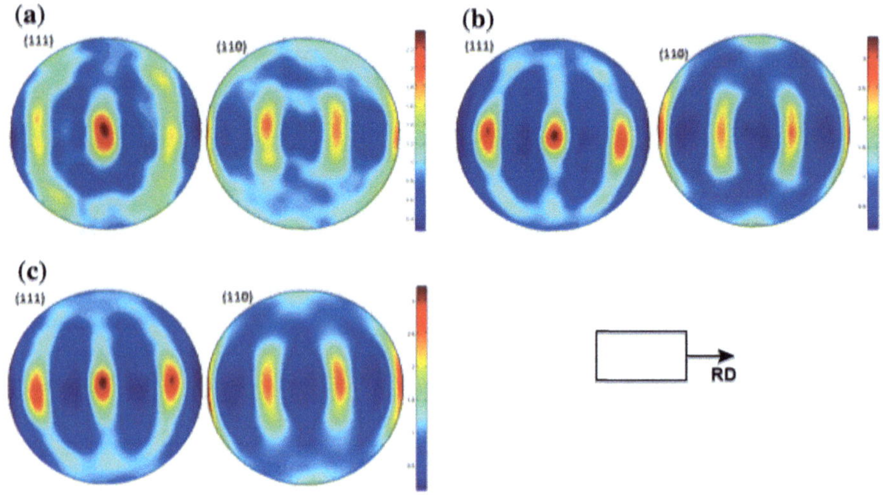

Fig. 3 Experimental pole figures {111} and {110} from initial materials. **a** DP500; **b** DP600; **c** DP780. RD—rolling direction of the sheet

to different phases: martensite having a high hardness and ferrite with a much lower hardness value.

Pole figures from ferritic grains obtained by EBSD (Fig. 3) show the existence of a γ fiber ({111} <uvw>) type crystallographic texture with a reinforcement of the {111} <110> component, which is typical of rolled steel sheets [6].

3 Mechanical Characterization

3.1 Uniaxial Tensile Test

Uniaxial tensile tests were performed at room temperature using a machine crosshead speed of 5 mm/s, which corresponds to an initial strain rate of 0.0016 s^{-1}. Specimens were machined according to ASTM E 8M-04 standard and the tests were performed for three different directions relative to the rolling direction (0°, 45° and 90°). Results were averaged from three experiments for each direction and material, in order to test and ensure repeatability. Figure 4 shows the corresponding stress-strain hardening curves for each DP steel and direction.

The obtained mechanical parameters extracted from curves along the rolling direction of the sheets are presented in Table 2, namely yield stress ($R_{p0,2}$), ultimate tensile strength (R_m), elongation at yield point (e_0), uniform elongation (e_u) and total elongation (e_t).

The anisotropy coefficient (r) for different angles to the rolling direction has been measured (Fig. 5). It is noticed the two different evolution behaviours presented by

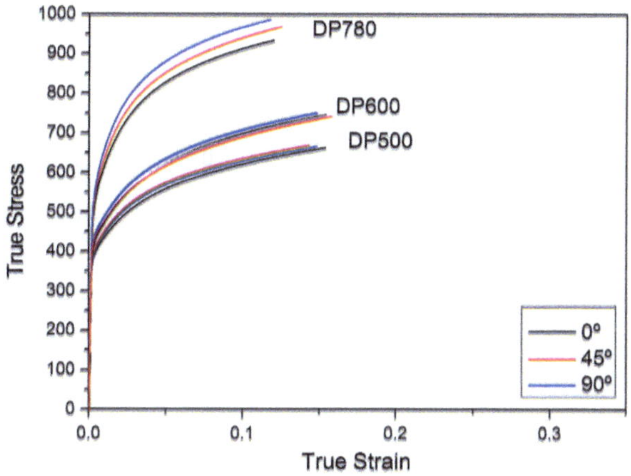

Fig. 4 Results from tensile test for true stress-strain curves of dual phase steels along three different directions 0°, 45° and 90° to rolling direction

Table 2 Mechanical properties for dual phase steels under current study, along rolling direction

Material	$R_{p0,2}$ (MPa)	R_m (MPa)	e_0 (%)	e_u (%)	e_t (%)
DP500	356.53	544.84	0.34	18.18	29.52
DP600	416.05	630.85	0.37	16.40	27.14
DP780	526.18	843.10	0.47	12.53	17.96

the three materials. The DP500 has the lowest r-value at 45° to the rolling direction, while the pair DP600/DP780 show the highest value for such direction. All of them have higher r-values for transverse direction (90º) when compared to rolling direction (0º).

By using the experimental crystallographic texture of initial materials as input of VPSC model [5] it was possible to predict the evolution of anisotropy coefficient (Fig. 6) for different angles to rolling direction. The obtained results showed that

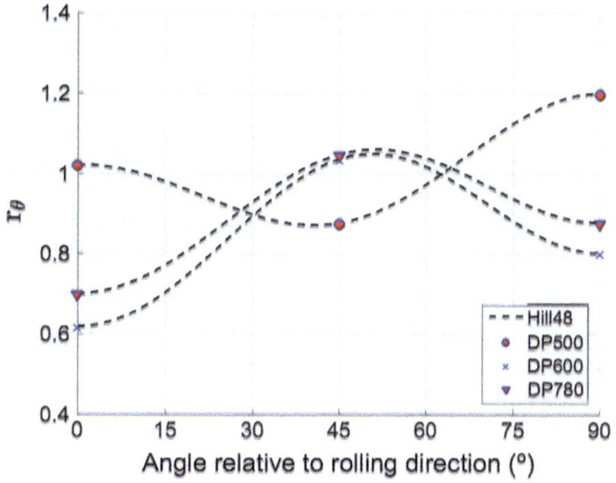

Fig. 5 Evolution of anisotropy coefficients along different angles to rolling direction

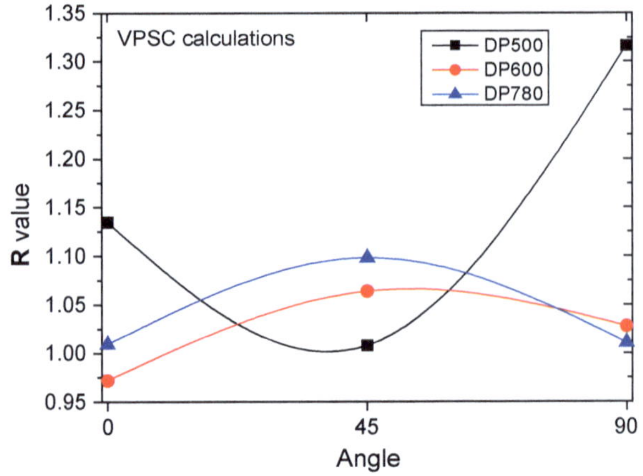

Fig. 6 Predicted anisotropy coefficients by VPSC model along different angles to rolling direction

both the simulated and experimental values present a similar evolution, thus giving evidence on the relation between initial crystallographic texture and initial anisotropy exhibited by materials.

3.2 *Hydraulic Bulge Test*

A biaxial hydraulic bulge test has been used with a tactile mechanical system (Fig. 7) to obtain biaxial stress-strain curve. The experimental system is composed by a set of tools, a hydraulic pump and a tactile mechanical device, which is in contact with sheet blank and permits to obtain data acquisition about specimen curvature and elongation at the pole, thus getting the relevant information for material characterization by coupling it with membrane theory.

Fig. 7 Experimental system used for biaxial hydraulic bulge test

Table 3 Mechanical properties for dual phase steels obtained from biaxial hydraulic bulge test

Material	Burst pressure (bar)	Total height at pole w/fracture (mm)
DP500	87.21	49.52
DP600	97.62	50.05
DP780	119.61	42.04

The set of tools contains a circular die, with a nominal diameter of 150 mm and die radius of 13 mm, as well as a blank holder with a drawbead, which restrains the sample and avoids any oil leakage during the test. The measuring system is calibrated before each test in order to ensure accuracy and reproducibility of data acquisition. The bulge test is performed with a pressure increment of 1 bar/s and the diameter of circular samples is 250 mm. The experimental system [7] allows the continuous acquisition of hydraulic pressure (p), as well as the variables provided by the measuring system: radius of curvature (ρ) and biaxial strain (ε). The ratio blank diameter/thickness permits the application of the membrane theory, which gives the evolution of material hardness behaviour for biaxial stress versus strain.

Table 3 presents the results obtained for maximum pressure (burst) obtained in biaxial hydraulic bulge test for each material, as well as the corresponding total height at the pole.

Results obtained for hardening biaxial stress-strain curves for each of DP steels from bulge test are shown in Fig. 8. Some oscillations (irregular behaviour) are observed for very low values of strain, which is due to variations for high levels of blank curvature (almost flat blank corresponds to an almost infinite radius of curvature) and low values of hydraulic pressure. Also, Fig. 8 shows that a similar initial yield stress for DP500 and DP600 exists, although DP600 has a higher strength evolution with continuing deformation, as expected.

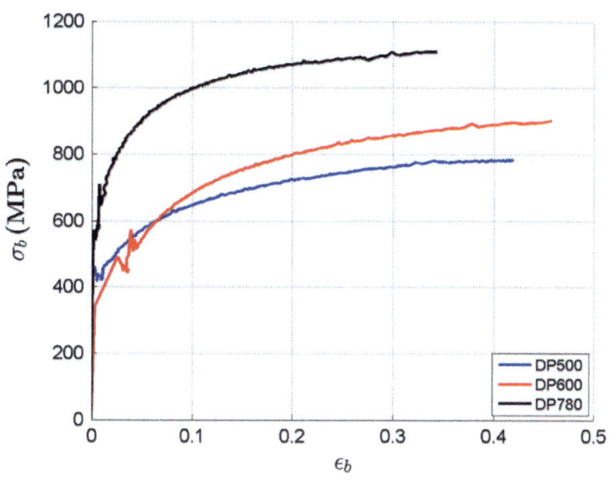

Fig. 8 Biaxial stress-strain curves for dual phase steels, obtained by hydraulic bulge test

3.2.1 Biaxial Stress-Strain Curve Conversion to Equivalent Stress-Strain Curve

By performing tensile and hydraulic bulge mechanical tests, one obtains two hardening curves for the same material, from each test. Defining that curve $\sigma = f(\varepsilon)$ corresponds to the uniaxial tensile test (along rolling direction) and curve $\sigma_b = f(\varepsilon_b)$ corresponds to hydraulic bulge test, one should have the same work performed to deform material, since the material is the same, therefore one should have a common hardening curve. However, the two hardening curves may not be coincident, since they don't belong to the same space of deformation and therefore the biaxial curve must be converted to the equivalent stress-strain curve.

The conversion of the biaxial stress-strain curve into the equivalent stress-strain curve, will be based on the principle of equivalent plastic work. In this way, assuming the incompressibility of the material and considering the state of stress at the pole as $\sigma_1 = \sigma_2 = \sigma_{b,}$ one can reshape the relationship of equivalent plastic work ($\sigma \cdot \varepsilon = \sigma_1 \cdot \varepsilon_1 + \sigma_2 \cdot \varepsilon_2$) by using Levy-von Mises and therefore Eq. 1 is obtained:

$$\frac{\sigma}{\sigma_b} + \frac{\varepsilon_b}{\varepsilon} = k \tag{1}$$

where k is a constant.

The equivalent plastic work is the link between the two curves, being W_u the plastic work per unit volume for the tensile test and W_b for the hydraulic bulge test. This methodology has been also used by other authors with corresponding satisfactory results [8–10].

Integrating $\sigma = f(\varepsilon)$ for the overall plastic domain one obtains the corresponding plastic work per volume unit for both tests, Eq. 2:

$$W(\varepsilon) = \int_{i=i(\varepsilon_i)}^{i=i(\varepsilon_f)-1} (\varepsilon_{i+1} - \varepsilon_i) \cdot \frac{\sigma_{i+1} + \sigma_i}{2} \tag{2}$$

When $W_u = W_b$ it is possible to establish a relationship between the stresses or strains for both tests, i.e.:

$$\begin{cases} \sigma \rightarrow W_u = W_b = \sigma_b \\ \varepsilon \rightarrow W_u = W_b = \varepsilon_b \end{cases} \tag{2}$$

To obtain k parameter, several methods can be considered [8, 9, 11]. In this chapter, it will be used a method based on the equivalent plastic work, by corresponding to the maximum values of stress and strain for the uniaxial tensile test ($\sigma_{max} \rightarrow W_u = W_b \rightarrow \sigma_b \Rightarrow k_1 = \sigma_{max}/\sigma_b$), as shown in Fig. 9 [9].

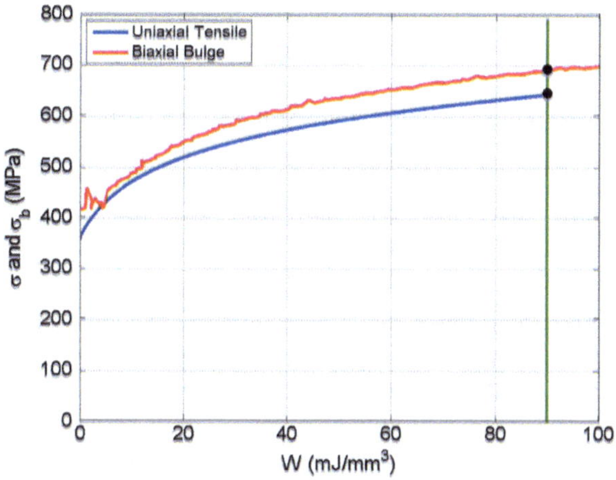

Fig. 9 Relation between plastic work and stress obtained from each test (tensile, bulge) by using the maximum stress from tensile test

In Table 4 it is shown the corresponding values for *k* parameter used to convert biaxial stress-strain curve into equivalent stress-strain curve, by using the defined method of maximum values of stress and strain from uniaxial tensile test.

The final result of converting biaxial data and including it in the equivalent hardening curve is presented in Fig. 10, for every dual phase steel. It is observed

Table 4 k parameter used to convert biaxial stress-strain curve into equivalent stress-strain curve, for each dual phase steel

Parameter	Material		
	DP500	DP600	DP780
k	0.9318	0.9645	0.9423

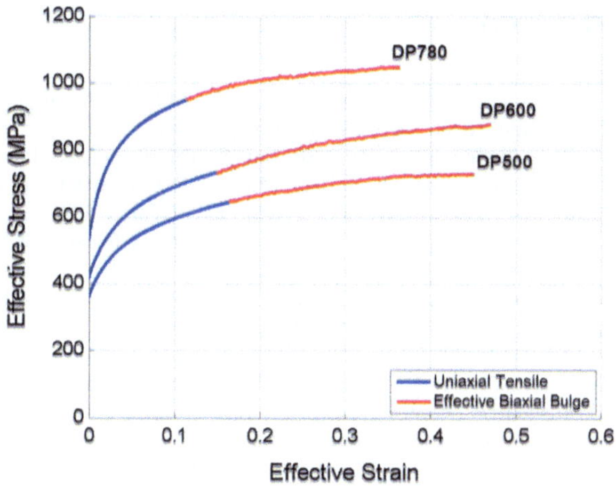

Fig. 10 Combination of stress-strain curve from tensile test with biaxial stress-strain curve, based on the principle of equivalent work and method of maximum values of stress and strain from uniaxial tensile test

that using the selected method there is a good first derivative continuity for the extension of hardening curve based on bulge test, both for DP500 and DP780. As for DP600 there is a less perfect continuity in the transition between tensile data and bulge data, although being still very smooth and having a good overall behaviour.

4 Conclusions

Three dual-phase steels with different amounts of martensite (DP500, DP600 and DP780) were selected and investigated for characterization and analysis of mechanical and microstructural analysis.

The incompatibilities of deformation between soft ferritic matrix and the hard martensite particles promotes a very inhomogeneous plastic deformation of the ferrite grains and the selected dual-phase steels sheets show an anisotropic behaviour for monotonic tensile tests, which is due and related to their initial crystallographic texture. The corresponding relation and prediction has been performed by a VPSC (viscoplastic self-consistent) model and its results show a good agreement with the experiments.

The conversion of bulge data to the equivalent stress-strain hardening curve by using the method of maximum values of stress and strain from uniaxial tensile test has shown to provide a good correlation, in order to obtain the extension of hardening information for selected dual phase steels, thus giving possibility of material characterization for higher strain values based on experimental data.

Acknowledgements The authors would like to acknowledge the support of this work by FEDER funds through the Operational Program for Competitiveness Factors (COMPETE) and by national funds through the FCT (Foundation for Science and Technology) through the projects PTDC/EMS-TEC/2404/2012, POCI-01-0145-FEDER-016876 and PTDC/EMS-TEC/6400/2014.

References

1. Kim, S.B., Huh, H., Bok, H.H., Moon, M.B.: Forming limit diagram of auto-body steel sheets for high-speed sheet metal forming. J. Mater. Proc. Technol. **211**, 851 (2011)
2. Cao, Y., Karlsson, B., Ahlström, J.: Temperature and strain rate effects on the mechanical behavior of dual phase steel. Mater. Sci. Eng. A **636**, 124–132 (2015)
3. Lopes, A.B., Barlat, F., Gracio, J.J., Duarte, J.F.F., Rauch, E.F.: Effect of texture and microstructure on strain hardening anisotropy for aluminum deformed in uniaxial tension and simple shear. Int. J. Plast. **19**, 1 (2003)
4. Vincze, G., Rauch, E.F., Gracio, J.J., Barlat, F., Lopes, A.B.: A comparison of the mechanical behaviour of an AA1050 and a low carbon steel deformed upon strain reversal. Acta Mater. **53**, 1005 (2005)
5. Lebensohn, R.A., Tome, C.N.: Acta Metall. Mater. **9**, 2611 (1993)
6. Rollett, A.D., Wright, S.I.: In: Kocks, U.F., Tome, C.N., Wenk, H.R. (eds.), Texture and Anisotropy: Preferred Orientations in Polycrystals and Their Effect on Materials Properties, p. 178. Cambridge University Press (1998)

7. Campos, H., Santos, A.D., Martins, B., Ito, K., Mori, N., Barlat, F.: Hydraulic bulge test for stress-strain curve determination and damage calibration for Ito-Goya model. In: 11th World Congress on Computational Mechanics, 5th European Conference on Computational Mechanics and 6th European Conference on Computational Fluid Dynamics, pp. 4223–4238 (2014)
8. Mutrux, A., Hochholdinger, B., Hora, P.: Procedure for evaluation and validation of hydraulic biaxial experiment. In: Numisheet 2008, Switzerland, pp. 67–71 (2008)
9. Sigvant, M., Mattiasson, K., Vegter, H., Thilderkvist, P.: A viscous pressure bulge test for determination of a plastic hardening curve and equibiaxial material data. Int. J. Mater. Form. **2**, 235–242 (2009)
10. Lazarescu, L., Nicodim, I., Ciobanu, I., Comsa, D.S., Banabic, D.: Determination of material parameters of sheet metals using the hydraulic bulge test. Acta Metall. Slovaca **19**(1), 4–12 (2013)
11. Lee, M.-G., Kim, D., Kim, C., Wenner, M.L., Wagoner, R.H., Chung, K.: Springback evaluation of automotive sheets based on isotropic-kinematic hardening laws and non-quadratic anisotropic yield functions—Part II: characterization of material properties. Int. J. Plast. **21**, 883–914 (2005)

Part II
Ceramics

Application of Cutting Edges with High Durability Made of Nanocrystalline Cemented Carbides

Maciej Jan Kupczyk, Piotr Siwak and Jedrzej Komolka

1 Introduction

The WC-Co type cemented carbides used to produce cutting tools are divided into three basic groups with respect to their application, i.e., S, H and U (according to PN) or P, M, K (according to ISO). Moreover, with respect to the particle size of the WC, they can be divided into the following groups [2, 4, 16]:

- Coarse—with an average particle diameter 3 ÷ 30 μm,
- Standard—with an average diameter 1.5 ÷ 3 μm,
- Fine—with an average diameter 0.5 ÷ 1.5 μm,
- Ultrafine—with an average diameter 0.2 ÷ 0.5 μm,
- Nanocrystalline—with an average diameter 20 ÷ 200 nm.

WC grain size has a significant influence on the properties of the cutting edges made of cemented carbides, and particularly on their bending strength and hardness [1, 3, 5, 7, 10–12]. When the diameter of the WC particles is greater than 1.5 μm, an increase in the bending strength and decrease in hardness of cutting edges made of cemented carbides is observed when the WC grain size increases. When the WC grain has a diameter less than 1.5 μm, simultaneous increase in bending strength and hardness of cemented carbides is observed as the WC grain size decreases. This observation leads to the overturning of the belief that the increase in hardness entails a reduction in ductility [5, 11, 13, 16].

M.J. Kupczyk (✉) · P. Siwak · J. Komolka
Poznan University of Technology, Piotrowo 3, 60-965 Poznan, Poland
e-mail: maciej.kupczyk@put.poznan.pl

P. Siwak
e-mail: piotr.siwak@put.poznan.pl

J. Komolka
e-mail: jedrzej.komolka@gmail.com

L.F.M. da Silva (ed.), *Materials Design and Applications*,
Advanced Structured Materials 65, DOI 10.1007/978-3-319-50784-2_4

The reduction in grain size is one way to modify the microstructure in order to obtain a significant improvement of mechanical properties. It should be noted, however, that in nanocrystalline materials with a grain size of less than 10–20 nm, a different mechanism of deformation associated with the slides at the grain boundaries gains importance, causing a reduction in hardness of the material with the nanocrystalline structure. Therefore the use of nanoparticles with a grain size smaller than 20 nm can be unprofitable. Elimination of this phenomenon is possible in suitably designed nanocrystalline composite materials in which amorphization of the structure of grain boundaries can weaken the activity of dislocations, making plastic deformation difficult [7, 11].

This chapter presents the results of comparative testing of wear and durability of nanocrystalline cemented carbides with and without the addition of Cr_3C_2. The addition of Cr_3C_2 is used in order to limit grain growth during the sintering of tungsten carbide nanopowder.

Modification of the chemical composition of the sintered nanocrystalline cemented carbides was performed in order to produce cutting edges with still greater wear resistance than nanocrystalline cemented carbides of the Nano_WC-5Co type without growth inhibitor, with a view to using them effectively in machining of hard machinable construction materials.

2 Experimental Details

Comparative studies used the cutting inserts made of:

- WC-5% wt. Co conventional cemented carbides,
- Nano_WC-5% wt. Co nanocrystalline cemented carbides,
- Nano_WC-5% wt. Co nanocrystalline cemented carbides with the addition of a given amount of Cr_3C_2:
  (a) 0.3 wt%,
  (b) 0.6 wt%,
  (c) 0.9 wt%.

Table 1 shows the chemical composition of the materials for insert cutting edges used in the study.

Table 1 Chemical analysis of the insert cutting edges

Material of cutting edges	Sintering method	Chemical analysis (wt%)		
		WC	Co	Cr_3C_2
Conventional WC-5Co	HP	95	5	–
Nanocrystalline nano_WC-5Co	PPS	95	5	–
Nanocrystaline nano_WC-5Co + Cr_3C_2	PPS	94.715	4.985	0.3
		94.430	4.970	0.6
		94.145	4.955	0.9

Samples made of conventional sintered carbides were produced by the Hot Pressing (HP) method in the Baildonit company.

Nanocrystalline cemented carbides of the Nano_WC-5Co and Nano_WC-5Co + Cr_3C_2 types were sintered using nanopowder with a mean grain diameter of 50 nm, applying the Pulse Plasma Sintering (PPS) method, in the Department of Materials Science and Engineering, Warsaw University of Technology [6, 8, 9, 14, 15, 17]. Samples made of nanocrystalline cemented carbides, whose characteristics are presented in this article, were prepared at the sintering temperature of about 1380, 1420, 1470, 1530 and 1560 K. The sintering time was 200 s. Rectangular cutting inserts (9.7 × 9.7 mm with a corner radius $r_\varepsilon = 0.8$ mm) were obtained from cylindrical samples with a diameter of $\phi = 20$ mm, using the wire EDM Agiecut Classic 2. The rake face of the cutting inserts was then ground to obtain the thickness of 3.18 mm.

The prepared cutting inserts made of nanocrystalline cemented carbides were subjected to wear and durability tests during the process of longitudinal turning, using the TK66 lathe.

After fixing the rectangular cutting inserts in CSRNR 252509-ID toolholder, the following geometry of the cutting insert was obtained:

- tool cutting edge angle $\kappa_r = 75°$,
- tool orthogonal clearance $\alpha_O = 6°$,
- tool included angle $\varepsilon_r = 90°$,
- tool orthogonal rake angle $\gamma_O = -6°$,
- tool cutting edge inclination $\lambda_s = -6°$.

The following conditions were assumed during wear and durability tests:

- cutting speed $v_c = 120$ m/min,
- feed $f = 0.2$ mm/rev,
- depth of cut $a_p = 1$ mm.
- workpiece: the EN 1.45.41 (PN/EN) austenitic stainless steel.

In order to interpret the results of the study of functional properties of the cutting inserts, the following tests of selected technological parameters of these inserts were performed:

- assessment of the microsections of sinters, using a scanning electron microscope Vega TESCAN, based on the secondary electrons (SE) images and backscattered electron (BSE) images at magnification 6000× and 15000×,
- measurement of hardness (HV30) using computer-aided FV-700 Vickers diamond testing machine with an automatic head and controlled load and with an FM AR-90 read-out automatic system.

3 Results and Discussion

Figures 1 and 2 present a graphical interpretation of the results of wear measurements of nanocrystalline cemented carbides cutting inserts with and without the addition of growth inhibitor, taken during machining of austenitic steel. The VB_c

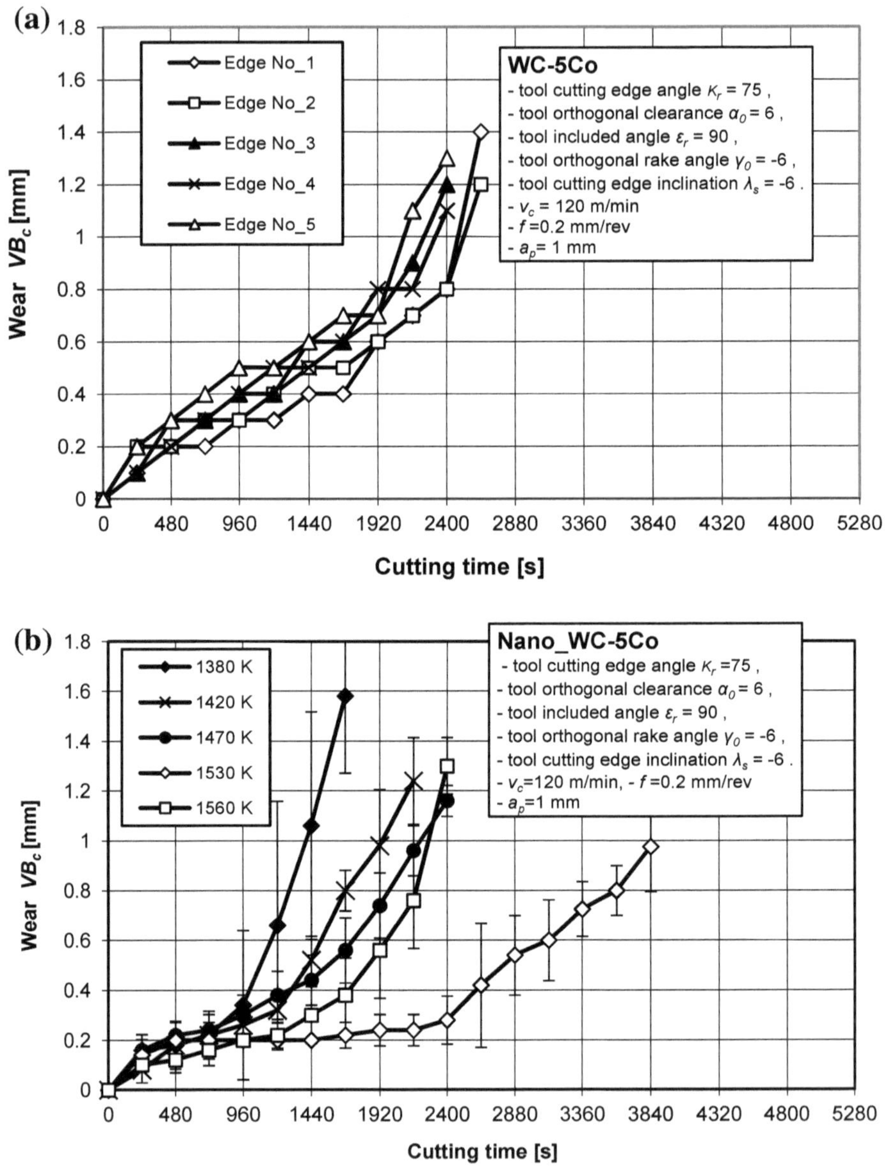

Fig. 1 Wear curves of cutting edges made of: **a** WC-5Co, **b** Nano_WC-5Co, obtained during turning of the EN 1.4544 (1H18N9T) austenitic steel

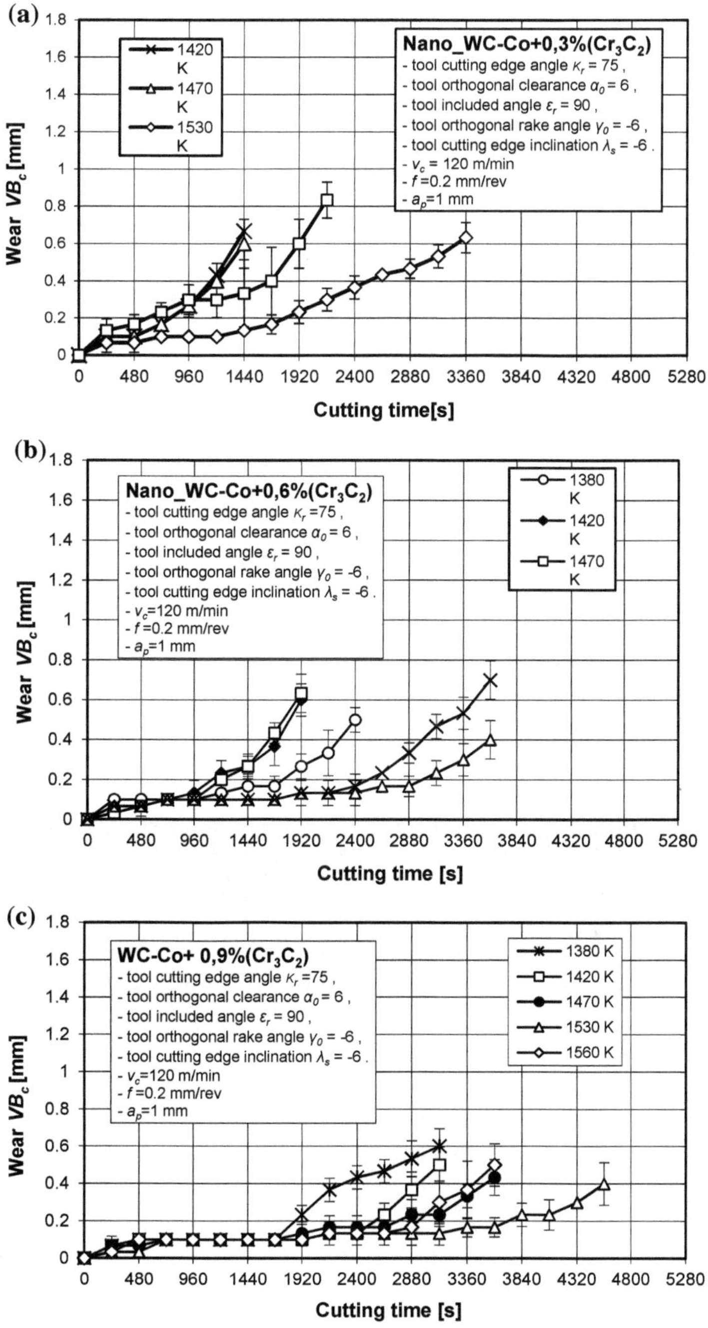

Fig. 2 Wear curves of cutting edges made of nanocrystalline cemented carbides with different amounts of Cr_3C_2: **a** Nano_WC-5Co + 0.3% (Cr_3C_2), **b** Nano_WC-5Co + 0.6% (Cr_3C_2), **c** Nano_WC-5Co + 0.9% (Cr_3C_2) obtained during turning of the EN 1.4544 (1H18N9T) austenitic steel

values (band width of the corner wear) given in these figures are the mean values of wear from the four corners of each of the cutting inserts.

Figure 3 shows the SE images of flank face of cutting edges of nanocrystalline cemented carbides at the end of their exploitation time.

Based on the results of wear measurements, the durability of cutting edges was determined for the wedge wear indicator $VB_C = 0.4$ mm (Fig. 4).

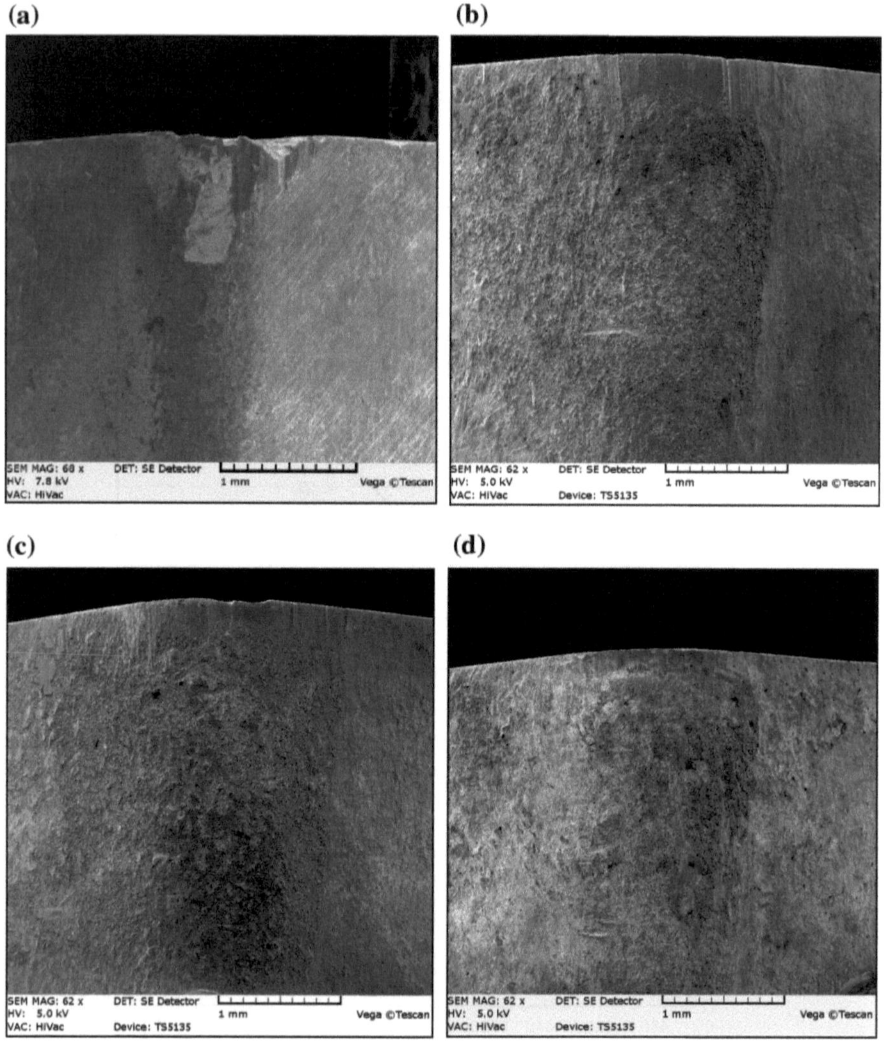

Fig. 3 SE images of cutting edges made of nanocrystalline cemented carbides with different amount of Cr_3C_2: **a** Nano_WC-5Co, **b** Nano_WC-5Co + 0.3% (Cr_3C_2), **c** Nano_WC-5Co + 0.6% (Cr_3C_2), **d** Nano_WC-5Co + 0.9% (Cr_3C_2) at the end of their exploitation period

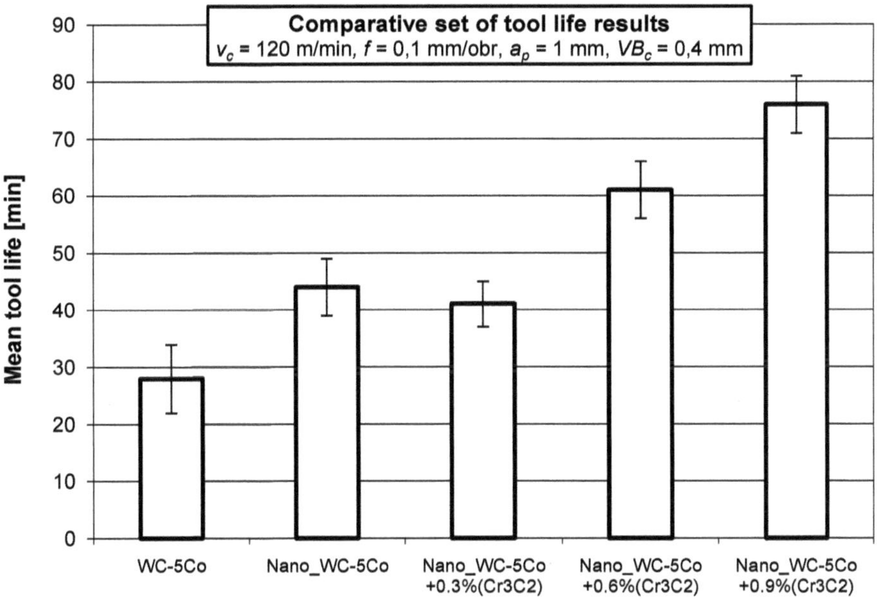

Fig. 4 Durability results of cutting edges made of: **a** WC-5Co, **b** Nano_WC-5Co, **c** Nano_WC-5Co + 0.3% (Cr_3C_2), **d** Nano_WC-5Co + 0.6% (Cr_3C_2), **e** Nano_WC-Co + 0.9% (Cr_3C_2) sintered in 1530 K

On the basis of earlier hardness measurements, it can be stated, that the highest hardness was obtained for the cutting inserts made of nanocrystalline cemented carbides containing 0.9% Cr_3C_2 by weight. This can be explained by the fact that 0.9% is a maximum quantity of growth inhibitor that can be used without exceeding the limit of Cr_3C_2 solubility in cobalt. For this amount of Cr_3C_2, after sintering, the cobalt binder phase is a solid solution of Cr and W. In the study, the higher content of Cr_3C_2 was not applied, because then during the cooling, the chromium carbide and cobalt carbide of the M_7C_3 type separate from the solution, when the ratio Cr/Co increases with the increase of Cr_3C_2. The presence of these cemented carbides leads to a decrease in fracture toughness and bending strength due to replacing of the plastic cobalt phase by brittle chromium and cobalt carbides of the M_7C_3 type [7, 8].

Figure 5 shows BSE images of microsections of nanocrystalline cemented carbides without and with the growth inhibitor Cr_3C_2 added in the amount of 0.3, 0.6 and 0.9% by weight (images were taken using a Vega TESCAN scanning microscope).

Table 2 presents a comparative set of the results of testing of the hardness, grain size and durability of cutting edges made of conventional and nanocrystalline cemented carbides with and without the inhibitor of growth. The values given in Table 2 are the average values of five measurements.

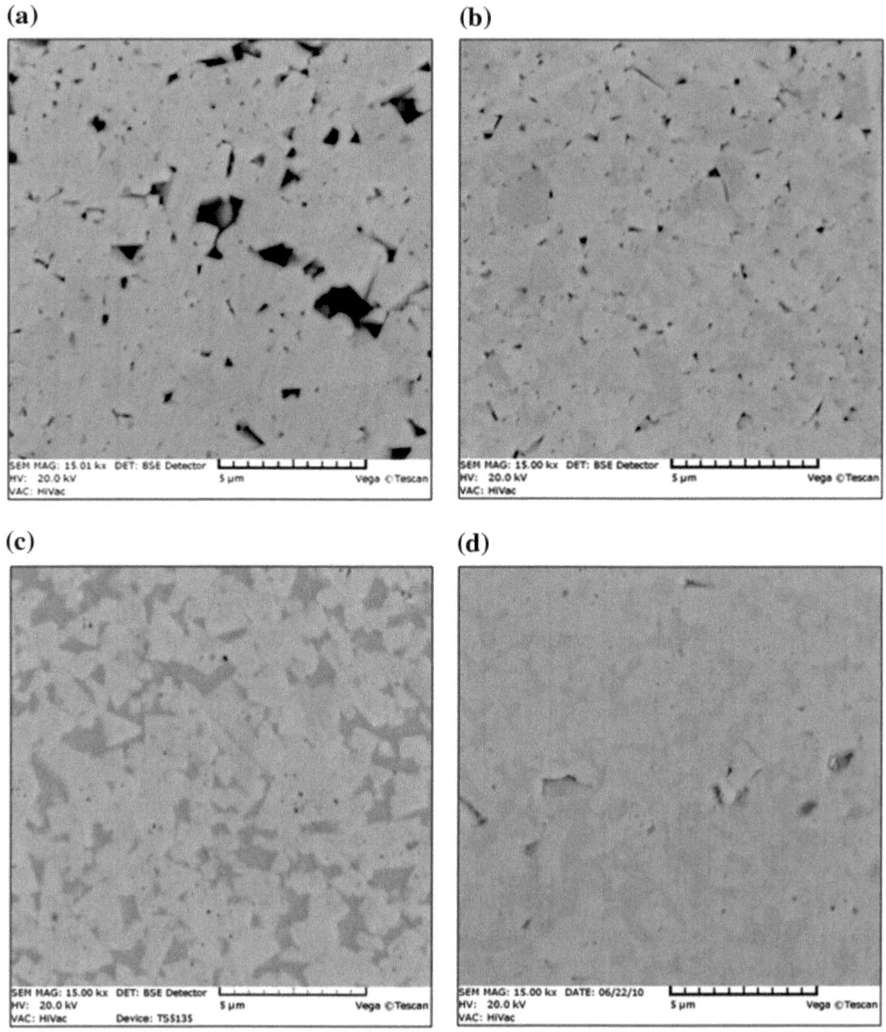

Fig. 5 BSE images of microsections of: **a** WC-5Co, **b** Nano_WC-5Co + 0.3% Cr_3C_2, **c** Nano_WC-5Co + 0.6% Cr_3C_2, **d** Nano_WC-5Co + 0.9% Cr_3C_2

The nanocrystalline cemented carbides have much higher hardness and smaller average grain size than standard cemented carbides with the same chemical composition and particularly with growth inhibitor. For these reasons, cutting inserts made of the nanocrystalline cemented carbides have significantly greater resistance to abrasive wear (the increase in hardness is equivalent to the increase in abrasive wear resistance—many researchers have demonstrated a linear relationship between

Table 2 Comparative set of investigation results of the hardness, granularity and durability of cutting edges made of conventional and nanocrystalline cemented carbides without and with the growth inhibitor

Material of cutting edges	Sintering method	Amount of Cr_3C_2 (wt%).	Grain size (nm)	Mean hardness (HV30)	Durability (min)
Conventional WC-5Co	HP	–	2000–3000	1436	32
Nanocrystalline nano_WC-5Co	PPS	–	300–500	1912	51
Nanocrystaline nano_WC-5Co + Cr_3C_2	PPS	0.3	200–300	1851	41
		0.6	150–200	2032	61
		0.9	60–150	2187	76

the increase of the relative abrasive wear resistance and hardness of the material subjected to friction [5, 16]) and greater durability during turning the EN 1.4544 (1H18N9T) austenitic steel.

4 Conclusions

The research results presented in the chapter concern wear and durability of nanocrystalline cemented carbides of the type Nano_WC-5Co without and with the growth inhibitor Cr_3C_2, produced with the PPS method. The results allow us to formulate the following conclusions:

- cutting inserts made of Nano_WC-5Co + 0.9% Cr_3C_2, produced using the PPS method at sintering temperature 1530 K and sintering time 200 s exhibit lower abrasive wear and more than 4 times greater durability during turning of austenitic steel compared to the cutting inserts made of cemented carbides without the growth inhibitor (for wedge wear indicator VBc = 0.4 mm);
- nanocrystalline cemented carbides with the addition of Cr_3C_2 have a much greater hardness (approximately 2180 HV30) than nanocrystalline cemented carbides without it (about 1800 HV30);
- the results of wear and durability studies of nanocrystalline cemented carbides of the type Nano_WC-5Co, containing 0.9% Cr_3C_2 by weight indicate the possibility of substantial enhancement of the cutting performance, e.g. increase in cutting speed,
- the use of a growth inhibitor in the form of chromium carbide, in particular in the amount of 0.9 wt%, ensures a considerable reduction in growth of the tungsten carbide grains in the process of sintering.

Acknowledgements The author gratefully acknowledges the financial support of the Ministry of Science and Higher Education (Polish State Committee for Scientific Research, contract No. NN503 147734) [8]. The author also thanks Prof. A. Michalski and Dr. M. Rosinski from Warsaw University of Technology and Dr. K. Jozwiak from Poznan University of Technology for access to the apparatus and preparation of samples.

References

1. Allen, C., Sheen, M., Williams, J., Pugsley, V.A.: The wear ultrafine WC-Co hard metals. Wear **250**, 604–610 (2001)
2. Dobrzanski, L.A.: The Principle of Material Science and Physical Metallurgy. WNT Publisker, Gliwice-Warsaw (2002)
3. Feldshtein, E.: Selected Problems of Machining. ISBN 978-83-7842-104-7. ZGU Publisher, Zielona Gora (2013)
4. Gordon, S., Hillery, M.T.: A review of the cutting of composite materials. Proc. Inst. Mech. Eng. Part L J. Mater. Design Appl. **217**(1), 35 (2003)
5. Kupczyk, M.J.: Manufacture and exploitation of cutting tools with wear-resistant coatings. PUT Publisher, Poznan (2009). ISBN 978-83-7143-864-6
6. Kupczyk, M.J.: Cutting edges with high hardness made of nanocrystalline cemented carbides. Int. J. Refractory Met. Hard Mater. **49**, 249–255 (2015)
7. Kupczyk, M.J. (chief of project), Michalski, A., Zurek, J., Jozwiak, K., Rosinski, M., Siwak, P.: Research project of the ministry of science and higher education, contract No. N N503 147734; Poznan (2010) (unpublished)
8. Michalski, A.: Pulse plasma sintering of ceramic materials. Ceramika **91**, 378–385 (2005)
9. Michalski, A.: Physical and Chemical Principles of Coating Obtained from Vapour. WUT Publisher, Warsaw (2000). ISBN 83-7207-171-3, ISBN 83-7207-202-7
10. Milman, Y.V., Luyckx, S., Northorp, I.T.: Influence of temperature, grain size and cobalt content on hardness of WC-Co alloys. Int. J. Refractory Met. Hard Mater. **17**, 39–44 (1999)
11. Olszyna, A.R.: Superhard Ceramics. WUT Publisher, Warsaw (2001). ISBN 83-7207-312-0
12. Richter, V., Ruthendorf, M.V.: On hardness of ultrafine nanocrystalline hard materials. Int. J. Refract. Metal Hard Mater. **17**, 141–152 (1999)
13. Sailer, T., Herr, M., Sockel, H.-G., Schulte, R., Feld, H., Praksh, L.J.: Microstructure and properties of ultrafine gradient hardmetals. Int. J. Refract. Metal Hard Mater. **19**, 553–559 (2001)
14. Siemiaszko, D.: Structure and properties of the WC-Co composites made of tungsten, carbon and kobalt powders sintered by high-electric current pulses. Doctor's thesis, Warsaw University of Technology, Warsaw (2006)
15. Sokolowska, A.: Unconventional Means of Materials Synthesis. PWN Publisher, Warsaw (1991). ISBN 83-01-10848-7
16. Wysiecki, M.: Modern Tool Materials. WNT Publisher, Warsaw (1997)
17. Zdunek, K.: Pulse Plasma in Surface Engineering. WUT Publisher, Warsaw (2004). ISBN 83-7207-476-3

Design of Nanocrystalline Cemented Carbides with High Hardness

Maciej Jan Kupczyk

1 Introduction

Investigations in the area of mechanical technology and machining are often stimulated by new achievements in the field of materials science, concerning the study and design of bulk materials in order to obtain the desirable technological and functional properties of these materials. This concerns construction materials as well as tool materials [1–4].

During creation of the concept of the investigations the author of this paper was guided by, among other things, the fact that in the most recent world literature there appear numerous examples of the use of nanocrystalline materials indicating that they have often much better functional properties than conventional construction materials [3–10]. This led the author to undertake complex research of tool materials with the nanocrystalline structure, so that manufacturers can be offered modern, high-performance, durable, reliable tools to obtain products of the highest quality.

Literature data indicates that the tool materials made of cemented carbides with a fine-grained structure (the average WC grain diameter: 0.5 ÷ 1.5 μm) and especially with a ultrafine grained structure (the average WC grain size: 0.2 ÷ 0.5 μm) have better mechanical properties than tool materials made of cemented carbides with standard grain size (now widely used—with the average WC diameter: 1.5 ÷ 3 μm) and especially the coarse structure (having an average WC grain diameter 3 ÷ 30 μm) [3, 4]. This is illustrated in Fig. 1, which shows, among other things, that the smaller the WC grain, the greater hardness of the WC-Co cemented carbides.

This has a very significant influence on the functional properties of tools made of cemented carbides because the increase in hardness is equivalent to the increase in abrasive wear resistance. This relationship has been experimentally confirmed by

M.J. Kupczyk (✉)
Poznan University of Technology, Piotrowo 3, 60-965 Poznan, Poland
e-mail: maciej.kupczyk@put.poznan.pl

L.F.M. da Silva (ed.), *Materials Design and Applications*,
Advanced Structured Materials 65, DOI 10.1007/978-3-319-50784-2_5

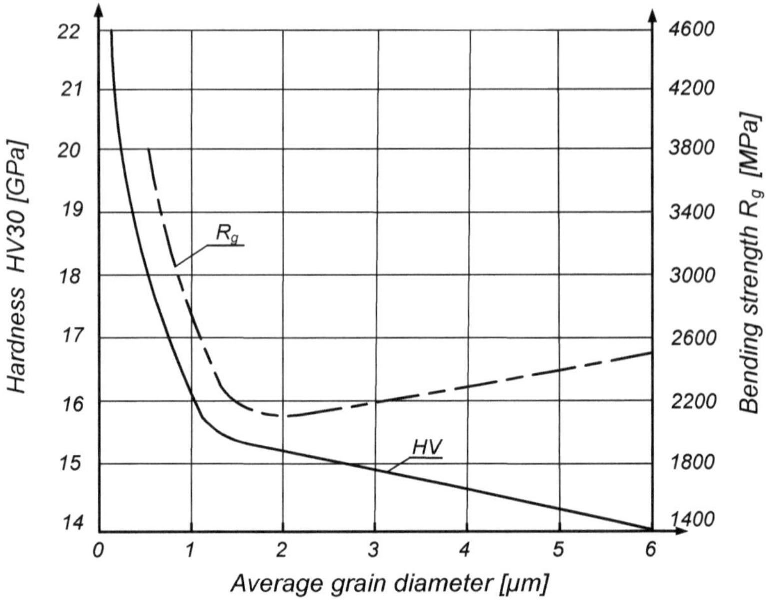

Fig. 1 Bending strength and hardness of cemented carbides versus the average grain diameter of the WC (elaborated on the basis of [3, 4])

Young, Khrushchev, Babiczew, Wachal and Janecki [11, 12]. They have demonstrated a linear relationship between the increase of the relative abrasive wear resistance and hardness of the material subjected to friction (metal, various steels, nitrides, oxides, silicides and borides of transition metals). The linear dependence of the relative abrasive wear resistance on the hardness is also included in the formulas given by the above authors [3, 11, 12]. Figures 2 and 3 illustrate the above relationship for different types of steel and for selected metals.

WC grain size also has a major influence on the bending strength of sintered carbides.

When the diameter of the WC particles is greater than 1.5 µm, an increase in the bending strength and a decrease in hardness of cutting edges made of cemented carbides with increase in the WC grain sizes is observed. When the WC grain has a diameter less than 1.5 µm, a simultaneous increase in bending strength and hardness of cemented carbides with the decrease in the WC grain size is observed. The smaller the grain size the higher the hardness and bending strength of cemented carbides. This observation, among others, led to the overturning of the belief that the increase in hardness entails a reduction in ductility [3, 4].

From the above examples it is clear that the reduction in the grain size is one way to modify the microstructure which allows for a significant improvement in mechanical properties of materials. This type of modification of the polycrystalline material makes it difficult to generate dislocations and limits their mobility; it also limits the spread of existing microcracks in the material.

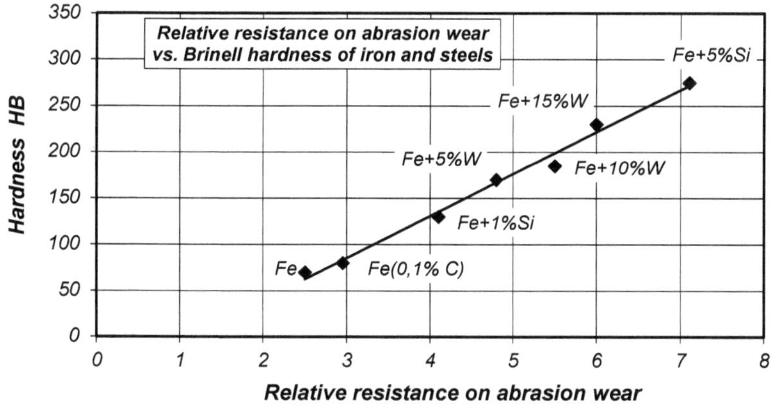

Fig. 2 Influence of hardness on relative abrasive wear resistance of iron and different types of steel (elaborated on the basis of [3, 11, 12])

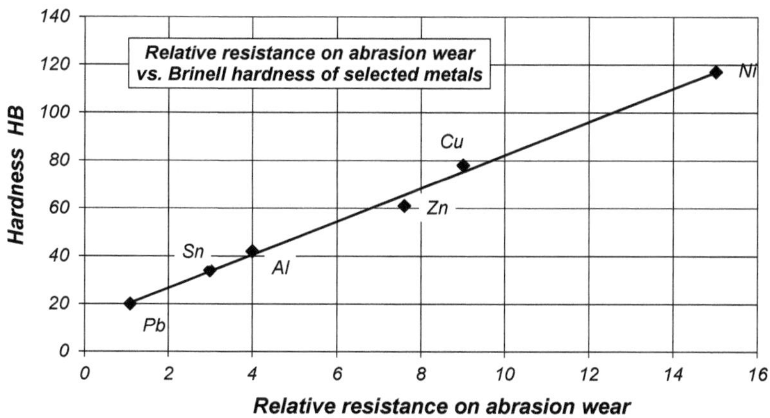

Fig. 3 Influence of hardness on relative abrasive wear resistance of selected metals (elaborated on the basis of [3, 11, 12])

Taking into account the above observations, the present author undertook innovative studies using nanoparticles to manufacture nanocrystalline cemented carbides (having an average grain diameter of not more than 200 nm), to form the cutting edges made of a material harder than the sintered carbides used so far.

The specific aims of the study were to confirm the applicability of the Pulse Plasma Sintering (PPS) method to the consolidation of nanopowders, and to determine the best parameters of the sintering process to obtain the cutting edges made of nanocrystalline cemented carbides with significantly higher abrasion wear resistance than current cutting edges made of cemented carbides.

2 Justification of the Use of the PPS Method

To describe the expected contribution of the proposed solution, it is necessary to present, the development of the methods of consolidation of powders, used so far, describe the characteristic features of the currently applied methods and identify the benefits of the proposed method of sintering carbides using pulse plasma.

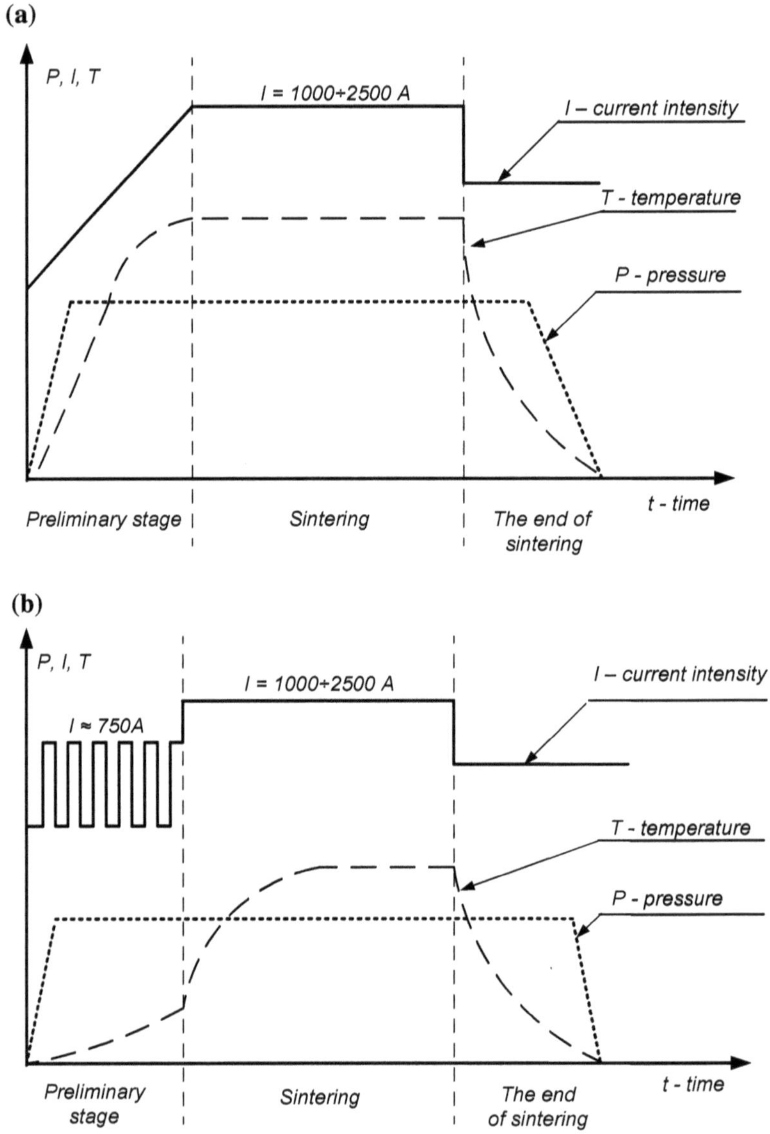

Fig. 4 Resistance heating powders: **a** conventional **b** SPS (elaborated on the basis of [6, 14, 26])

Resistance sintering of cemented carbides using DC electric current was first applied by Taylor in 1933 [13]. In the 1950s Taylor applied resistance heating with simultaneous pressing. This solution is now widely used. In this process the voltage is in the range of 5–40 V and current does not exceed 25 kA [13]. After 1960 new methods were created, in which the sintered powder was heated and activated by pulse current. One of these methods was the SPS (Spark Plasma Sintering) method. The use of pulse current made it possible to significantly reduce the sintering time and lower the sintering temperature compared to earlier methods [14–19]. Figure 4 shows the heating of sintered powders in the conventional process and in the SPS process. In the SPS method the pulse current was applied only in the initial stage of sintering.

In the PPS method the consolidation process phenomena occur with increased intensity, because the intensity of the pulse current is several times greater than in the SPS method. Previous studies, which used different materials, have shown that the PPS method can be used to sinter nanocrystalline composites [3, 20–25]. In the case of the SPS method, it is not possible to obtain a nanocrystalline structure—at most, it is possible to obtain a fine-grained structure [21–23]. An additional advantage of the PPS method is the fact that it is energy efficient resulting, among others, from the 40% decrease in sintering time compared to the time required in the SPS method [6, 15, 19, 21, 26]. Using the PPS method, it is possible to obtain high density sintered materials [6, 21–23, 26].

To summarize the advantages of the PPS method, it should be noted that with regard to the possibility of obtaining a nanocrystalline structure the cutting inserts made of cemented carbides manufactured in this process should have greater hardness. Therefore highly improved anti abrasive properties of obtained materials are likely to be achieved.

3 Conditions of Sintering Nanocrystalline Powders Using the PPS Method

Sintered nanocrystalline carbides were prepared using the following powders:

- a mixture of tungsten carbide (WC), 95% by weight and cobalt (Co), 5% by weight (WC-5 wt% Co),
- a mixture of tungsten carbide (WC), cobalt (Co) and varying amounts of the growth inhibitor in the form of chromium carbide Cr_3C_2 (WC-5 wt% Co + Cr_3C_2).

The percentage of the added inhibitor Cr_3C_2 was 0.3, 0.6 and 0.9%, where the greatest value was determined by the solubility of Cr_3C_2 in cobalt.

The use of growth inhibitors is related to the fact that during the sintering there occurs unfavorable grain growth of tungsten carbide. The higher the sintering temperature, the greater the grain growth. On the other hand, high temperature is desirable to obtain cemented carbides with a compact structure and high density.

Fig. 5 SE images of the WC-5Co nanopowder agglomerates [3]

In the studies, there was used a mixture of the WC-5Co nanopowder with the purity of 99.9% and the Cr_3C_2 nanopowder with the purity of 99.6%. The powders were produced and supplied by Inframat Advanced Materials, USA.

The particle size of the nanopowder particles, according to the manufacturer, was 40 ÷ 80 nm. The SE (secondary electron) image of the WC-5Co nanopowder is shown in Fig. 5. The picture taken with a scanning microscope, shows groups of nanopowder grains in the form of agglomerates.

In order to verify the manufacturer's data, the diameter of nanocrystalline powder grains was selectively measured, using an atomic force microscope (AFM) Quesant Q-Scope 250, equipped with a camera with an optical system for precise positioning of the laser beam on the lever of the probe. This microscope uses a pointed probe to examine the surface of a sample. The length of the probe is several microns and the diameter of the tip is less than 10 nm. The probe is located at the free end of the lever, whose length ranges from 100 to 200 μm. The force which occurs between the probe and the surface of the tested powder causes the twisting or the deflection of the lever. At the end of the lever there is mounted a sensitive detector, which measures the deflection as the probe moves over the surface of the powder. The measurement of the deflection of the lever allows a computer to create a topographical map of the sample surface. The change in the position of the lever is mainly caused by the van der Waals forces in no-contact mode of measurement.

On the basis of measurements made with the AFM microscope (an average of 12 measurements) it was found that the average grain size of the tested powder is 48 nm.

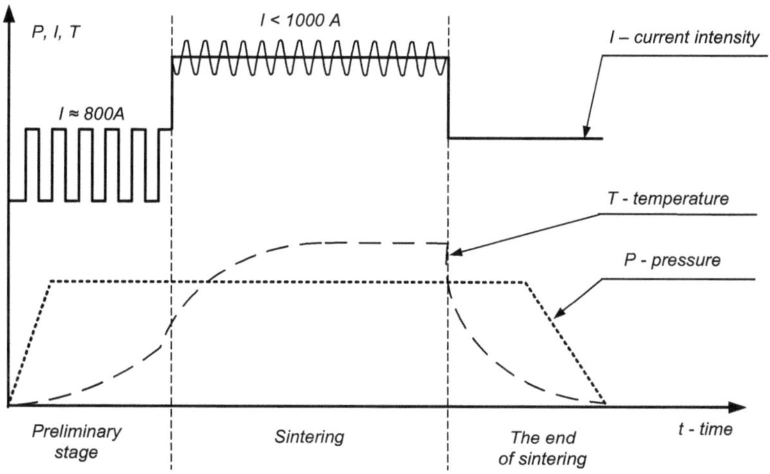

Fig. 6 Sintering of nanocrystalline carbides with the PPS method (elaborated on the basis of [3, 6, 21, 23, 26])

The WC-5Co nanopowder with a suitable amount of additives limiting the growth of the WC grain during sintering was placed in a graphite mold between punches and sealed in the working chamber. Then the pressure of 60 MPa was applied and the chamber was degassed to 4×10^{-2} Pa. After degassing of the chamber the process of sintering was started. As a result of the process there were obtained samples made of nanocrystalline sintered carbides in the form of discs with a diameter of 20 mm and a height of 3.5 mm. The flow chart of the nanopowder sintering process is presented in Fig. 6.

Nanocrystalline powders were sintered at different sintering temperatures: 1380, 1420, 1470, 1530 and 1560 K. The sintering time was 200 s. At the outset of the process, the powder was heated to a predetermined sintering temperature and held for a preset sintering time. After the sintering process, the sample was cooled under pressure of 60 MPa in vacuum of 4×10^{-2} Pa to ambient temperature.

4 Results of the Microstructure and Hardness

Microstructure and chemical composition of the sintered carbides were studied using the VEGA—TS 5135 scanning microscope. Particularly useful for the research were secondary electrons (SE) images and the backscattered electrons (BSE) images. The microscope, working at the Institute of Materials Science, Poznan University of Technology, was produced by an English company TESCAN—Digital Microscopy Imaging and is equipped with an X-ray microanalyser.

The investigations involved three kind of samples made of:

- conventional cemented carbides of the WC-5Co type,
- nanocrystalline cemented carbides of the Nano_WC-5Co type,
- nanocrystalline cemented carbides of the Nano_WC-5Co + Cr_3C_2 type with a different content of growth inhibitor.

Samples made of conventional sintered carbides were produced by the Hot Pressing (HP) method in the Baildonit company. The processes of adding, mixing and sintering powders of nanocrystalline cemented carbides were carried out at the Department of Materials Science and Engineering, Warsaw University of Technology.

4.1 *Study of Conventional WC-5Co Cemented Carbides Microstructure*

Figure 7 shows example BSE images of microsections of the WC-5Co conventional sintered carbides produced by the Hot Pressing (HP) method in the Baildonit company.

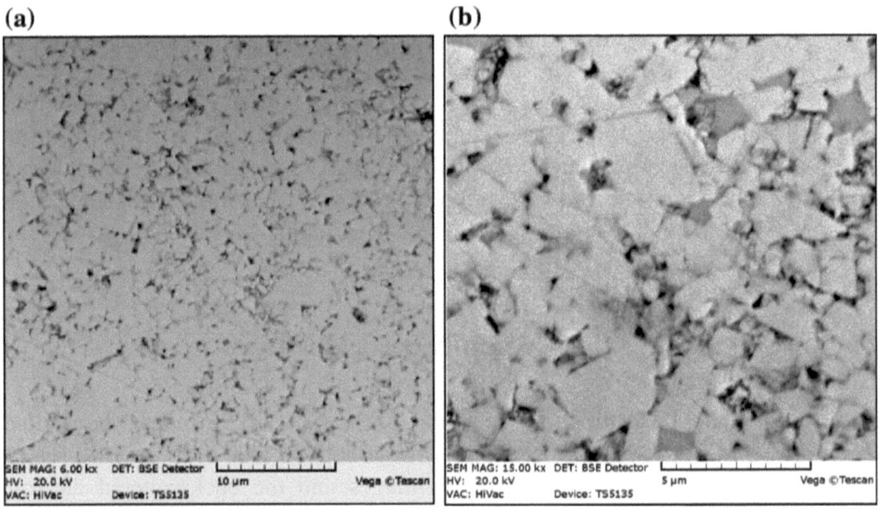

Fig. 7 BSE images of microsections of the WC-5Co conventional cemented carbides at a magnification of: **a** 6000x, **b** 15 000x

4.2 Microstructure Study of Nanocrystalline Cemented Carbides of the Nano_WC-5Co Type

In the study, nanocrystalline cemented carbides were formed at various values of sintering temperature (1380 K, 1420 K, 1470 K, 1530 K, 1560 K) in sintering time 200 s.

Figure 8 presents images of microsections to illustrate the effect of sintering temperature on the structure of nanocrystalline cemented carbides.

The observations made on the basis of the images from the scanning electron microscope indicate that the sinters produced at a temperature of 1530 K have a most compact structure. The sintering of nanocrystalline cemented carbides at higher temperatures, i.e., at 1560 K, leads to substantial grain growth, which can be disadvantageous with respect to the technological and functional characteristics of the cemented carbides.

4.3 Microstructure Study of Nanocrystalline Cemented Carbides of the Nano_WC-5Co + Cr_3C_2 Type—with the Addition of Growth Inhibitor

Figure 9 presents examples of BSE images of microsections of nanocrystalline cemented carbides of the Nano_WC-5Co + Cr_3C_2 type obtained using the most advantageous temperature (1530 K) and sintering time (200 s) with a different content of Cr_3C_2 (0.3, 0.6 and 0.9%).

4.4 Results of Hardness Measurements

Measurements of hardness of samples made of nanocrystalline cemented carbides were performed in the Laboratory of Hard and Superhard Materials of the Institute of Mechanical Technology, Poznan University of Technology [27].

Nanocrystalline cemented carbides, before the testing of hardness and microstructure, were subjected to the procedure of surface preparation. The surfaces underwent rough and finishing grinding and polishing, according to the PN-EN 23878: 1996 standard.

Polishing was carried out using buffing wheels, with the addition of diamond grains in the emulsion.

Next, the samples underwent the purification process, which involved:

- pre-treatment cleaning,
- final cleaning.

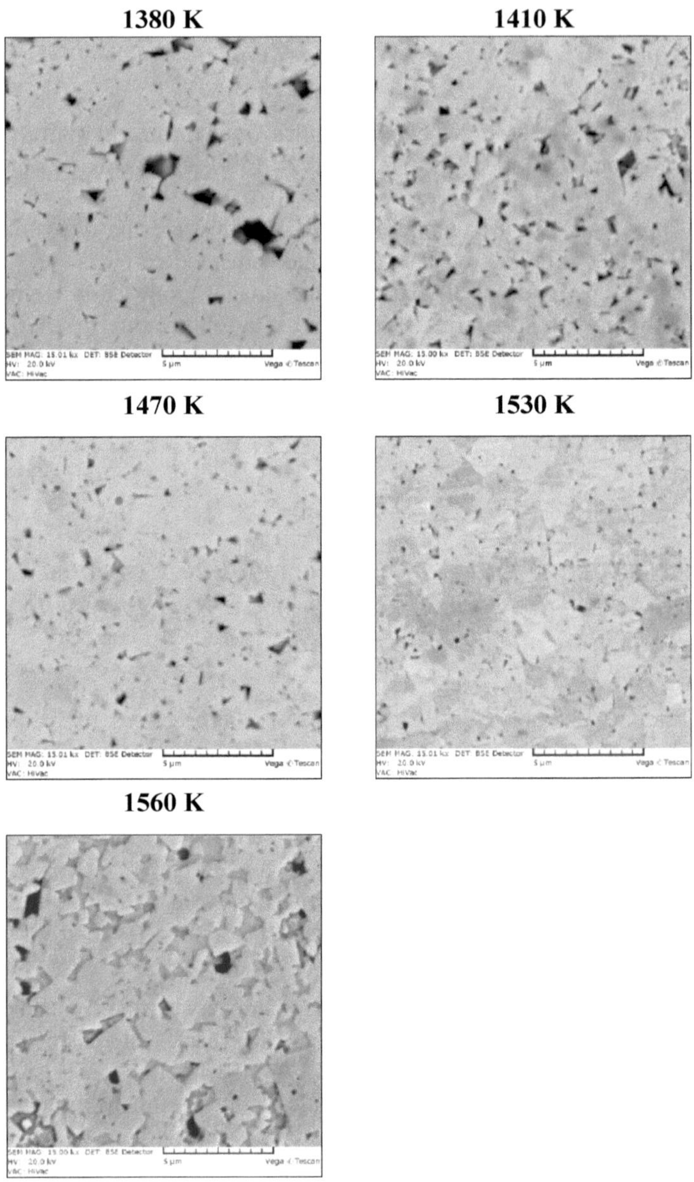

Fig. 8 BSE images of microsections of the Nano_WC-5Co nanocrystalline cemented carbides obtained at various sintering temperatures

Purification consists in removal of all types of contaminants (dirt, liquids, dust, grease, etc.) from the cutting inserts surfaces.

The hardness was measured using a digital Vickers hardness tester FV 700 equipped with an LCD screen, automatic head, ARS-90 control and an automatic

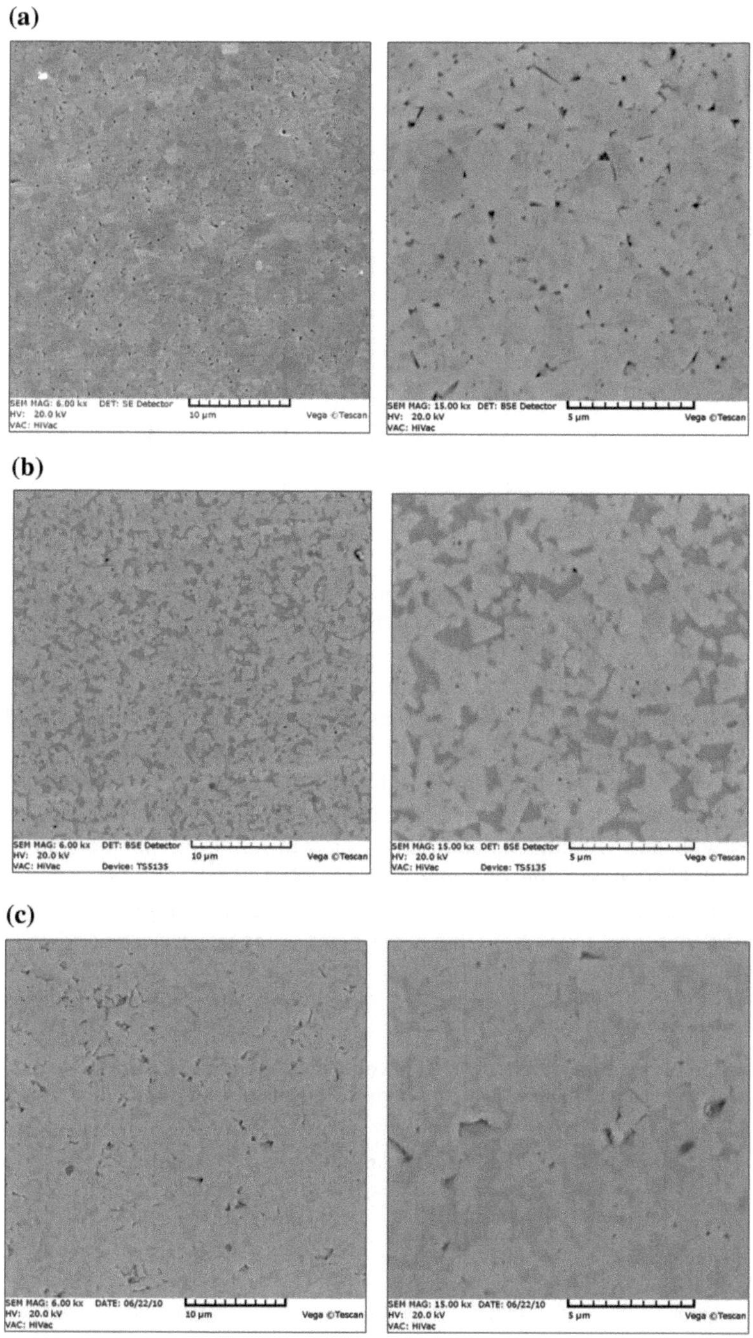

Fig. 9 Comparative set of BSE images of microsections of the Nano_WC-5Co + Cr_3C_2 nanocrystalline cemented carbides obtained for 1530 K and 200 s and for different amount of Cr_3C_2: **a** 0.3%, **b** 0.6%, **c** 0.9%

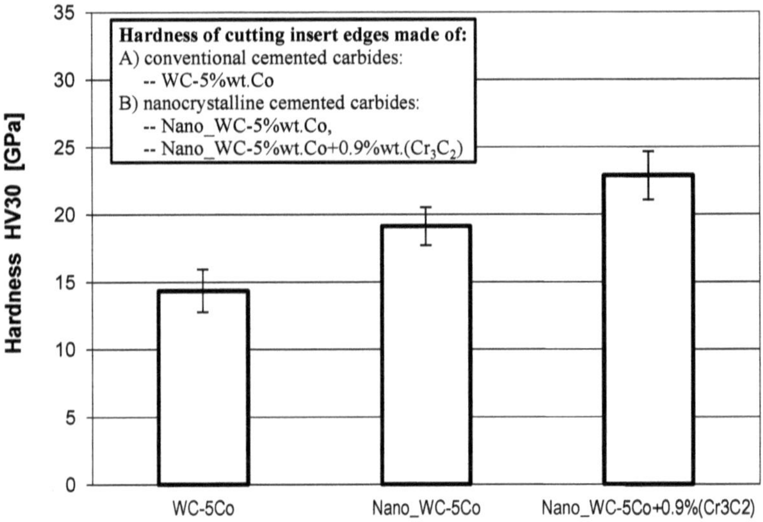

Fig. 10 Results of hardness measurements of the WC-5Co conventional cemented carbides obtained using the HP method and nanocrystalline cemented carbides, with and without of the addition of Cr_3C_2, obtained using the PPS method (1530 K, 200 s)

reading system. In the tests, the recommended load of 294 N (30 kg) was applied for 7 s. The hardness of each sample was measured 9 times. The final hardness of each sample was taken as the arithmetic mean of the nine measurements.

Figure 10 compares the results of hardness measurements for the WC-5Co conventional cemented carbide and the Nano_WC-5Co nanocrystalline cemented carbides with and without the addition of growth inhibitor, for the sintering temperature of 1530 K and the sintering time of 200 s (Table 1).

Nanocrystalline cemented carbides obtained in the sintering temperature 1530 K and at the sintering time 200 s have the most compact structure and the highest hardness.

Table 1 Results of hardness measurements of conventional cemented carbides obtained using the HP method and nanocrystalline cemented carbides obtained using the PPS method (1530 K, 200 s)

Material of cutting edges	Sintering method	Amount of Cr_3C_2 (wt%)	Grain size (nm)	Mean hardness (GPa)
Conventional WC-5Co	HP	–	2000–3000	14.36
Nanocrystalline nano_WC-5Co	PPS	–	300–500	19.12
Nanocrystaline nano_WC-5Co + Cr_3C_2	PPS	0.3	200–300	18.51
		0.6	150–200	20.32
		0.9	60–150	21.87

From the above results it is clear that the reduction in the grain size is one way to modify the microstructure which allows for a significant improvement in mechanical properties of cemented carbides (increase of hardness), because this type of modification of the polycrystalline tool material makes it difficult to generate dislocations and limits their mobility [8].

The study confirmed the usefulness of the Pulse Plasma Sintering (PPS) method to obtain the nanocrystalline cemented carbides.

5 Conclusions

Microscopic observations of microsections revealed a significant impact of temperature on the hardness of the cutting inserts made of nanocrystalline cemented carbides produced with the PPS method. The most advantageous temperature in this regard was the sintering temperature of 1530 K, which allowed us to achieve nanocrystalline cemented carbides with the most compact structure. In higher sintering temperature there was observed significant grain growth and a decrease of hardness.

Based on the results presented in the chapter it should be noted that:

- nanocrystalline cemented carbides with the addition of Cr_3C_2 have a much greater hardness (approximately 2180 HV30) than nanocrystalline cemented carbides without Cr_3C_2 (about 1800 HV30), and than the conventional cemented carbides (about 1470 HV30).
- use of a growth inhibitor in the form of Cr_3C_2, in particular in the amount of 0.9%, allows a considerable reduction of growth of the WC grain in the nanocrystalline cemented carbides (from 150–300 to 60–150 nm).
- the best results were obtained for the hardness of the cutting inserts of nanocrystalline cemented carbides containing 0.9% of Cr_3C_2. This can be explained by the fact that this is a maximum quantity of growth inhibitor not exceeding the limit of solubility in cobalt. For this amount of Cr_3C_2, after sintering, the cobalt binder phase is a solid solution of Cr and W. In the study, the higher content of Cr_3C_2 was not applied, because then during the cooling, the chromium carbide and cobalt carbide of the M_7C_3 type separate from the solution, when the ratio Cr/Co increases with the increase in Cr_3C_2. The presence of these cemented carbides leads to a decrease in fracture toughness and bending strength due to replacing of the plastic cobalt phase by brittle chromium and cobalt carbides type M_7C_3.

Acknowledgements The author gratefully acknowledges the financial support of the Ministry of Science and Higher Education (Polish State Committee for Scientific Research, contract No. NN503 147734) [20]. The author also thanks Prof. A. Michalski and Dr. M. Rosinski from Warsaw University of Technology and Dr. K. Jozwiak and Dr. P. Siwak from Poznan University of Technology for preparation of the samples and for access to the apparatus.

References

1. Burakowski, T., Wierzchoń, T.: Surface Engineering of Metals—Principles, Equipment, Technologies. CRC PRESS LLC, Catalog No. 8225 (1999)
2. Feldshtein, E.: Selected Problems of Machining. ISBN 978-83-7842-104-7. ZGU Publisher, Zielona, Gora (2013)
3. Kupczyk, M.J.: Manufacture and Exploitation of Cutting Tools with Wear-Resistant Coatings. PUT Publisher, Poznan (2009), ISBN 978-83-7143-864-6
4. Wysiecki, M.: Modern tool materials. WNT Publisher, Warsaw (1997)
5. Allen, C., Sheen, M., Williams, J., Pugsley, V.A.: The wear ultrafine WC-Co hard metals. Wear **250**, 604–610 (2001)
6. Kupczyk, M.J., Komolka, J.: High durability of cutting insert edges made of nanocrystalline cemented carbides. Int. J. Refract. Metal Hard Mater. **49**, 225–231 (2015)
7. Milman, Y.V., Luyckx, S., Northorp, I.T.: Influence of temperature, grain size and cobalt content on hardness of WC-Co alloys. Int. J. Refract. Metal Hard Mater. **39–44**, 17 (1999)
8. Olszyna, A.R.: Superhard ceramics. WUT Publisher, Warsaw (2001). ISBN 83-7207-312-0
9. Razavi, M., Zamani, S., Rahimipour, M.R.: Effect of heating on the properties of nano-crystalline titanium carbide during mechanical alloying process. Proc. Inst. Mech. Eng. Part L: J Mater. Design Appl. **228**(3), 147 (2014)
10. Sailer, T., Herr, M., Sockel, H.-G., Schulte, R., Feld, H., Praksh, L.J.: Microstructure and properties of ultrafine gradient hardmetals. Int. J. Refract. Metal Hard Mater. **19**, 553–559 (2001)
11. Hebda, M., Wachal, A.: Tribology. WNT Publisher, Warsaw (1980)
12. Janecki, J., Hebda, M.: Friction, Lubrication and Wear of Machine Pieces. WNT Publisher, Warsaw (1972)
13. Taylor, G.F.: US Patent No. 1896 854 (1933)
14. Groza, J.R., Zavaliangos, A.: Sintering activation by external electrical field. Mater. Sci. Eng. A **287**, 171–177 (2000)
15. Kim, H.C., Shon, I.J., Yoon, J.K., Doh, J.M.: Consolidation of ultra fine WC and WC-Co hard metals by pulsed current activated sintering and its mechanical properties. Int. J. Refract. Metal Hard Mater. **25**, 46–52 (2007)
16. Kim, H.C., Jeong, I.K., Shon, I.J., Ko, I.Y., Doh, J.M.: Fabrication of WC-8 wt.% Co hard materials by two rapid sintering processes. Int. J. Refract. Metal Hard Mater. 25, 336-340 (2007)
17. Kim, H.C., Oh, D.Y., Shon, I.J.: Sintering of nanophase WC-15 vol.% Co hard metals by rapid sintering process. Int. J. Refract. Metal Hard Mater. **22**, 197–203 (2004)
18. Risbud, S.H., Shan, ChH: Fast consolidation of ceramics powders. Mater. Sci. Eng., A **204**, 146–155 (1995)
19. Zhu, L.H., Huang, Q.W., Zhao, H.F.: Preparation of nanocrystalline WC-10Co-0.8VC by spark plasma sintering. Int. J. Refract. Metal Hard Mater. **22**, 1631–1633 (2003)
20. Kupczyk, M.J. (chief of project), Michalski, A., Zurek, J., Jozwiak, K., Rosinski, M., Siwak, P.: Research Project of the Ministry of Science and Higher Education; contract No. NN503 147734; Poznan (2010) (unpublished)
21. Michalski, A.: Pulse plasma sintering of ceramic materials. Ceramika **91**, 378–385 (2005). (in Polish)
22. Michalski, A.: Physical and chemical principles of coating obtained from vapour. WUT Publisher, Warsaw (2000). ISBN 83-7207-171-3, ISBN 83-7207-202-7
23. Siemiaszko, D.: Structure and properties of the WC-Co composites made of tungsten, carbon and kobalt powders sintered by high-electric current pulses. Doctor's thesis, Warsaw University of Technology (2006)
24. Sokolowska, A.: Unconventional Means of Materials Synthesis. PWN Publisher, Warsaw (1991). ISBN 83-01-10848-7

25. Zdunek, K.: Pulse plasma in surface engineering. WUT Publisher, Warsaw (2004). ISBN 83-7207-476-3
26. Kupczyk, M.J.: Cutting edges with high hardness made of nanocrystalline cemented carbides. Int. J. Refract. Metal Hard Mater. **49**, 249–255 (2015)
27. Kupczyk, M.J.: Hardness of nanocrystalline cemented carbides. Elaboration of the Laboratory of Hard and Superhard Materials of the Institute of Mechanical Technology of PUT; Poznan (2013) (unpublished)

Probing Oxygen Vacancies in $BaTiO_3$ Powders and Single Crystals by Micro-Raman Scattering

Hiroaki Fukushima, Hirotaka Oka, Hiroki Moriwake, Hiroshi Uchida, Hiroshi Funakubo and Ken Nishida

1 Introduction

Barium titanate ($BaTiO_3$) is an important ferroelectric perovskite material with excellent ferroelectric properties at room temperature. Due to its interesting electrical properties, $BaTiO_3$ is a prevalent dielectric material for electronic applications such as thermistors, multilayer ceramic capacitors (MLCCs), pressure sensors, high-density dynamic random access memories (DRAMs), infrared detectors, electro-optic switches, and surface acoustic wave devices [1–3]. One of the most important

H. Fukushima (✉) · H. Oka · K. Nishida
Department of Communications Engineering, National Defense Academy, 1-10-20, Hashirimizu, Yokosuka, Kanagawa 239-8686, Japan
e-mail: ed14001@nda.ac.jp

H. Oka
e-mail: challchall0526@gmail.com

K. Nishida
e-mail: nishida@nda.ac.jp

H. Moriwake
Japan Fine Ceramics Center, 2-4-1, Mutsuno, Atsuta-ku, Nagoya 456-8587, Japan
e-mail: moriwake@jfcc.or.jp

H. Uchida
Department of Materials and Life Sciences, Sophia University, 7-1, Kioi-cho, Chiyoda-ku, Tokyo 102-8554, Japan
e-mail: uchidah@sophia.ac.jp

H. Funakubo
Department of Innovative and Engineered Materials, Tokyo Institute of Technology, 4259-J2-43, Nagatsuta-cho, Midori-ku, Yokohama 226-8502, Japan
e-mail: funakubo.h.aa@m.titech.ac.jp

L.F.M. da Silva (ed.), *Materials Design and Applications*,
Advanced Structured Materials 65, DOI 10.1007/978-3-319-50784-2_6

applications of $BaTiO_3$ is MLCCs. Because $BaTiO_3$ is fired in a reductive atmosphere or in a vacuum during MLCCs device processing, oxygen vacancies are generated [4–6]. These oxygen vacancies deteriorate device performance and limit the lifetime [7–10]. As devices become highly integrated and miniaturized, it is crucial to evaluate the small amount of oxygen vacancies using highly sensitive techniques.

Both the transition electron microscopy electron energy loss spectroscopy (TEM–EELS) method [5, 11, 12] and the thermally stimulated current (TSC) method [5, 13–15] have been reported to evaluate oxygen vacancies in $BaTiO_3$. However, TEM–EELS is destructive and sample preparation is time consuming. On the other hand, TSC only provides information about the amount of defect charges and the polarity of defects. Currently, a highly sensitive method with a high spatial resolution does not exist to evaluate the oxygen vacancies. Therefore, a novel evaluation method for oxygen vacancies is required.

Raman spectroscopy is sensitive to the crystal symmetry and local structural changes. Advantages of Raman spectroscopy include a relatively high spatial resolution, noncontact and nondestructive measurements, no sample preparation, and measurements independent of the atmosphere. Although our group has investigated the oxygen vacancies in $PbTiO_3$ thin films by Raman spectroscopy [16–18], it is difficult to determine the amount of oxygen vacancies because oxygen is a light element.

In this study, we evaluate the amount of oxygen vacancies in $BaTiO_3$ using Raman spectroscopy. It has been reported that an additional mode related to the oxygen vacancies occurs on the lower frequency side of the B_1 phonon for $PbTiO_3$ [16, 17]. Since the intensity for the additional mode is weak, we apply a resonant Raman technique.

The crystal structure of $BaTiO_3$ depends on temperature. At room temperature, $BaTiO_3$ has a tetragonal perovskite structure belonging to the $P4mm$ space group (C_{4v}) [19], which yields eight Raman-active modes ($3A_1 + B_1 + 4E$). The B_1 phonon mode consists of only the oxygen ion vibration [20, 21].

2 Experimental Details

The samples were either (100)-oriented $BaTiO_3$ single crystals or $BaTiO_3$ powders. To vary the amount of oxygen vacancies, $BaTiO_3$ singles crystals and powders were reduced from 700 to 900 °C for 120 min in a hydrogen atmosphere. The crystal structure after the thermal treatment was measured by X-ray diffraction (PA Nalytical X'pert MRD for single crystals and Rigaku MiniFlex300 for powders).

Micro-Raman spectroscopy was performed in the spectral range from 70 to 600 cm^{-1} with a micro-Raman system (Photon Design, NFR-800-NSD) and a wavenumber resolution of about 0.2 cm^{-1}. An Ar/Kr mixed gas ion laser (488.0, 514.5, 568.2, 647.1 nm) was used for the excitation. The laser spot size on the sample surface was 1–2 μm through an objective lens. The laser power of the sample surface was approximately 5 mW. Polarized Raman spectroscopy was carried out under a backscattering geometry with a measurement time of 20–300 s at room temperature. The backward scattering light was collected and dispersed by a spectrometer.

The amount of oxygen vacancies was estimated using the following procedure. For a single crystal, the weight of the reduced $BaTiO_3$ was measured using an analytical balance, and the mass decrement was equivalent to the amount of oxygen vacancies. For a powder, the reduced $BaTiO_3$ was thermally treated in air (re-oxidized) using thermal gravimetric and differential thermal analysis (TG-DTA). A BRUKER WS003 system was used to measure the TG-DAT data. Then the $BaTiO_3$ powders were heated from room temperature up to 1200 °C at a rate of 10 °C/min, and the mass increment of $BaTiO_3$ was equivalent to the amount of oxygen vacancies.

3 Results and Discussion

Figure 1 shows optical photographs of the $BaTiO_3$ single crystals (Fig. 1a) and powders (Fig. 1b) before and after reduction at various temperatures in a hydrogen atmosphere. Prior to the thermal treatment, the $BaTiO_3$ single crystals are

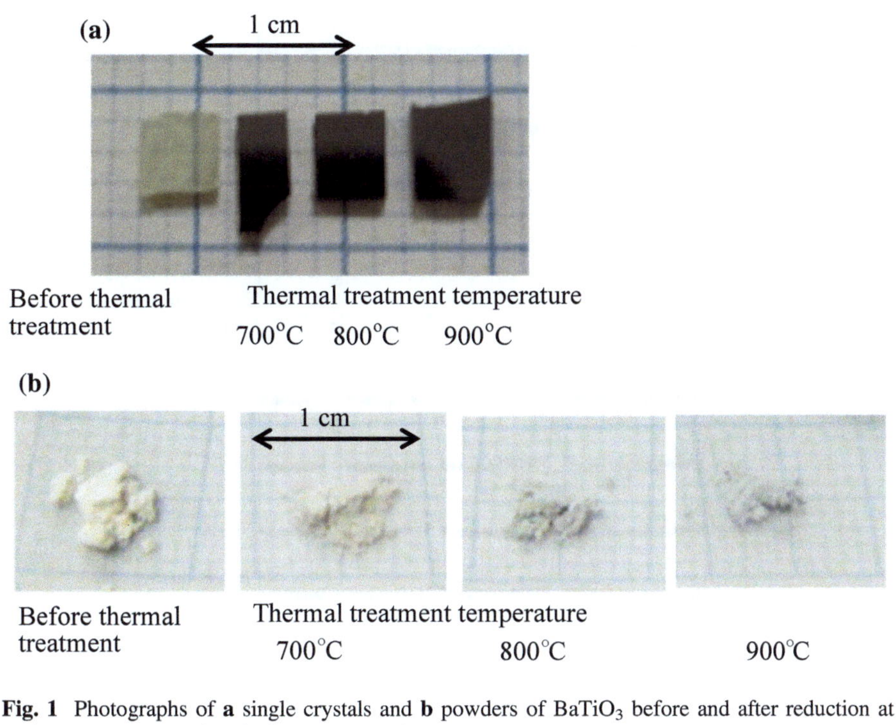

Fig. 1 Photographs of **a** single crystals and **b** powders of $BaTiO_3$ before and after reduction at various temperatures in a hydrogen atmosphere

transparent, but become black and translucent after the treatment. Although the $BaTiO_3$ powder is initially white, it turns slightly dark blue after thermal treatment in a reduced atmosphere. As the thermal treatment temperature increases, the $BaTiO_3$ single crystals and powders become more blackish in color because the absorption of reduced $BaTiO_3$ increases at shorter wavelengths in the visible light range [22].

Figure 2 shows the 2θ-ω XRD pattern of the $BaTiO_3$ single crystals (Fig. 2a) and powders (Fig. 2b) before and after the thermal treatment in a hydrogen atmosphere at 700–900 °C for 120 min. Typical tetragonal $BaTiO_3$ XRD spectra are observed. The thermal treatment does not alter the spectral features and the heterogeneous phase is not confirmed.

Figure 3 shows the Raman spectra of a $BaTiO_3$ single crystal before the thermal treatment with unpolarization [$z(un, un)z$], parallel polarization [$z(xx)z$], and cross

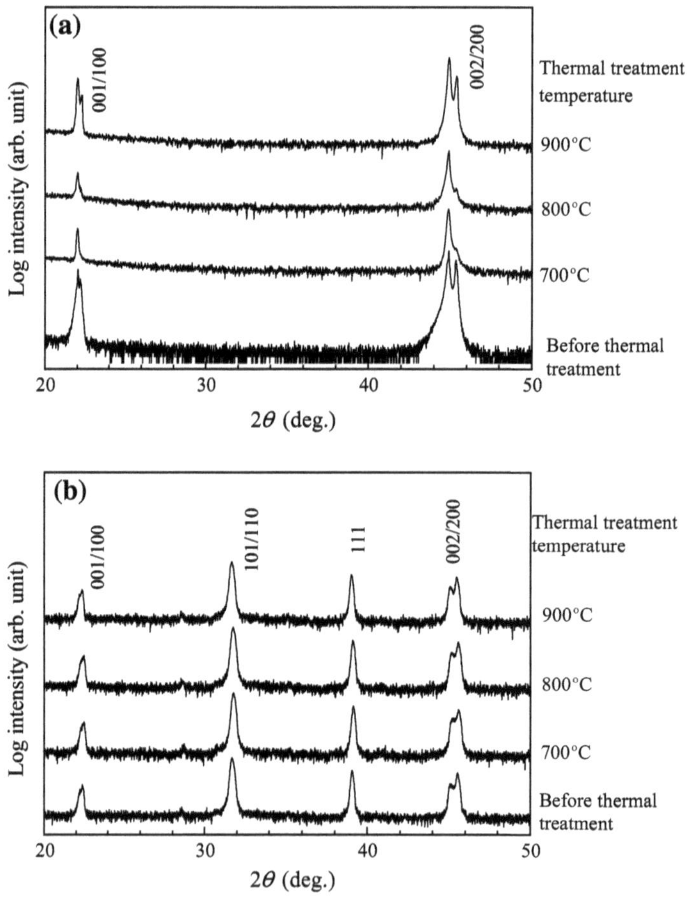

Fig. 2 XRD patterns of **a** single crystals and **b** powders of $BaTiO_3$ before and after thermal treatment at various temperatures in a hydrogen atmosphere

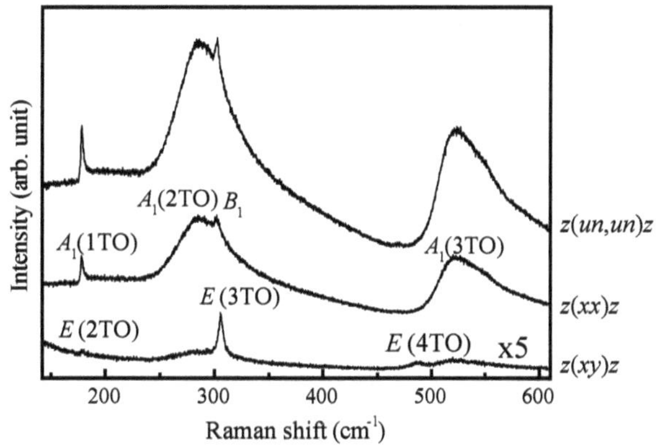

Fig. 3 Raman spectra of a $BaTiO_3$ single crystals corrected with unpolarized, parallel, and cross configurations

polarization [$z(xy)z$] at an excitation wavelength of 514.5 nm. The Raman spectra show the typical tetragonal $BaTiO_3$ phase for a perovskite structure. These spectral features and observed phonon frequencies of $BaTiO_3$ agree well with those previously reported [23, 24].

According to the Raman selection rules, the A_1-symmetry and B_1-symmetry modes can be observed in the parallel polarization configuration [$z(xx)z$], while the E-symmetry mode can be observed in the cross polarization configuration [$z(xy)z$]. The A_1 and B_1 modes appear in the parallel polarized spectra, while the E mode only appears in the cross polarized spectra, demonstrating that the spectra are clearly separated according to the Raman selection rules. Most importantly, the E(3TO) mode is separated from the B_1 mode. Consequently, the peak position of the E(3TO) and B_1 modes can be accurately determined. The Raman peaks of the A_1(1TO), A_1(2TO), B_1, and A_1(3TO) modes are observed around 145, 260, 305, and 510 cm^{-1} in the parallel polarization configuration [$z(xx)z$], respectively. The Raman peaks of the E(2TO), E(3TO), and E(4TO) modes are observed around 180, 307, and 485 cm^{-1} in the cross polarization configuration [$z(xy)z$], respectively.

Figure 4a shows the parallel polarized Raman spectra of $BaTiO_3$ single crystals thermally treated at 800 °C in a hydrogen atmosphere with various wavelengths of incident light from 488.0 to 647.1 nm. The Raman peak corresponding the B_1 phonon mode, which consists of only an oxygen ion vibration [20], occurs around 305 cm^{-1}. An additional mode (302 cm^{-1}) related to the oxygen vacancies occurs on the lower frequency side of the B_1 phonon mode (Fig. 4b). The peaks of the additional phonon mode and the B_1 phonon mode are separated by a Lorentz function. Figure 4c shows the dependence of the Raman intensity of the additional mode on the incident laser wavelength. The Raman peak intensity of the additional mode is maximized when the excitation is 568.2 nm. The excitation energy of 2.22 eV (568.2 nm) is close to the energy level for oxygen vacancies (2.2 eV) [25]. Due to this coincidence, the incident laser wavelength (energy) dependence by Raman scattering in resonance with transitions is involved in the recombination process of the oxygen vacancy band. It should be noted that the resonance effect can

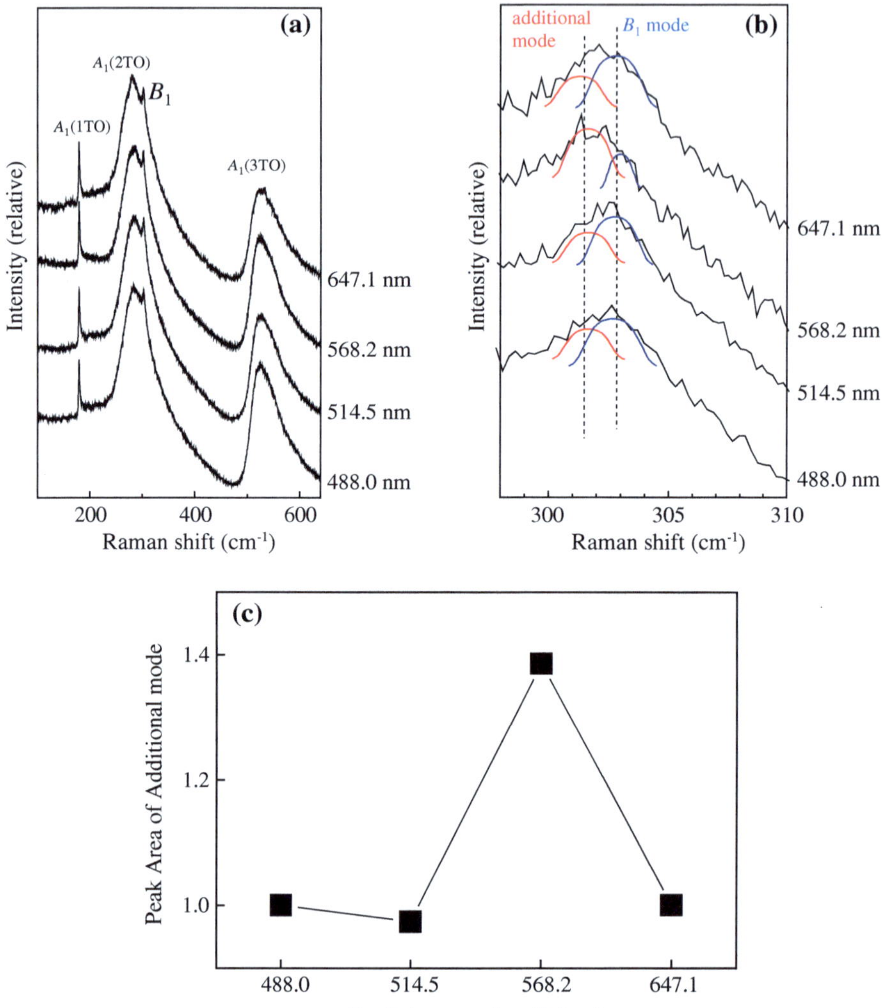

Fig. 4 **a** Parallel polarized Raman spectra of $BaTiO_3$ single crystals thermally treated at 800 °C for 120 min in a hydrogen atmosphere. Wavelength of the excitation laser changed from 488.0 to 647.1 nm. **b** Enlarged view of the spectra around 305 cm^{-1}. **c** Raman intensity of the additional mode as a function of the excitation wavelength

be used to evaluate the oxygen vacancies. First-principles calculations of a $BaTiO_3$ model with oxygen vacancies support the assignment of the additional mode.

Figure 5a shows the parallel polarized [*z*(*xx*)*z*] Raman spectra of the $BaTiO_3$ single crystals before and after the thermal treatment at 700–900 °C for 120 min in a hydrogen atmosphere. Based on the results shown in Fig. 4c, the wavelength of

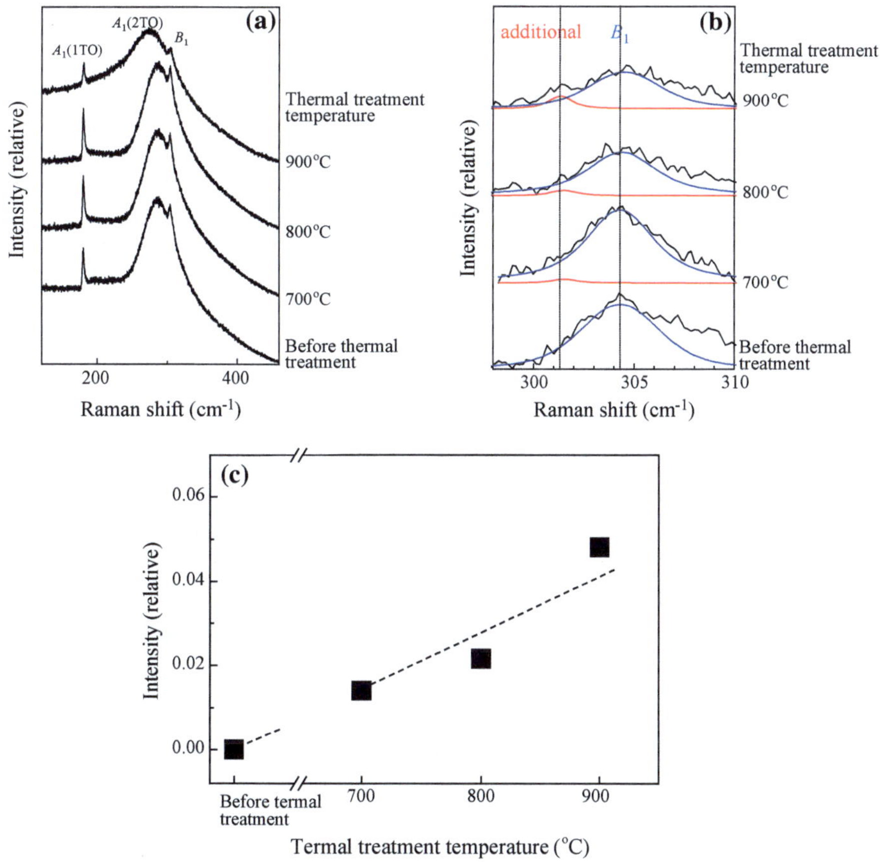

Fig. 5 **a** Parallel polarized Raman spectra of the $BaTiO_3$ single crystals thermally treated at various temperatures for 120 min in a hydrogen atmosphere. **b** Enlarged view of the spectra around 305 cm^{-1}. **c** Raman intensity of the additional mode for single crystals as a function of the thermal treatment temperature

the excitation laser is fixed at 568.2 nm, which is the resonant condition for the oxygen vacancies. The additional phonon mode is predominant. As shown in Fig. 5b, the Raman peaks around 305 cm^{-1} shift to the lower side as the thermal treatment temperature increase because the additional phonon mode occurs at a lower wavelength than the B_1 mode and the B_1 mode decreases. The peaks of the additional and B_1 modes are separated by a Lorentz function. The Raman intensity of the additional mode increases as the thermal treatment temperature increases (Fig. 5c). On the other hand, the Raman intensity of the B_1 mode decreases as the thermal treatment temperature increases because this mode consists of only the oxygen ion vibration.

Figure 6a shows the Raman spectra of the $BaTiO_3$ powders before and after the thermal treatment at 700–900 °C for 120 min in a hydrogen atmosphere. Typical

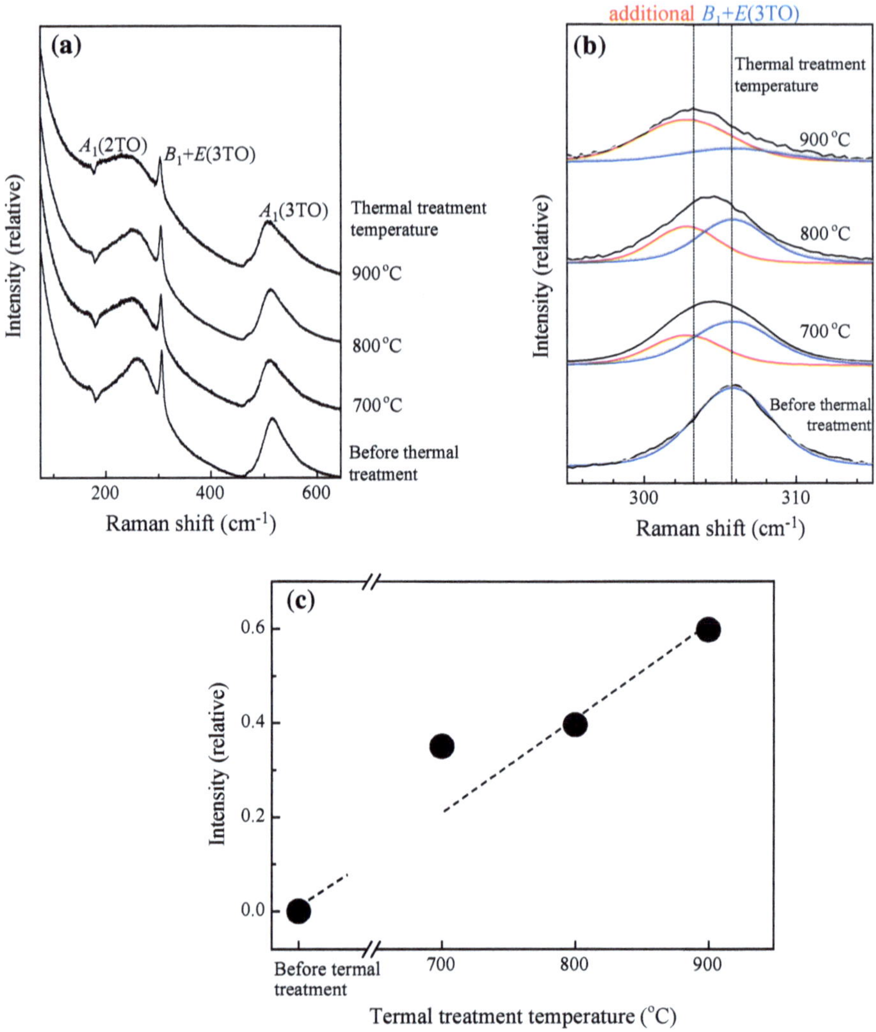

Fig. 6 **a** Raman spectra of the $BaTiO_3$ powders thermally treated at various temperatures for 120 min in a hydrogen atmosphere. **b** Enlarged view of the spectra around 305 cm^{-1}. **c** Raman intensity of the additional mode for powders as a function of the thermal treatment temperature

tetragonal $BaTiO_3$ Raman spectra are observed. The Raman peaks of the A_1(2TO), B_1 + E(3TO), and A_1(3TO) modes are observed around 206, 305, and 510 cm^{-1}, respectively. The thermal treatment does not alter the spectral features and the heterogeneous phase is not confirmed. The Raman peaks around 305 cm^{-1} shift to the lower side as the thermal treatment temperature increases (Fig. 6b). Similar to that for a single crystal, the peaks of the additional and B_1 + E(3TO) modes are separated by a Lorentz function. As the thermal treatment temperature increases, the Raman intensity of the additional mode increases (Fig. 6c). On the other hand, the Raman intensity of the $B_1 + E(3TO)$ mode decreases.

Figure 7 shows the amount of oxygen vacancies in $BaTiO_3$ single crystals (Fig. 7a) and powders (Fig. 7b) as a function of the thermal treatment temperature. For both sample types, the amount of oxygen vacancies increases as the thermal treatment temperature increases. For a given thermal treatment temperature, the single crystal has fewer oxygen vacancies than the powder, implying that the single crystal is harder to reduce because it has less surface area than the powder.

Figure 8 shows the relationship between the Raman intensity of the additional phonon mode and the ratio of oxygen vacancies in $BaTiO_3$ single crystals and powders. As the oxygen vacancies increase, the Raman intensity of the additional phonon mode increases linearly. Because a good correlation is observed between the Raman intensity of the additional phonon mode and the amount of oxygen vacancies in $BaTiO_3$, the Raman intensity of the additional phonon mode can be used to evaluate the oxygen vacancies in $BaTiO_3$.

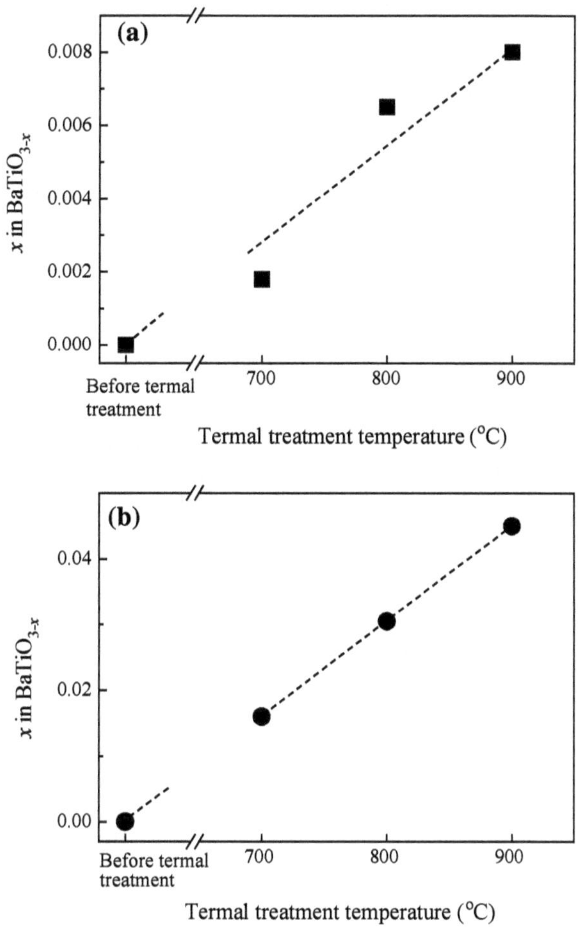

Fig. 7 Amount of oxygen vacancies in **a** single crystals and **b** powders of $BaTiO_3$ as a function of the thermal treatment temperature

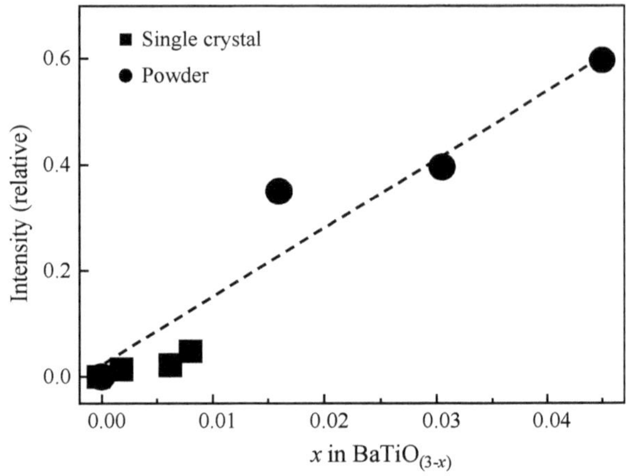

Fig. 8 Raman intensity of the additional phonon mode for single crystals and powders of $BaTiO_3$ as a function of the oxygen vacancies ratio

4 Conclusions

An additional mode, which is related to oxygen vacancies, is dominant in the resonant Raman measurement conditions. The intensity of this additional mode increases as the oxygen vacancy concentration increases in $BaTiO_3$ single crystals and powders. The relationship between the Raman intensity of the additional phonon mode and that of oxygen vacancies is used to estimate the oxygen content in $BaTiO_3$, demonstrating that resonant Raman spectroscopy is a highly sensitive and accurate method to evaluate oxygen vacancies in $BaTiO_3$.

References

1. He, Y.: Heat capacity, thermal conductivity, and thermal expansion of barium titanate-based ceramics. Thermochim. Acta **419**(1), 135–141 (2004)
2. Preda, L., Courselle, L., Despax, B., Bandet, J., Ianculescu, A.: Structural characteristics of RF-sputtered $BaTiO_3$ thin films. Thin Solid Films **389**(1), 43–50 (2001)
3. Wang, B., Zhang, L.D., Zhang, L., Yan, Y., Zhang, S.L.: Raman scattering from $BaTiO_3$ thin film prepared on silicon substrate by r.f. sputtering. Thin Solid Films **354**(1), 262–266 (1999)
4. Kishi, H., Mizuno, Y., Chazono, H.: Base-metal electrode-multilayer ceramic capacitors: past, present and future perspectives. Jpn. J. Appl. Phys. **42**(1), 1–15 (2003)
5. Nishida, K., Kishi, H., Osada, M., Funakubo, H., Nishide, M., Takeuchi, H., Katoda, T., Yamamoto, T.: Raman spectroscopy evaluation of oxygen vacancy migration by electrical field in multilayer ceramic capacitors. Jpn. J. Appl. Phys. **48**(9S1), 09KF11 (2009)
6. Waser, R., Baiatu, T., Härdtl, K.H.: DC electrical degradation of perovskite-type titanates: I Ceramics. J. Am. Ceram. Soc. **73**(6), 1645–1653 (1990)
7. Woodward, D.I., Reaney, I.M., Yang, G.Y., Dickey, E.C., Randall, C.A.: Vacancy ordering in reduced barium titanate. Appl. Phys. Lett. **84**(23), 4650–4652 (2004)
8. Nishida, K., Kishi, H., Funakubo, H., Takeuchi, H., Katoda, T., Yamamoto, T.: Evaluation of residual strain and oxygen vacancy in multilayer ceramic capacitor using laser raman spectroscopy. Jpn. J. Appl. Phys. **46**(10S), 7005–7007 (2007)

9. Chazono, H., Kishi, H.: DC-electrical degradation of the BT-based material for multilayer ceramic capacitor with Ni internal electrode: impedance analysis and microstructure. Jpn. J. Appl. Phys. **40**(9S), 5624–5629 (2001)
10. Morita, K., Mizuno, Y., Chazono, H., Kishi, H.: Effect of Mn addition on dc-electrical degradation of multilayer ceramic capacitor with Ni internal electrode. Jpn. J. Appl. Phys. **41** (11S), 6957–6961 (2002)
11. Yang, G.Y., Dickey, E.C., Randall, C.A., Barber, D.E., Pinceloup, P., Henderson, M.A., Hill, R.A., Beeson, J.J., Skamser, D.J.: Oxygen nonstoichiometry and dielectric evolution of $BaTiO_3$. Part I—improvement of insulation resistance with reoxidation. J. Appl. Phys. **96**(12), 7492–7499 (2004)
12. Yang, G.Y., Lian, G.D., Dickey, E.C., Randall, C.A., Barber, D.E., Pinceloup, P., Henderson, M.A., Hill, R.A., Beeson, J.J., Skamser, D.J.: Oxygen nonstoichiometry and dielectric evolution of $BaTiO_3$. Part II—insulation resistance degradation under applied DC bias. J. Appl. Phys. **96**(12), 7500–7508 (2004)
13. Fukami, T., Kusunoki, M., Tsucniya, H.: TSC study on Fe-doped barium-strontium titanate ceramics. Jpn. J. Appl. Phys. **26**(S2), 46–49 (1987)
14. Nishida, T., Matsuoka, M., Okamura, S., Shiosaki, T.: Thermally stimulated current measurement for degraded Pb(Ti, Zr)O3 thin-film capacitors. Jpn. J. Appl. Phys. **42**(9S), 5947–5951 (2003)
15. Takeoka, S., Morita, K., Mizuno, Y., Kishi, H.: Thermally Stimulated Current (TSC) studies on resistance degradation of Ni-MLCC. Ferroelectrics **356**(1), 78–84 (2007)
16. Nishida, K., Osada, M., Sakai, J., Ito, N., Katoda, T., Ikariyama, R., Funakubo, H., Moriwake, H., Yamamoto, T.: Oxygen vacancies in $PbTiO_3$ thin films probed by resonant Raman spectroscopy. J. Ceram. Soc. Jpn. **121**(8), 598–601 (2013)
17. Nishida, K., Osada, M., Moriwake, H., Sakai, J., Ito, N., Nishide, M., Oka, H., Matsuoka, M., Ikariyama, R., Funakubo, H., Katoda, T., Yamamoto, T.: Investigation of $PbTiO_3$ thin films with reduced and re-oxidized treatment using Raman spectroscopy. J. Ceram. Soc. Jpn. **121**(9), 859–862 (2013)
18. Nishida, K., Osada, M., Takeuchi, H., Yosiaki, I., Sakai, J., Ito, N., Ikariyama, R., Kamo, T., Fujisawa, T., Funakubo, H., Katoda, T., Yamamoto, T.: Raman spectroscopy study of oxygen vacancies in $PbTiO_3$ thin films generated heat-treated in hydrogen atmosphere. Jpn. J. Appl. Phys. **47**(9), 7510–7513 (2008)
19. Zalar, B., Lebar, A., Seliger, J., Blinc, R., Laguta, V.V., Itoh, M.: NMR study of disorder in $BaTiO_3$ and $SrTiO_3$. Phys. Rev. B **71**(6), 064107 (2005)
20. Freire, J.D., Katiyar, R.S.: Lattice dynamics of crystals with tetragonal $BaTiO_3$ structure. Phys. Rev. B **37**(4), 2074–2085 (1988)
21. Ikegami, S.: Raman spectrum of $BaTiO_3$. J. Phys. Soc. Jpn. **19**(1), 46–51 (1964)
22. Yoo, H.I., Chang, M.W., Oh, T.S., Lee, C.E., Becker, K.D.: Electrocoloration and oxygen vacancy mobility of $BaTiO_3$. J. Appl. Phys. **102**(9), 093701 (2007)
23. Das, R.R., Yuzyuk, Y.I., Bhattacharya, P., Gupta, V., Katiyar, R.S.: Folded acoustic phonons and soft mode dynamics in $BaTiO_3/SrTiO_3$ superlattices. Phys. Rev. B **69**(13), 132302 (2004)
24. Burns, G.: Lattice modes in ferroelectric perovskites. II. $Pb_{1-x}Ba_xTiO_3$ including $BaTiO_3$. Phys. Rev. B **10**(5), 1951–1959 (1974)
25. Moreira, M.L., Mambrini, G.P., Volanti, D.P., Leite, E.R., Orlandi, M.O., Pizani, P.S., Mastelaro, V.R., Paiva-Santos, C.O., Longo, E., Varela, J.A.: Hydrothermal Microwave: a new route to obtain photoluminescent crystalline $BaTiO_3$ nanoparticles. Chem. Mater. **20**(16), 5381–5387 (2008)

Part III
Composites

Investigation of Film Formation and Electrical Properties of PS Latex/MWCNT Nanocomposites

Saziye Ugur, Okan Yildiz and Selim Kara

1 Introduction

Polymer latexes have been important to industry for many decades due to their performance and in a number of applications in a wide variety of products in daily life. They have many applications including functioning as a film former or a binder in paints and coatings. For example, in a typical paint formulation, a binder is the main factor that determines the physical, mechanical, and chemical properties of the coating film [1]. Obviously, film formation is a critical aspect of all latex applications, and excellent film formation is required to obtain good film properties. Latex film formation is a complicated, multistage phenomenon and depends strongly on the characteristics of latex particles. Traditionally, the film formation process of polymer latex is considered in terms of three sequential steps: (i) Water evaporation and subsequent packing of polymer particles. (ii) Deformation of the particles and close contact between the particles if their glass transition temperature (Tg) is less than or close to the drying temperature (soft or low T_g latex). Latex with a T_g above the drying temperature (hard or high T_g latex) stays undeformed at this stage. In the annealing of a hard latex system, deformation of particles first leads to void closure [2, 3] and then, after the voids disappear, diffusion across particle–particle boundaries starts, i.e., the mechanical properties of hard latex films evolve during annealing, after all solvent has evaporated and all voids have disappeared.

S. Ugur (✉) · O. Yildiz
Department of Physics, Istanbul Technical University, 34469 Maslak, Istanbul, Turkey
e-mail: saziye@itu.edu.tr

O. Yildiz
e-mail: yildizokann@gmail.com

S. Kara
Department of Physics, Trakya University, 22030 Edirne, Turkey
e-mail: skara@trakya.edu.tr

L.F.M. da Silva (ed.), *Materials Design and Applications*,
Advanced Structured Materials 65, DOI 10.1007/978-3-319-50784-2_7

(iii) Coalescence of the deformed particles to form a homogeneous film [3] where macromolecules belonging to different particles mixed by interdiffusion [4, 5].

Polymeric materials are widely used in industry, e.g., for transparent film and paper coating [6–8]. In some cases, the properties of neat polymers are not quite sufficient for certain high demand applications. Appropriately adding high-quality fillers into a polymer matrix can significantly improve its performance and even endow some new properties to the polymer. Nanocomposite materials are interesting since they have potential for dramatically improved materials performance, and can also provide multifunctional characteristics. The precise tuning of the physical properties is a challenge in these materials. In the past two decades, the extension of the range of fillers for polymer reinforcement to nanometer-sized filler materials has lead to progress in both the design of innovative materials and understanding of their properties [9]. Apart from inorganic fillers, functional carbon-based nanoparticles are of interest. The field of polymer nanocomposites with carbon nanomaterial has highly diversified; mainly carbon nanotubes (CNT) based polymer composites have been widely explored in many aspects in recent times [10]. Researchers have effectively utilized the advantages of CNT for the development of various composite systems from renewable resource-based polymeric materials; CNT reinforcement not only provides improvement in mechanical and thermal properties but also creates additional functional properties such as flame retardancy, conductive properties, and EMI shielding [11].

In order to successfully introduce such materials in large-scale industrial applications, several challenges need to be addressed. Perhaps the most important one is to disperse nanoparticles homogeneously on small scale. Another challenge is that many nanoparticles are quite expensive. For the incorporation of CNTs in polymer matrices, several methods have been proposed, including melt-processing [12] and extrusion [13], mechanical stretching [14], spin coating [15], the use of latex technology [16, 17] or magnetic fields [18], a coagulation method [19] and in situ polymerization [20, 21]. Application of latex technology is a relatively new approach to incorporate CNTs inside any kind of highly viscous polymers, which can be synthesized by emulsion polymerization such as PS, PMMA and polyethylene [22], or which can be brought into artificial latex (e.g., by applying high shear conditions) [23]. The advantages of this technique are obvious: it is easy, versatile, reproducible, and reliable and allows a good incorporation of predominantly individual CNTs into a highly viscous polymer matrix. It does not require the use of toxic and inflammable solvents, which is safe and environmentally friendly. Besides, no difficult synthesis of special polymers has to be carried out. Basically, it also exhibits a great degree of flexibility with respect to the choice of the matrix: a homogeneously dispersed CNT network can be obtained for any kind of polymer, which can be produced by emulsion polymerization or which can be brought into a latex form. For example, amorphous polymer latex such as polystyrene PS, PMMA, or even semicrystalline polymer (e.g., polyethylene) can be used [24]. Furthermore, since the CNT walls are not chemically modified, their properties are preserved. Low percolation thresholds of typically 0.3 wt% SWNTs have been determined for highly viscous composites based on PS or PMMA prepared with the latex-based process [25]. Polymer latex is industrially produced at large scale and this type of industry is mature. Since the process basically consists of a simple mixing of two aqueous components and counts a limited number of steps, it offers the prospect of facile scale-up for the future.

A large number of these CNT-based polymer composites exploit CNTs as conductive filler in insulating hosts for applications ranging from electronics to the automobile and aerospace sectors, such as printable circuits and transparent conductive coatings. Since the filler size is below the wavelength of visible light, the optimized reinforced polymer may remain transparent. The aims are to develop easily processed materials for future applications in which metals and/or semiconductors are currently preferred. Basically, for CNT–polymer composites, a highly conductive filler is added to the polymer matrix in order to provide a three-dimensional conductive network of the filler through the composite. This situation is generally described by the percolation theory [26]. At a critical filler volume or mass fraction, called percolation threshold, the conductivity drastically increases by many orders of magnitude with very little increase in the filler loading. Since they have a high conductivity, as well as a very high aspect ratio (length: diameter ratio), the percolation threshold for CNTs is extremely low, and hence CNTs are particularly convenient to use as filler in polymer matrices. Fillers such as carbon black or synthetic graphite, which have aspect ratios varying from 2 to less than 500 on average, require filler concentrations typically ranging from 18 to 2 wt %, respectively, to get conductive materials, depending on the polymer and the level of conductivity needed [27]. Such filler concentrations make the polymer brittle. On the other hand, while using CNTs as filler, one gets the same conductivity for 1–3 wt% loading due to their high conductivity, combined with a very high aspect ratio (about 1000). Low filler loadings generally do not significantly affect the other physical properties of the host polymer and usually only result in a minimal increase in melt viscosity, which still allows easy processing. Furthermore, from a commercial point of view, it is highly relevant to take this low required loading into account since CNTs are nowadays still very expensive. Carbon nanotubes (CNTs) are the third allotropic carbon form in addition to diamond and graphite and have attracted a great deal of interest since the study published by Iijima [28]. In general, CNTs can be seen as one-dimensional conductors [29, 30]. Singlewall nanotubes (SWNTs) consist of single layers of graphite lattice rolled into perfect cylinders with a diameter usually in the range of 0.7–2 nm, whereas multiwall nanotubes (MWNTs) consist of sets of concentric cylindrical shells, each of which resembles a SWNT, although generally with larger diameters. The electronic properties of perfect MWNTs are rather similar to those of perfect SWNTs. However, due to weak coupling between the concentric cylinders of the MWNTs, only the outer shell contributes to the electron transport and thus to the final electronic properties. However, MWNTs are less expensive than SWNTs and polymer—MWNT composites may be more acceptable than polymer—SWNT composites in industrial application. Furthermore, MWNTs are easier to disperse in the polymer matrix compared to SWNTs.

In recent study, we prepared polymer/MWCNT composites applying latex technology which is easy, versatile, reliable, environmentally friendly, and allows the production of conductive composites based on exfoliated individual CNTs or bundles of a few CNTs in a polymer matrix of choice, with low percolation threshold. We investigated the film formation behavior and electrical conductivity properties of these composites depending on the MWCNT content using the photon transmission technique, electrical conductivity measurements and scanning electron microscopy (SEM), respectively. Polystyrene (PS) latex was used as the polymer matrix because its properties are well-known; it is easy to process, it is soluble in a broad range of solvents, and its clarity allows dispersion of MWCNTs to be optically observed at the micron scale. Films were prepared by mixing PS latex with MWCNT particles in the presence of surfactant PVP in various compositions and annealing them at temperatures above the glass transition temperature (T_g) of PS. After each annealing step, the transmitted light intensity, I_{tr}, was monitored to observe the film formation process. The increase in I_{tr} above minimum film formation temperature (T_0) up to the healing temperature (T_h), and above T_h during annealing was explained by void closure and interdiffusion processes, respectively. From the measurements of the electrical conductivities of the composites, the percolation threshold of conductivity was found to be 1.5 wt% MWCNT. The morphological results obtained are correlated with the conductivity of the composites.

2 Experimental

2.1 Materials

2.1.1 Preparation of Latex Dispersion

Polystyrene (PS) latexes were synthesized by using emulsion polymerization technique [31]. The polymerization was performed batch-wisely using a thermostatted reactor equipped with a condenser, thermocouple, mechanical stirring paddle and nitrogen inlet. Water (50 ml), Styrene monomer (3 g; 99% pure from Janssen) and 0.014 g of fluorescent 1-Pyrenylmethyl methacrylate (PolyFluorVR 394) were first mixed in the polymerization reactor where the temperature was kept constant (at 70 °C). The water soluble radical initiator potassium persulfate (KPS) (1.6% wt/wt over styrene) dissolved in small amount of water (2 ml) was then introduced in order to induce styrene polymerization. Surfactant sodium dodecyl sulfate (SDS) (0.12% wt/vol) was added in the polymerization recipe. The polymerization was conducted under 400 rpm agitation during 12 h under nitrogen atmosphere at 70 °C. The particle size was measured using Malven Instrument NanoZS. The mean diameter of these particles is 382 nm (see Fig. 1a). The weight-average molecular weights (M_w of individual PS chains (Mw) were

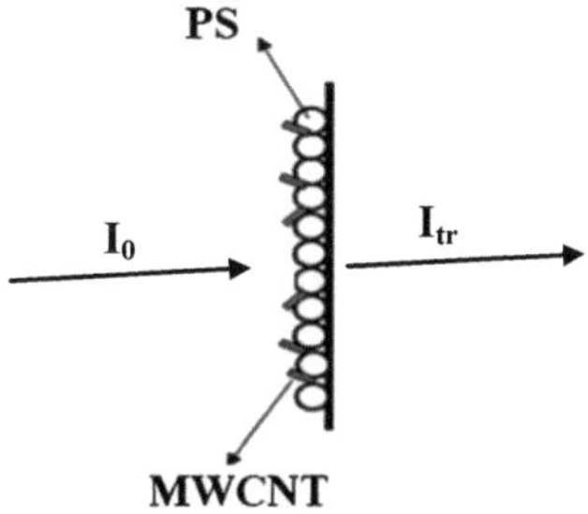

Fig. 1 A schematic illustration of sample position and transmitted light intensity (I_{tr})

measured by gel permeation chromatography (GPC) and found as 99.3×10^3 g/mol. Glass transition temperature (T_g) of the PS latexes were determined with the use of differential scanning calorimeter (DSC) and found to be around 105 °C.

2.1.2 Multiwalled Carbon Nanotubes (MWCNTs)

Commercially available HDplas MWCNTs (Cheap Tubes Inc., VT, USA, length: 1–12 µm, outer diameter (OD): 13–18 nm, the density is approximately 2.1 g/cm^3, and purity is higher than 99 wt%) were used as supplied in black powder form without further purification. The MWCNTs are plasma purified and exfoliated, so they disperse easier in water. However, they still require the use of a surfactant and polyvinyl pyrrolidone (PVP) is a good stabilizing agent for dispersions of carbon nanotubes, enabling preparation of composites from dispersions of MWCNTs in PS solution. A stock solution of MWCNTs was prepared following the manufacturers regulations: nanotubes were dispersed in deionized (DI) water with the aid of PVP in the proportions of 10 parts MWCNTs, 1–2 parts PVP, 2.000 parts DI water stirring for 3 h. Figure 1b shows the SEM image of MWCNTs used in this study (http://www.cheaptubesinc.com).

2.2 *Preparation of PS/MWCNT Composite Films*

A 15 g/L solution of polystyrene (PS) in water was prepared separately. The dispersion of MWCNTs in water was mixed with the solution of PS yielding the required ratio, R, of MWCNTs in PS latex by using the relation $R(wt\%) = \frac{M_{MWCNT}}{M_{MWCNT} + M_{PS}} \times 100$, where M_{PS} and M_{MWCNT} represent the weight of PS and MWCNTs in the mixture, respectively. Seventeen different mixtures were prepared with 0.0, 0.17, 0.35, 0.50, 0.75, 1.0, 1.15, 1.30, 1.40, 1.50, 2.50, 3.50, 5.0, 7.50, 10.0, 15.0 and 20 wt% MWCNTs by using this relation. Each mixture was stirred for 1 h at room temperature. By placing the same number of drops on glass plates with similar surface areas (0.8×2.5 cm^2) and allowing the water to evaporate at 45 °C in the oven, dry films were obtained. After drying, samples were separately annealed above the T_g of PS for 10 min at temperatures ranging from 100 to 250 °C. The temperature was maintained within ±2 °C during annealing. After each annealing step, films were removed from the oven and cooled down to room temperature. The thickness of the films was determined from the weight and the density of samples and ranged from 8 to 12 μm.

2.3 Measurements

Photon Transmission Measurements. Photon transmission experiments were carried out using model Carry 100 bio UV–visible (UVV) spectrometer from Varian. The transmittances of the films were detected at 550 nm at which the composite's spectra is almost flat, i.e., at this wavelength polystyrene and carbon nanotubes have no specific absorption. A glass plate was used as a standard for all UVV experiments. The sample position and the transmitted light intensity, I_{tr}, are presented in Fig. 2.

SEM Measurements. Scanning electron micrographs of the PS/MWCNT films were taken at 10–20 kV in a JEOL 6335F microscope. A thin film of gold (10 nm) was sputtered onto the surface of samples using a Hummer-600 sputtering system to help image the PS/MWCNT films against the glass background.

Electrical Conductivity Measurements. Electrical conductivity was measured by a two-probe method using a Keithley Model 6517a Electrometer with an ultrahigh resistance meter. For the surface resistance measurements, the samples were coated onto thin rectangular glass slabs with typical dimensions of 2.5 × 2.5 cm^2. Electrical resistivities of the composite films were measured by alternating polarity technique with electrometer and a test fixture. The composite films were placed in the text fixture, which have disk shaped electrodes, then their surface resistivities, R_s (Ω), were measured for 15 s under 100 V alternating potential. All the resistivities of the composite films were determined for four

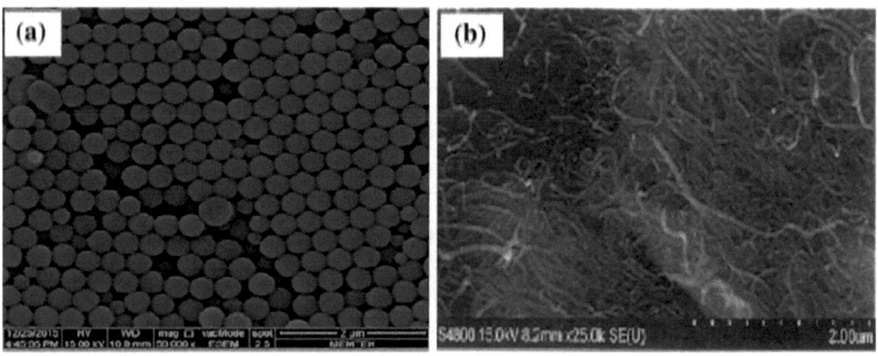

Fig. 2 SEM images of **a** pure PS latex and **b** HDplas (plasma purified) multiwalled carbon nanotubes (MWCNTs) (http://www.cheaptubesinc.com) used in this study

different orientations and measurements were repeated many times to lower the error level. The surface resistivity **values were** converted into surface conductivity **values**.

3 Results and Discussions

3.1 Film Formation Process of PS/MWCNT Composites

Transmitted light intensities, I_{tr}, versus annealing temperatures are plotted in Fig. 3 for the films with 0, 1.0, 1.5, 3.5, 7.5 and 10 wt% MWCNT content. Upon annealing, the transmitted light intensity, I_{tr}, started to increase above a certain onset temperature, called the minimum film-formation temperature T_0, for all film samples except for 10 wt% MCWNT content film. The increase in I_{tr} with annealing

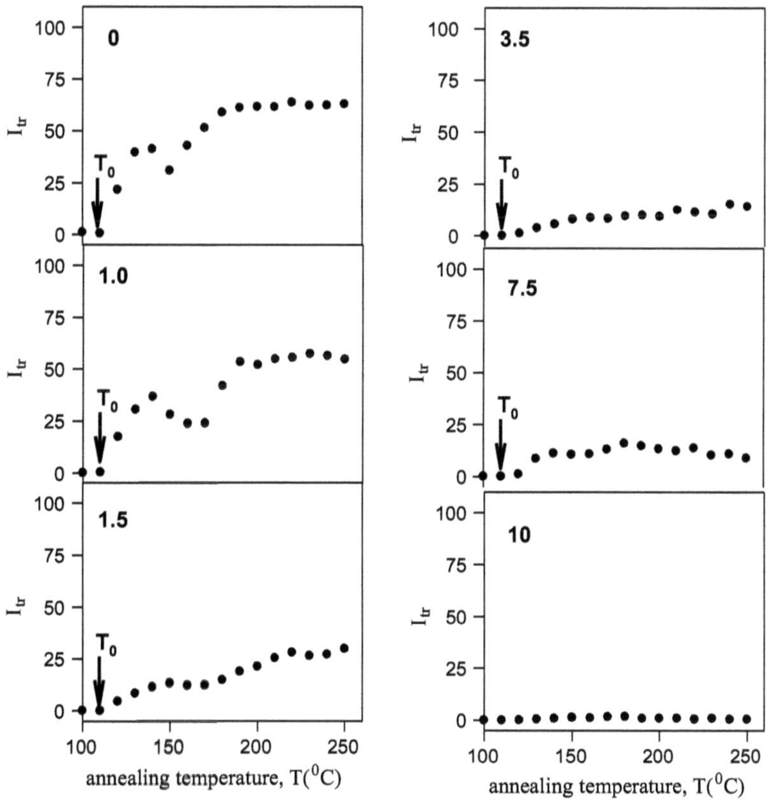

Fig. 3 Plots of transmitted photon intensities, I_{tr}, versus annealing temperatures depending on MWCNT content in the films. The *numbers* on each figure show the MWCNT content in the film. T_0 is the minimum film formation temperature

can be explained by the evaluation of the transparency of the films and surface smoothing upon annealing. Most probably, the increase in I_{tr} above T_0 up to T_h (see Fig. 8) corresponds to the void closure process [32, 33] i.e., the polystyrene starts to flow upon annealing and voids between particles can be filled. On the other hand, the increase in I_{tr} above T_h corresponds to the interdiffusion process. However, for 10 wt% MWCNT content film, I_{tr} almost doesn't change with annealing, which means that no film formation process occurs, and light transmission is completely blocked by dispersion of the MWCNTs in the composite film. On the other hand, I_{tr} decreases with increasing MWCNT content in films at all annealing temperatures, predicting that less transparency occurs at high MWCNT content films. The plots of the maximum values of $(I_{tr})_m$ versus MWCNT content in Fig. 4a also confirms this picture, i.e., as the MWCNT content is increased, $(I_{tr})_m$ almost decreases continuously from 63% to its minimum value (0%) around 10 wt% MWCNT.

Minimum film formation (T_0) and healing (T_h) temperatures are important characteristics related to the film-formation properties of latexes. The T_0 and T_h values measured for composite films reported in Table 1 support these findings. T_0

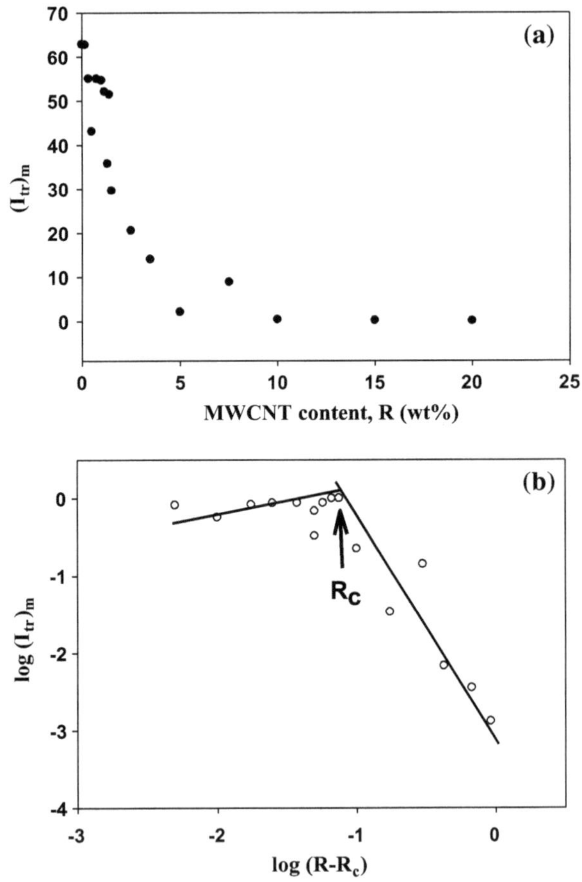

Fig. 4 **a** A plot of the maxima of transmitted light intensities, $(I_{tr})_m$, in Fig. 3 versus MWCNT content, *R* (wt%); **b** log–log plot of $(I_{tr})_m$ versus ($|R - R_c|$)

Table 1 Experimentally determined activation energy (ΔH and ΔE) values and minimum film formation (T_0) and healing (T_h) temperatures

MWCNT (%)	ΔH (kcal.mol^{-1})	ΔE (kcal.mol^{-1})	T_0 (°C)	T_h (°C)
0.0	14.8	4.4	110	130
0.17	20.1	7.9	110	130
0.35	10.4	2.0	100	130
0.50	5.1	3.4	110	140
0.75	14.6	5.9	110	130
1.00	19.6	4.5	110	130
1.15	16.3	5.4	110	130
1.30	11.8	8.8	110	130
1.40	20.3	1.4	110	130
1.50	21.5	8.8	110	130
2.50	11.8	5.2	110	140
3.50	17.4	6.0	110	140
5.00	21.2	10.6	110	130
7.50	26.0	7.3	110	130

is often used to indicate the lowest possible temperature for particle deformation sufficient to decrease interstitial void diameters to sizes well below the wavelength of light [34]. Below this critical temperature, the dry latex is opaque and powdery. However, at and/or above this temperature, a latex cast film becomes **a** continuous and clear film [35]. Therefore, T_0 has been considered in this study as the temperature above which the I_{tr} starts to increase. The healing temperature (T_h) is the minimum temperature at which the latex film becomes continuous and free of voids. The healing point indicates the onset of the particle–particle adhesion [35]. Here, T_h is defined as the intersection of two broken lines from the logarithmic plot of I_{tr} curves (see Fig. 8) versus temperature. From Table 1, it is seen that T_0 and T_h do not change so much indicating that film formation of PS latexes are not affected with MWCNT content in the films in the range of (0–7.5) wt% MWCNT.

3.2 Morphology of the Composites

To gain more information on the organization of MWCNTs in the PS matrix during annealing, the SEM images of the PS/MWCNT nanocomposites containing 0, 1, 2.5 and 10 wt% MWCNT annealed at 100 °C and 250 °C are compared in Figs. 5, 6 and 7, respectively. After annealing at 100 °C (see Fig. 5), no deformation in PS particles is observed and PS particles keep their original spherical shapes for all film samples. However, some depressions are seen in 2.5 and 10 wt% MWCNT content film samples. In order to understand these depressions, the higher magnification images of these films were taken and given in Fig. 6. From these images, it is understood that these depressions are due to the MWCNT bundles subsided in PS powder films. When the latex mixture was coated on the substrate and dried, the

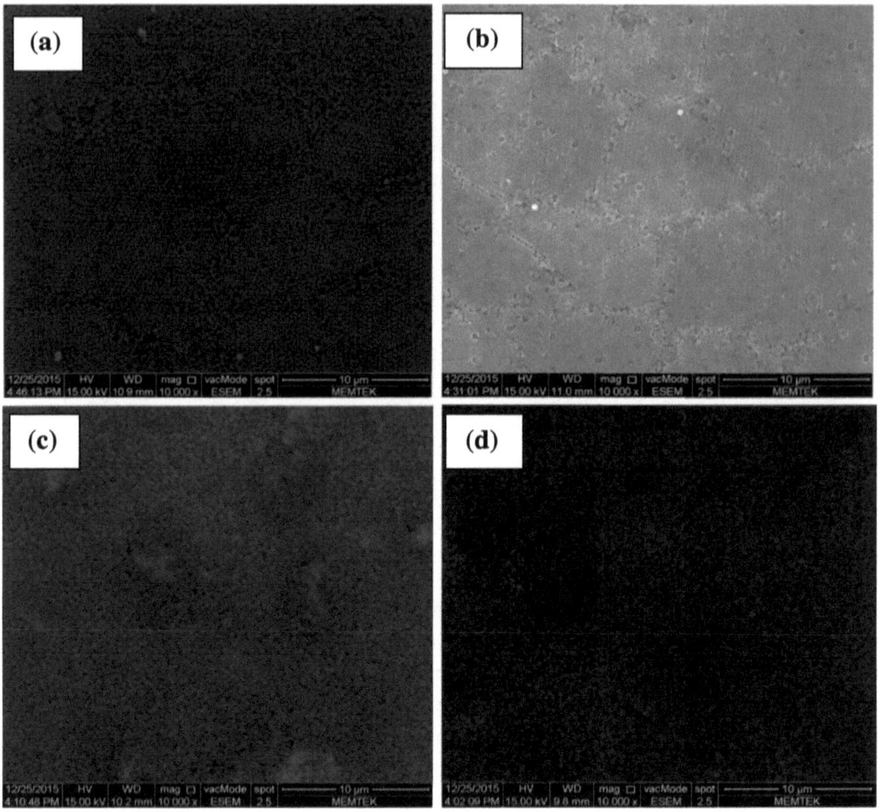

Fig. 5 SEM images of composite films prepared with **a** 0, **b** 1, **c** 2.5 and **d** 10 wt% MWCNT content annealed at 100 °C for 10 min

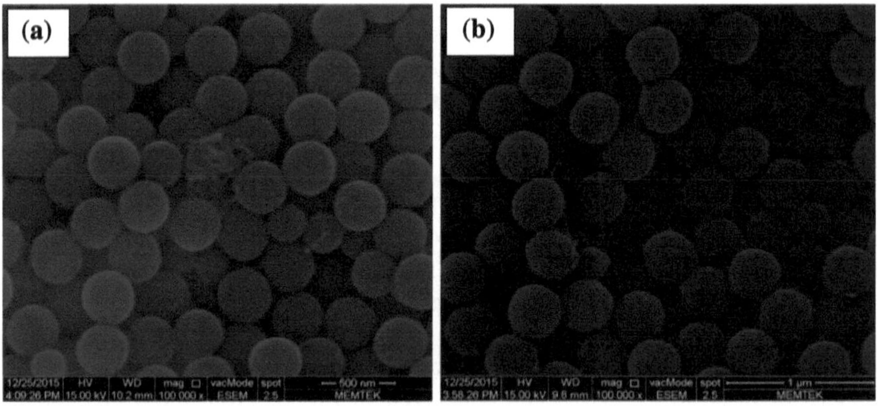

Fig. 6 SEM images of composite films taken in higher magnification (x100000) for **a** 2.5 and **b** 10 wt% MWCNT content films annealed at 100 °C for 10 min. MWCNT bundles are clearly seen in these films

MWCNTs would subside to the bottom layer during the evaporation of water due to their high density. After annealing treatment at 250 °C (Fig. 7), SEM images show that complete particle coalescence has been achieved in all composite films. It is seen that all memory of the original particles are lost and a continuous latex film formed due to the interdiffusion of polymer chains. SEM image of pure PS (0 wt% MWCNT) film surface appear relatively smooth and flat, revealing that a continuous and homogenous film is formed (see Fig. 7a). Figure 7b shows a SEM image of the surface of a PS/MWCNT composite film with 1 wt% MWCNT content. Here the bright areas represent conductive MWCNT bundles in the insulating PS polymer matrix. These small and large MWCNT bundles are seen to disperse in the PS network with or without very few contacts between each other in the film. Most of them just lay close to each others without contacting each others. However, the size and number of contacts between the MWCNT bundles increased with increasing MWCNT content (Fig. 7c and d) forming a continuous conductive network in the polymer matrix. Here, due to annealing at 250 °C, PS particles deformed and

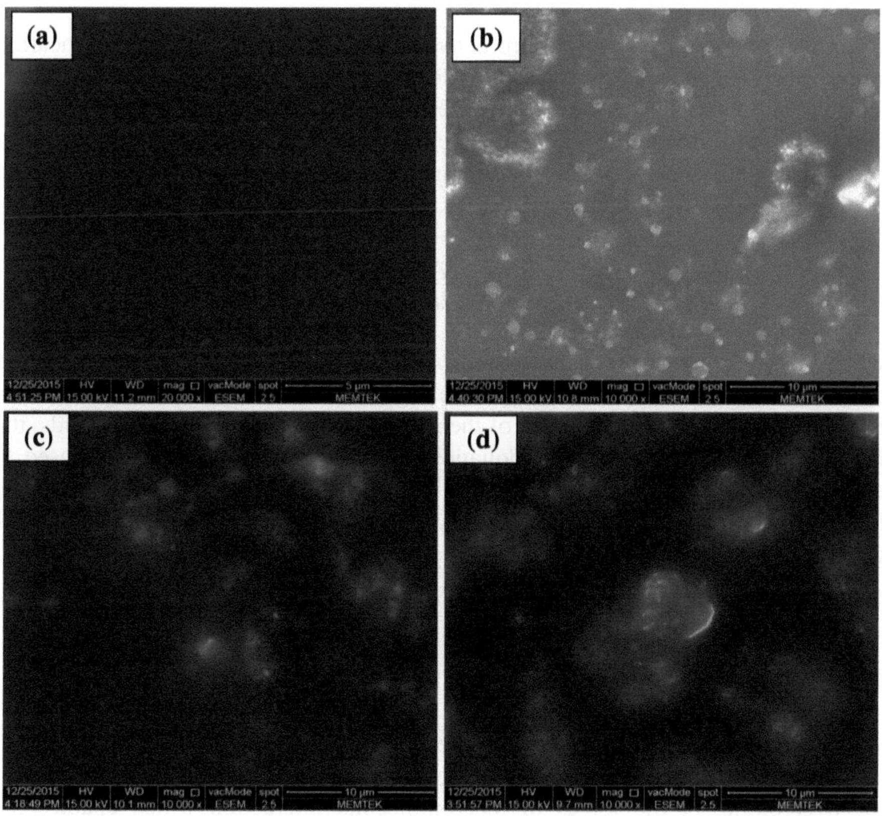

Fig. 7 SEM images of composite films prepared with **a** 0, **b** 1, **c** 2.5 and **d** 10 wt% MWCNT content annealed at 250 °C for 10 min

became a continuous phase in the top layer, and played an essential role in the connection of conductive MWCNTs bundles in the bottom layer. As a result, a conductive network was constructed after the film-forming process. In addition, these bundles also cause strong light scattering resulting in low I_{tr} values which confirms the optical transmission results versus MWCNT given in Fig. 4a.

3.3 *Film Formation Mechanisms*

The increase in I_{tr} intensity below and above the T_h point in the 0–7.5 wt% MWCNT range can be explained by void closure and interdiffusion processes, respectively [36]. To understand these phenomena, the following mechanisms and their formulations are proposed.

3.3.1 Voids Closure

Latex deformation and void closure between particles can be induced by shearing stress which is generated by surface tension of polymer i.e., polymer-air interfacial tension. The void closure kinetics can determine the time for optical transparency and latex film formation [37]. In order to relate the shrinkage of spherical void of radius, r to the viscosity of surrounding medium, η an expression was derived and given by the following relation [37].

$$\frac{dr}{dt} = -\frac{\gamma}{2\eta}\left(\frac{1}{\rho(r)}\right) \tag{1}$$

where γ is surface energy, t is time and $\rho(r)$ is the relative density. It has to be noted that here surface energy causes a decrease in void size and the term $\rho(r)$ varies with the micro structural characteristics of the material, such as the number of voids, the initial particle size and packing. Equation 1 is similar to one, which was used to explain the time dependence of the minimum film formation temperature during latex film formation [38]. If the viscosity is constant in time, integration of Eq. 1 gives the relation as

$$t = -\frac{2\eta}{\gamma}\int_{r_0}^{r}\rho(r)dr \tag{2}$$

where r_0 is the initial void radius at time t = 0.

The dependence of the viscosity of polymer melt on temperature is affected by the overcoming of the forces of macromolecular interaction which enables the segments of polymer chain to jump over from one equilibration position to another. This process happens at temperatures at which free volume becomes large enough and is connected with the overcoming of the potential barrier. Frenkel-Eyring theory produces the following relation for the temperature dependence of viscosity [39–41].

$$\eta = (N_0 h/V)\exp(\Delta G/kT) \tag{3}$$

where N_O is Avagadro's number, h is Planck's constant, V is molar volume and k is Boltzmann constant. It is known that $\Delta G = \Delta H - T\Delta S$, then Eq. 3 can be written as

$$\eta = A\ \exp(\Delta H/kT) \tag{4}$$

where ΔH is the activation energy of viscous flow i.e., the amount of heat which must be given to one mole of material for creating the act of a jump during viscous flow. ΔS is the entropy of activation of viscous flow. Here A represents a constant for the related parameters which do not depend on temperature. Combining Eqs. 2 and 4 the following useful equation is obtained

$$t = -\frac{2A}{\gamma}\exp\left(\frac{\Delta H}{kT}\right)\int_{r_0}^{r}\rho(r)dr \tag{5}$$

In order to quantify the above results, Eq. 5 can be employed by assuming that the interparticle voids are in equal size and number of voids stays constant during film formation (i.e., $\rho(r) \infty r^{-3}$), then integration of Eq. 5 gives the relation

$$t = \frac{2AC}{\gamma}\exp\left(\frac{\Delta H}{kT}\right) - \left(\frac{1}{r^2} - \frac{1}{r_0^2}\right) \tag{6}$$

where, C is a constant related to relative density $\rho(r)$.

In order to quantify the behavior of I_{tr} curves below T_h presented in Fig. 3, void closure model can be applied, where decrease in void size (r) causes an increase in $I_{tr}/(I_{tr})_m$ ratios and vice versa. If the assumption is made that $I_{tr}/(I_{tr})_m$ ratio is inversely proportional to the 6th power of void radius, r then Eq. 6 can be written as

$$t = \frac{2AC}{\gamma}\exp\left(\frac{\Delta H}{kT}\right)\left(\frac{I_{tr}}{(I_{tr})_m}\right)^{1/3} \tag{7}$$

Here, r_0^{-2} is omitted from the relation since it is very small compared to r^{-2} values after a void closure process is started. Equation 7 can be solved for $I_{tr}/(I_{tr})_m$

$$I_{tr}(T) = S(t)\exp\left(-\frac{3\Delta H}{kT}\right) \tag{8}$$

where $S(t) = (\gamma t/2AC)^3$, and $I_{tr} = I_{tr}/(I_{tr})_m$. For a given time the logarithmic form of Eq. 8 can be written as follows

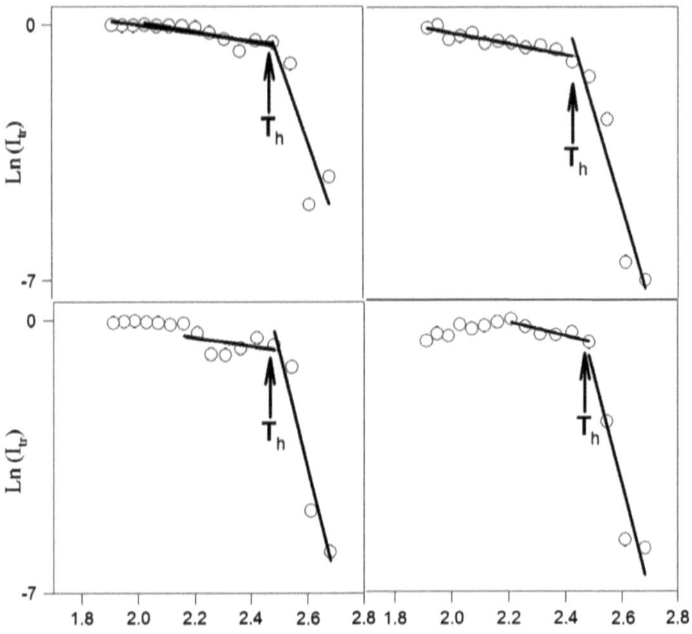

Fig. 8 The ln(I_{tr}) versus T^{-1} plots of the data in Fig. 3. The *slope* of the *straight lines* produces ΔH and ΔE activation energies, which are listed in Table 1

$$LnI_{tr}(T) = LnS(t) - \left(\frac{3\Delta H}{k_B T}\right) \tag{9}$$

Equation (9) can now be used to produce viscous flow activation energies, ΔH. LnI_{tr} versus T^{-1} plots and their fits to Eq. (9) are presented in Fig. 8 (right hand side of the curves) from which ΔH activation energies were obtained. The measured void closure, *ΔH* activation energies are listed in Table 1. It is seen that *ΔH* energy values do not change much by increasing the MWCNT content, i.e., the amount of heat which was required by one mole of polymeric material to accomplish a jump during viscous flow do not change by varying the MWCNT content. From here it is understood that energy need for viscous flow for PS particles is not affected by MWCNT content. Since MWCNTs subside in PS matrix as bundles, they do not affect the void closure process.

3.3.2 Healing and Interdiffusion

The increase in I_{tr} was already explained in the previous section, by the increase in transparency of latex film due to disappearing of deformed particle-particle interfaces. As the annealing temperature is increased above healing temperature, T_h, some part of the polymer chains may cross the junction surface and particle boundaries start to disappear, as a result I_{tr} increases due to the shorter optical and long mean free paths of a photon [42–46]. In order to quantify these results, the Prager-Tirrell (PT) model [47] for the chain crossing density can be employed. These authors used de Gennes's "reptation" model to explain configurational relaxation at the polymer-polymer junction where each polymer chain is considered to be confined to a tube in which executes a random back and forth motion. A homopolymer chain with N freely jointed segments of length L was considered by PT, which moves back and forth by one segment with a frequency, ν. In time the chain displaces down the tube by a number of segments, m. Here, $\nu/2$ is called the "diffusion coefficient" of m in one dimensional motion. PT calculated the probability of the net displacement with m during time t in the range of $n - \Delta$ to $n - (\Delta + d\Delta)$ segments. A Gaussian probability density was obtained for small times and large N. The total "crossing density" $\sigma(t)$ (chains per unit area) at junction surface then was calculated from the contributions $\sigma_1(t)$ due to chains still retaining some portion of their initial tubes, plus a remainder, $\sigma_2(t)$. Here the $\sigma_2(t)$ contribution comes from chains which have relaxed at least once. In terms of reduced time $\tau = 2\nu t/N^2$ the total crossing density can be written as

$$\sigma(\tau)/\sigma(\infty) = 2\pi^{-1/2}\left[\tau^{1/2} + 2\sum_{k=0}^{\infty}(-1)^n\left[\tau^{1/2}\exp(-k^2/\tau) - \pi^{-1/2}erfc(k/\tau^{1/2})\right]\right] \tag{10}$$

For small τ values the summation term of the above equation is very small and can be neglected, which then results in

$$\sigma(\tau)/\sigma(\infty) = 2\pi^{-1/2}\tau^{1/2} \tag{11}$$

This was predicted by de Gennes on the basis of scaling arguments [48]. Here it should be mentioned that the dependence on time, t in Eq. 10 goes as $t^{1/4}$ at early times of healing [48]. In order to compare our results with the crossing density of the PT model, the temperature dependence of $\sigma(\tau)/\sigma(\infty)$ can be modeled by taking into account the following Arrhenius relation for the linear diffusion coefficient

$$\nu = \nu_o \exp(-\Delta E_b/kT) \tag{12}$$

Here ΔE_b is defined as the activation energy for backbone motion depending on the temperature interval. Combining Eqs. 11 and 12 a useful relation is obtained as

$$\sigma(\tau)/\sigma(\infty) = A_o \exp(-\Delta E_b/2kT) \tag{13}$$

where $A_o = (8\nu_o t/\pi N^2)^{1/2}$ is a temperature independent coefficient.

The increase in I_{tr} above T_h is already related to the disappearance of particle-particle interfaces i.e., as annealing temperature is increased, more chains relaxed across the junction surface and as a result the crossing density increases.

Now, it can be assumed that I_{tr} is proportional to the crossing density $\sigma(\tau)$ in Eq. 13 and then the phenomenological equation can be written as

$$I_{tr}(T)/I_{tr}(\infty) = A\ \exp(-\Delta E/2kT) \tag{14}$$

Logarithmic plots of I_{tr} versus T^{-1} are presented in Fig. 8 (left hand side of the curves) for various MWCNT content. The activation energies, ΔE are produced by least squares fitting the data to Eq. 14 and are listed in Table 1, where it is seen that ΔE values not change much with increasing MWCNT indicating that MWCNT content does not affect the backbone motion of the polymer chains across the junction surfaces. The calculated activation energies were found roughly independent of MWCNT content in the film, since the film formation occurred on **the** top surface of composites during annealing.

3.4 Electrical Conductivity of Composites

The electrical conductivity (resistivity) of a composite is strongly dependent on the volume and/or mass fraction of the conductive phase. At low filler concentrations, the conductivity remains very close to the conductivity of the pure, electrically insulating polymer matrix since the fillers are dispersed individually or in small clusters in the matrix. When a critical filler volume or mass fraction (the percolation threshold) is reached, the conductivity drastically increases by many orders of magnitude with very little increase in the filler loading. It coincides with the formation of a conductive, three-dimensional network of the filler in the continuous polymer phase. Finally, the conductivity levels off at a certain value, the maximum conductivity of the composite.

Figure 9 shows the effect of MWCNT concentration on the surface conductivity of PS/MWCNT composite films annealed at 250 °C. The surface conductivity properties of the films were measured at room temperature by using a two probe technique. Figure 9a shows the electrical conductivity (σ) of PS/MWCNT composite films and its best fit as a function of the MWCNTs content (R). As the MWCNT content was in the range of (0–1.5) wt% a substantially low surface conductivity was obtained (10^{-15}–10^{-14} S) because of insufficient connection of the conductive network. A drastic increase in the electrical conductivity is obtained for the composites when the MWCNT content above 1.5 wt%, indicating that the percolation threshold, $R_c = 0.015$, for formation of a conductive MWCNT network in the PS matrix is reached. The conductivity of pure PS is in the order of 10^{-15} S, and for a MWCNT concentration of 2.5 wt% conductivity of 1.1×10^{-10} S can be achieved. With the increasing content of MWCNT, the surface conductivity increased due to the formation of the more uniform and continuous conductive network. When the content of MWCNT was 20 wt%, the conductivity reached 1.23×10^{-5} S. The incorporation of MWCNT increased the conductivity by 10

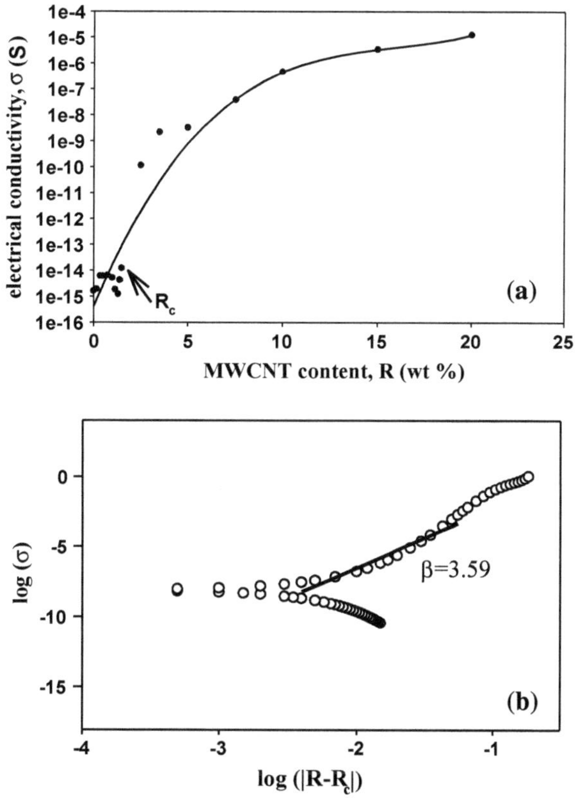

Fig. 9 **a** Conductivity, σ, and its best fitted curve versus MWCNT content, R (wt%); **b** log–log plot of the best fitted curve of σ versus (|R − R_c|)

orders of magnitude, from 1.56 × 10^{-15} S for the pure PS polymer to 1.23 × 10^{-5} S at 20 wt% MWCNTS.

Depending on the matrix, the processing technique and the CNT type used, percolation thresholds ranging from 0.001 wt% to more than 10 wt% have been reported [49].

Percolation theory has been used to interpret the behavior in a mixture of conducting and nonconducting components [50]. The sudden transition in such materials from insulator to conductor is evidence of a percolation threshold. The conductivity, σ of a percolative system is generally described as a function of mass fraction, R by the scaling law in the vicinity of the percolation threshold (R_c):

$$\sigma = \sigma_0 (R - R_c)^{\beta} \quad (15)$$

where σ is composite conductivity (in Siemens), σ_0 is the self-conductivity of MWCNTs film and is equal to 1. R presents weight fraction of MWCNTs, R_c presents percolation threshold of conductivity and β_σ is critical exponent. This equation is valid at concentrations above the percolation threshold, i.e., when

$R > R_c$,. The value of the critical exponents, β, is dependent on the dimensions of the lattice [51].

In order to calculate the percolation threshold, Eq. 15 was transformed into the logarithmic form:

$$\log(s) = \log(\sigma_0) + \beta \log|R - R_c| \quad (16)$$

Then to produce an estimated value for R_c, and the critical exponent β, we fitted the $\log(\sigma) - log|R - R_c|$ data in Fig. 9b for $R > R_c$, to Eq. 16. Here, due to the limitation on getting more data in critical region, we used the continuous function fitted best to the experimental points in Fig. 9a. The goodness of the fit was around $r^2 \approx 0.94$. The value of β_σ has been determined from the slope of the linear relation of $\log(\sigma) - log|R - R_c|$ plot of the best fitted curve in Fig. 9b and found to be as 3.59. In three dimensional lattice systems [52] β_σ values change from 1.3 to 3 and this value agrees well with the universal scaling value of $\beta_\sigma = 3.0$. The difference between the theoretical and experimental values (~%20) might be explained with the errors originating from the fitting. In our previous study [53], we used latex technology to disperse MWCNTs into polymer matrix, and the percolation threshold was found about 4 wt% with maximum conductivity of 2.94×10^{-6} S. It was observed that the bundles are not separated by polymer barrier and the polymer coated the MWCNT network as whole. Dufresne et al. [54] used a similar latex-based process to disperse MWCNTs into a polymer matrix. The conductivity behavior was found to be characteristic for a very anisotropic shape or aggregate formation of the conducting filler, with a percolation threshold around 3 wt% and a maximum conductivity of about 1 S/m.

On the other hand as seen in Fig. 4a that inclusion of MWCNTs into PS lattice, strongly decreases the transmitted light intensity. This finding can be rationalized by first assuming that σ is proportional to the scattered light intensity, $I_{sc} = 1 - (I_{tr})_m$ by obeying the following relation

$$I_{sc}(R) = (R - R_c)^\beta \quad (17)$$

Here it should be realized that inclusions of MWCNT into PS lattice create a two phase heterogeneous structure, which cause light scattering from the composite film surface, while the conductivity increases in a same fashion. Equation 17 describes the percolation model for MWNTs distribution in PS lattice where R_c was produced in Fig. 4b from the intersection of two broken straight lines. When R approaches R_c, the largest cluster of MWCNTs appears by connecting the left and right edges to the bottom edge of the MWCNTs. The scattered light intensity, I_{sc} versus MWCNTs content and its best fit are plotted in Fig. 10a, where it is seen that I_{sc} has increased to large values for all samples above 1.5 wt% MWCNTs content. This behavior of scattering light intensity can be explained by percolating MWCNTs particles in PS lattice. In Fig. 10b, log–log plot of Eq. 17 is fitted to the data in Fig. 10a, where the slope of the straight line produced the critical exponent, $\beta = 0.15$, above $R_c = 0.015$.

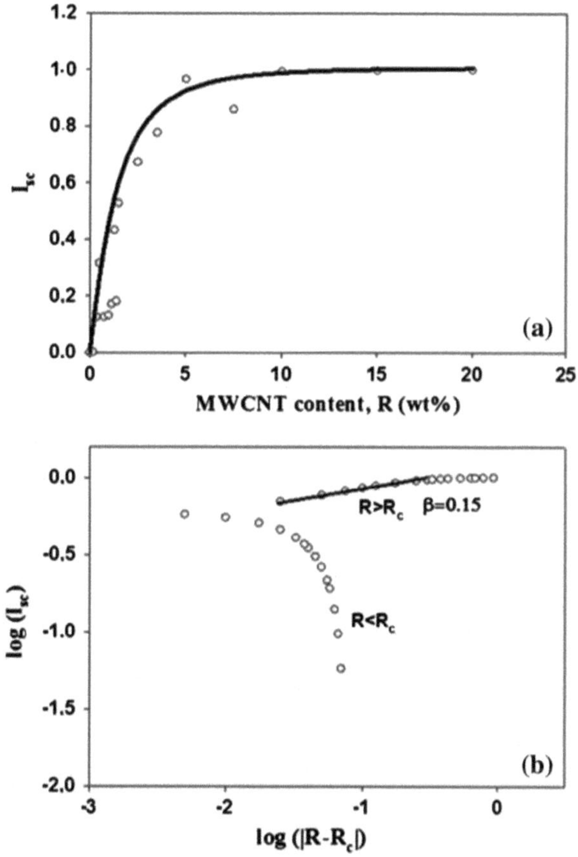

Fig. 10 **a** Scattering light intensity, I_{sc}, and its best fitted curve versus MWNT content, R (wt%); **b** log–log plot of the best fitted curve of I_{sc} versus ($|R - R_c|$)

4 Conclusions

We have reported an investigation of the film formation and electrical conductivity of PS/MWCNT composites. A simple and environmentally friendly method based on latex technology was used for preparation of lightweight and electrically conductive nanocomposites. PS latexes were mixed with MWCNT in the presence of PVP surfactant to form a stable aqueous suspension. Below 7.5 wt% MWCNT content, two distinct film formation stages which are named as void closure and interdiffusion were observed. However, for MWCNT concentrations above 7.5 wt% MWCNT no film formation can be achieved. On the other hand, sample conductivities were observed to depend strongly on the MWCNT contents which are drastically changed with **the** increase of the MWCNT content above the percolation threshold (R_c) of 1.5 wt% MWCNT. The electrical conductivity of these PS/MWCNT nanocomposites increased about 10 orders of magnitude compared with neat PS film, since a continuous percolated MWCNT phase was formed. These composites consist of MWCNT bundles in PS polymer matrix, with very low percolation threshold ($R_c = 1.5$ wt%). SEM images in Fig. 7 showed that the MWCNT bundles are separated by polymer tunneling barriers, the polymer coats individual bundles and prevent intimate contact between bundles at junction sites resulting in diminished electrical properties ($\sim 10^{-5}$ S). Thus, further investigation is underway in our laboratory to improve the electrical properties of the composite films by incorporating exfoliated individual CNTs, or at least relatively thin CNT bundles inside PS matrix with lower R_c.

References

1. Wicks, Z.W., Jones, F.N., Pappas, P.S., Wicks, D.A.: Organic Coatings: Science and Technology, 3rd edn. Wiley, Hoboken, New Jersey (2007)
2. Mazur, S.: Coalescence of polymer particles. In: Rosenweig, N. (ed.) Polymer Powder Processing. Wiley, New York (1995)
3. Mackenzie, J.K., Shutlewort, R.: A Phenomenological theory of sintering. Proc. Phys. Soc. **62**, 833–852 (1949)
4. Yoo, J.N., Sperling, L.H., Glinka, C.J., Klein, A.: Characterization of film formation from polystyrene latex particles via SANS. 2. High molecular weight. Macromolecules **24**(10), 2868–2876 (1991)
5. Pekcan, O.: Interdiffusion during latex film formation. Trends Polym. Sci. **2**, 236 (1994)
6. Paul, D.R., Robeson, L.M.: Polymer nanotechnology: nanocomposites. Polymer **49**, 3187–3204 (2008)
7. Dong, S.J., Yan, J.T., Xu, N., Xu, J., Wang, H.Y.: Molecular dynamics simulation on surface modification of carbon black with polyvinyl alcohol. Surf. Sci. **605**, 868–874 (2011)
8. Rozenberg, B.A., Tenne, R.: Polymer-assisted fabrication of nanoparticles and nanocomposites. Prog. Polym. Sci. **33**, 40–112 (2008)
9. Tjong, S.C.: Structural and mechanical properties of polymer nanocomposites. Mater. Sci. Eng. **R53**, 73–197 (2006)
10. Valentini, L., Puglia, D., Kenny, J.M.: In: Tasis, D. (ed.) Carbon Nanotube-Polymer Composites, vol. 234. RSC Publishing, Cambridge (2013)
11. Ramoa, S.D., Barra, G.M., Oliveira, R.V., De Oliveira, M.G., Cossa, M., Soares, B.G.: Electrical, rheological and electromagnetic interference shielding properties of thermoplastic polyurethane/carbon nanotube composites. Polym. Int. **62**, 1477–1484 (2013)
12. Manchado, M.A.L., Valentini, L., Biagiotti, J., Kenny, J.M.: Thermal and mechanical properties of single-walled carbon nanotubes–polypropylene composites prepared by melt processing. Carbon **43,** 1499–505 (2005)
13. Cooper, C.A., Ravich, D., Lips, D., Mayer, J., Wagner, H.D.: Distribution and alignment of carbon nanotubes and nanofibrils in a polymer matrix. Compos. Sci. Technol. **62**, 1105–1112 (2002)
14. Jin, L., Bower, C., Zhou, O.: Alignment of carbon nanotubes in a polymer matrix by mechanical stretching. Appl. Phys. Lett. **73**, 1197–1199 (1998)
15. Xu, X.J., Thwe, M.M., Shearwood, C., Liao, K.: Mechanical properties and interfacial characteristics of carbon-nanotube-reinforced epoxy thin films. Appl. Phys. Lett. **81**, 2833–2835 (2002)
16. Regev, O., ElKati, P.N.B., Loos, J., Koning, C.E.: Preparation of conductive nanotube-polymer composites using latex technology. Adv. Mater. **16**, 248–251 (2004)
17. Grossiord, N., Loos, J., Koning, C.E.: Strategies for dispersing carbon nanotubes in highly viscous polymers. J. Mater. Chem. **15**, 2349–2352 (2005)

18. Kimura, T., Ago, H., Tobita, M., Ohshima, S., Kyotani, M., Yumura, M.: Polymer composites of carbon nanotubes aligned by a magnetic field. Adv. Mater. **14**, 1380–1383 (2002)
19. Du, F.M., Fischer, J.E., Winey, K.I.: Coagulation method for preparing single-walled carbon nanotube/poly(methyl methacrylate) composites and their modulus, electrical conductivity, and thermal stability. J. Polym. Sci. Part Polym. Phys. **41**, 3333–3338 (2003)
20. Kumar, S., Dang, T.D., Arnold, F.E., Bhattacharyya, A.R., Min, B.G., Zhang, X.: Synthesis, structure, and properties of PBO/SWNT composites. Macromolecules **35**, 9039–9043 (2002)
21. Barraza, H.J., Pompeo, F., O'Rear, E.A., Resasco, D.E.: SWNT-filled thermoplastic and elastomeric composites prepared by miniemulsion polymerization. Nano Lett. **2**, 797–802 (2002)
22. Grossiord, N., Loos, J., Regev, O., Koning, C.E.: Toolbox for dispersing carbon nanotubes into polymers to get conductive nanocomposites. Chem. Mater. **18**, 1089–1099 (2006)
23. Grunlan, J.C., Mehrabi, A.R., Bannon, M.V., Bahr, J.L.: Water-based single-walled-nanotube-filled polymer composite with an exceptionally low percolation threshold. Adv. Mater. **16**, 150–153 (2004)
24. Bauers, F.M., Mecking, S.: aqueous homo- and copolymerization of ethylene by neutral nickel(II) complexes. Macromolecules **34**, 1165–1171 (2001)
25. Grossiord, N., Loos, J., Koning, C.E.: Strategies for dispersing carbon nanotubes in highly viscous polymers. J. Mater. Chem. **15**, 2349–2352 (2005)
26. Kirkpatrick, S.: Percolation and conduction. Rev. Mod. Phys. **45**, 574–588 (1973)
27. Clingerman, M.L., Weber, E.H., King, J.A., Schulz, K.H.: Development of an additive equation for predicting the electrical conductivity of carbon-filled composites. J. Appl. Polym. Sci. **88**, 2280–2299 (2003)
28. Iijima, S.: Helical microtubules of graphitic carbon. Nature **354**, 56–58 (1991)
29. McEuen, P.L., Bockrath, M., Cobden, D.H., Lu, J.G.: One dimensional transport in carbon nanotubes. Microelectron. Eng. **47**, 417–420 (1999)
30. Hamada, N., Sawada, S., Oshiyama, A.: New one-dimensional conductors: Graphitic microtubules. Phys. Rev. Lett. **68**, 1579–1581 (1992)
31. Liu, J.S., Feng, J., Winnik, M.A.: Study of polymer diffusion across the interface in latex films through direct energy transfer experiments. J. Chem. Phys. **101**, 9096–9103 (1994)
32. Du, F.M., Fischer, J.E., Winey, K.I.: Effect of nanotube alignment on percolation conductivity in carbon nanotube/polymer composites. Phys. Rev. B **72**(12), 121404 (2005)
33. Pekcan, O., Arda, E.: Encyclopedia of Surface and Colloid Science, p. 2691. Marcel and Dekker, New York (2002)
34. Keddie, J.L., Meredith, P., Jones, R.A.L., Donald, A.M.: Kinetics of film formation in acrylic latices studied with multiple-angle-of-incidence ellipsometry and environmental SEM. Macromolecules **28**(8), 2673–2682 (1995)
35. Eckersley, S.T., Rudin, A.: Mechanism of film formation from polymer latexes. J. Coat. Technol. **62**(780), 89–99 (1990)
36. Ugur, S., Yargi, O., Pekcan, O.: Film formation from PS latex doped PNIPAM hydrogels at various heating and cooling rates. Compos. Interfaces **15**(4), 411–424 (2008)
37. Keddie, J.L., Meredith, P., Jones, R.A.L., Donald, A.M.: Film Formation in Waterborne Coatings. In: Provder, T., Winnik, M.A., Urban, M.W. (eds.) ACS Symposium Series, vol. 648, pp. 332–348. American Chemical Society (1996)
38. Mc Kenna, G.B.: Comprehensive Polymer Science, vol. 2. In: Booth, C., Price, C. (eds.). Pergamon Pres, Oxford UK (1989)
39. Vogel, H.: Das Temperaturabhaengigkeitsgesetz der Viskositaet von Fluessigkeiten. Phys. Z. **22**, 645–646 (1925)
40. Fulcher, G.S.: Analysis of recent measurements of the viscosity of glasses. J. Am. Ceram. Soc. **8**, 339–355 (1925)
41. Frenkel, J.: Viscous flow of crystalline bodies under the action of surface tension. J. Phys. USSR **9**, 385–391 (1945)
42. Pekcan, O., Arda, E.: Encyclopedia of Surface and Colloid Science, p. 2691. Marcel and Dekker (2002)

43. Pekcan, O., Arda, E., Bulmus, V., Piskin, E.: Photon transmission method for studying film formation from polstyrene latexes with different molecular weights. J. Appl. Polym. Sci. **77**, 866–874 (2000)
44. Arda, E., Ozer, F., Piskin, E., Pekcan, O.: Film formation from nanosized copolymeric latex particles: a photon transmission study. J. Coll. Interface Sci. **233**, 271–279 (2001)
45. Arda, E., Pekcan, O.: Time and temperature dependence of void closure, healing and interdiffusion during latex film formation. Polymer **42**, 7419–7428 (2001)
46. Ugur, S., Yargi, O., Pekcan, O.: Film formation from PS latex doped PNIPAM hydrogels at various heating and cooling rates. Compos. Interface **15**, 411–424 (2008)
47. Prager, S., Tirrell, M.: The healing process at polymer-polymer interfaces. J. Chem. Phys. **75**, 5194–5198 (1981)
48. de Gennes, P.G.: Kinetics of diffusion-controlled processes in dense polymer systems. II. Effects of entanglements. J. Chem. Phys. **76**, 3322–3326 (1982)
49. Li, C., Thostenson, E.T., Chou, T.W.: Sensors and actuators based on carbon nanotubes and their composites: a review. Compos. Sci. Technol. **68**(6), 1227–1249 (2008)
50. Patrick, S.K.: Percolation and conduction. Rev. Mod. Phys. **45**, 574–588 (1973)
51. Broadbent, S.R., Hammersley, J.M.: Percolation processes. I crystals and mazes. Proc. Camb. Philos. Soc. **53**, 629–645 (1957)
52. Sahimi, M.: Applications of Percolation Theory. Taylor and Francis, London (1994)
53. Ugur, S., Yargi, O., Pekcan, O.: Conductivity percolation of carbon nanotubes (CNT) in polystyrene (PS) latex film. Can. J. Chem. **88**, 267–276 (2010)
54. Dufresne, A., Paillet, M., Putaux, J.L., Canet, R., Carmona, F., Delhaes, P., Cui, S.: Processing and characterization of carbon nanotube/poly(styrene-co-butyl acrylate) nanocomposites. J. Mater. Sci. **37**, 3915–3923 (2002)

Cellulose Nanowhiskers Obtained from Waste Recycling of Paper Industry

Alana G. Souza, Daniel B. Rocha and Derval S. Rosa

1 Introduction

The paper industry is one of the major consumers of wood and produces residues during paper processing and the waste material is usually just incinerated for disposal. This primary residue is also called sludge and contains the lignocellulosic components: lignin, hemicellulose, and type II cellulose (with stable monoclinic structure) [1]. There are also pollutants resulting from the chemical process, including heavy metals at levels lower than found in domestic sewage. These are partially dissolved during the pulp cellulose production and are currently incinerated with the final sludge waste.

The cellulose is one of the components found in wood, natural fibers, and other lignocellulosic materials. Annually, nearly 7.5×10^{10} tons of cellulose are processed [2]. The complex structure of the cellulose is being researched for its structural, morphological, and supramolecular properties, as well as the molecular arrangements of the cell wall, fibrillar matrix, micro- and nanofibrils, and other cellulose molecules [3]. Cellulose has gained scientific and industrial interest because of its mechanical properties, such as a high elastic modulus associated with its nanostructure properties, a high superficial area, and a high aspect ratio [3]. Marine biomass, rice husk, potatoes, and sisal fibers are some examples of sources used to prepare cellulose nanocrystals (CNCs) [4–6].

A.G. Souza · D.B. Rocha · D.S. Rosa (✉)
Universidade Federal do ABC, Av. dos Estados 5001, Santo André, SP 09210-580, Brazil
e-mail: dervalrosa@yahoo.com.br; derval.rosa@ufabc.edu.br

A.G. Souza
e-mail: alana_gs@live.com

D.B. Rocha
e-mail: danielb_rocha@outlook.com

L.F.M. da Silva (ed.), *Materials Design and Applications*,
Advanced Structured Materials 65, DOI 10.1007/978-3-319-50784-2_8

To obtain CNCs, it is necessary to isolate the cellulose through a chemical or mechanical process [2]. The high organization induced by hydrogen bounds and van der Waals forces creates a barrier that protects lignocellulosic structures and their components, making it difficult to access the cellulose structures. The most common isolation method is chemical processing using acid hydrolysis, which involves strong acid for the CNC isolation, generally H_2SO_4. For the isolation, pretreatments are necessary to make it easier the attack of H^+ ions to break the linkages, and degrade the amorphous region present in the material. The pretreatment involves the removal of oils and waxes for better exposure of the fiber surface [7]. Step 1 is mercerization, which involves alkaline chemical reagents such as sodium hydroxide. This treatment is based on an industrial system used for cotton fibers [8]. Step 2 is bleaching, which consists of lignin removal to increase the efficiency of the hydrolysis process. Studies have performed delignification using oxidative agents, [9] ionic solutions, [10] and alkaline solutions [11].

In this study, CNCs were isolated from sludge from the paper industry. This was done through sulfuric acid hydrolysis with pretreatment using sodium hydroxide as a mercerizing agent and hydrogen peroxide as a bleaching agent. The goal of this method is to reduce the amount of incinerated waste and increase the value of this lignocellulosic material.

2 Materials and Methods

2.1 Materials and Preparation of Nanocrystals

The raw material is primary paper residue provided by Multiverde Papeis Especiais Ltda. (Mogi das Cruzes, Brazil). CNC isolation was done with two different acid concentrations. First, mercerization was done using 1 g of residue in 20 ml of an aqueous solution of 2% (wt./vol.) sodium hydroxide (NaOH) under constant agitation at 50 °C for 1 h. After this process, the samples were washed with distilled water, filtered, and dried at 100 °C for 3 h. Bleaching was done with a solution of 34% (vol./vol.) hydrogen peroxide (H_2O_2) and 4% NaOH (wt./vol.) under constant agitation at 50 °C for 1 h. The samples were washed and filtered again and then dried at 80 °C for 2 h. The hydrolysis was carried out with 40 and 50% (vol./vol.) sulfuric acid (H_2SO_4) with agitation at 50 °C for 1 h and 30 min. The solution was centrifuged at 6000 rpm for 10 min and neutralized in a dialysis membrane for five days. The samples obtained after each step are identified in Table 1.

Table 1 Designation and description of the obtained samples

Designation	Samples
RP—V	Virgin paper residue
RP—M	Mercerized paper residue
RP—B	Bleached paper residue
RP—H40	Hydrolyzed paper residue (40% H_2SO_4) (method 1a)
RP—H50	Hydrolyzed paper residue (50% H_2SO_4) (method 1b)

2.2 *Characterization*

2.2.1 Compositional Analysis

The percentage composition of lignin and extractives was determined using Klason lignin (soluble and insoluble) and the TAPPI T 222 om-22 (2002c) methodology with modifications.

2.2.2 Fourier Transform Infrared Spectroscopy (FTIR)

FTIR analysis was used to identify the chemical structure of the lignocellulosic components present in the samples. The equipment was a Frontier 94942 (PerkinElmer, USA). Using attenuated total reflectance, the spectra were recorded in the range of 400–4000 cm^{-1}, and 64 scans were carried out.

2.2.3 Thermogravimetric Analysis (TGA)

Thermal stability was tested with progressive temperature increases in a thermogravimetric analyzer (STA 6000, PerkinElmer, USA). The samples were heated from 20 to 600 °C with a heating rate of 20 °C/min under a nitrogen flow.

2.2.4 Scanning Electron Microscopy (SEM)

The surface morphology was analyzed by SEM (JEOL—JCM 600, 20 kV). The samples were covered with gold by scattering equipment (SCANCOAT, PIRANI 501) at a pressure of 0.3 mbar at 1.5 kV for 35 s.

2.2.5 Dynamic Light Scattering (DLS)

The hydrodynamic radius was obtained as an approximation for the crystal size of the nanocellulose dispersion in water. The information was obtained using a dynamic light scattering system (ALV-CGS3) with $\Theta = 90°$. A HeNe polarized laser (22 mW) with a wavelength of 633 nm was used.

3 Results

3.1 *Compositional Analysis*

The virgin paper residue, mercerized, and bleached samples compositions are listed in Table 2. The virgin samples contained 1.8% wt. extractives, 34.8% wt. insoluble

Table 2 Extractives, insoluble lignin, and soluble lignin quantification

Sample	Extractives (%)	Insoluble lignin (%)	Soluble lignin (%)
RP-V	1.8 ± 0.4	34.8 ± 10.5	0.1 ± 0.1
RP-M	0.4 ± 0.2	26.6 ± 13.6	0.1 ± 0.1
RP-B	0.3 ± 0.1	19.4 ± 4.1	0.1 ± 0.1

lignin, and a low percentage of soluble lignin, as expected. These values are similar to the lignocellulose composition of softwood, which is between 20 and 35% [12, 13]. After mercerizing the waste, the extractives and a part of the lignin were removed. This is important for producing pure and high-strength cellulose nanofibers by removing unbound components, such as waxes, fats, resins, gums, and sugars.

After bleaching, the lignin content decreased from 26.6 to 19.4% wt. This result is not ideal for CNC isolation due the high chemical stability of lignin in acids, which creates a barrier for cellulose hydrolysis. Good values reported for lignin content are between 1 and 3% [14, 15]. The removal could be improved by repeating the process to remove the remaining lignin [14]. However, the presence of lignin can produce a good interface effect for application in composites due the possibility of adjusting the surface polarity of the cellulose portion and increasing its compatibility with polymeric matrices [16].

3.2 *Fourier Transform Infrared Spectroscopy (FTIR)*

The chemical structure of the samples was examined after each process using FTIR absorption spectroscopy to evaluate the presence of alkanes, esters, aromatics, ketones, and alcohol groups in the chemical structure. Figure 1a, b show the spectra after the pretreatment processes and hydrolysis, respectively. The typical bands of lignocellulosic compounds are observed in the spectra.

The similarity of the spectra show that the samples do not have huge structural modifications after the different chemical processes. For the RP-V and RP-M samples (Fig. 1a), the peaks from the lignin groups were expected. After the bleaching and hydrolysis (Fig. 1b), the total removal of lignin was expected, but the peaks at 1425 and 1330 cm^{-1} indicate the presence of residual lignin [17]. For the hydrolyzed samples (H40 and H50), Fig. 1b shows an increase in the intensity of the peak at 910 cm^{-1}, which represents cellulose monomer linkages. This result indicates better exposure of the cellulose structures. The spectra of commercial cellulose show a similar structure to the processed sample, indicating that the isolation methodology was efficient for CNC isolation. The assignments of the peaks are shown in Table 3.

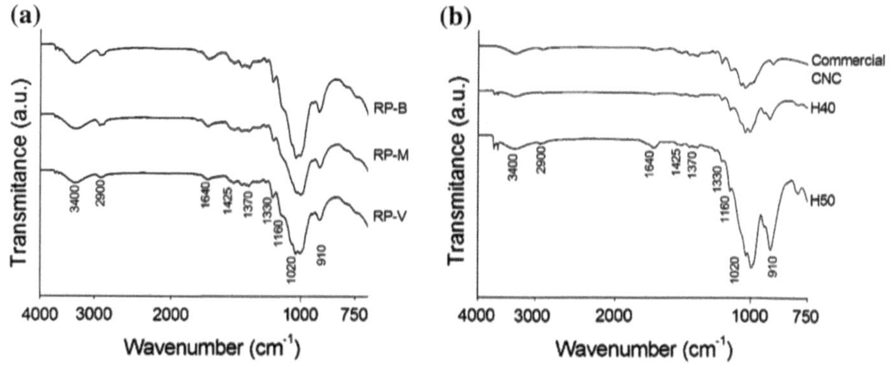

Fig. 1 **a** FTIR spectra of pretreated samples, **b** FTIR spectra of hydrolyzed samples and commercial nanocellulose

Table 3 Assignments of the transmittance spectra peaks

Wavenumber (cm^{-1})	Group	Component	References
3400	O-H stretching	Cellulose, lignin and hemicellulose	[17]
2900	C-H stretching	Cellulose, lignin and hemicellulose	[17]
1640	H-O-H stretching	Moisture	[17, 21]
1425	C-H deformation	Cellulose and lignin	[17, 21, 22]
1370	C-H deformation	Hemicellulose	[17, 21, 22]
1330	Aromatic skeletal vibration	Lignin	[23]
1160	C-O-C stretching	Hemicellulose and lignin	[24]
1020	C-O-C stretching	Lignin	[25]
910	β-glycosidic bond Deformation	Cellulose	[17, 24]

3.3 *Thermogravimetry Analysis (TGA)*

The TGA results are shown in Fig. 2. For all samples, the first weight loss was observed at 100–120 °C, which was mainly due to the evaporation of absorbed and intermolecular H-bonded water in the fiber. Significant weight loss was observed around 200–400 °C, which was mainly due to the thermal degradation of the cellulosic materials [18]. For the pretreated samples, the onset temperatures (T_{onset}) of the thermal decomposition of RP-V, RP-M, and RP-B were 310 °C, 290 °C, and 250 °C, respectively (Fig. 2a). The decrease of T_{onset} was due the increased exposure of less stable components like hemicellulose.

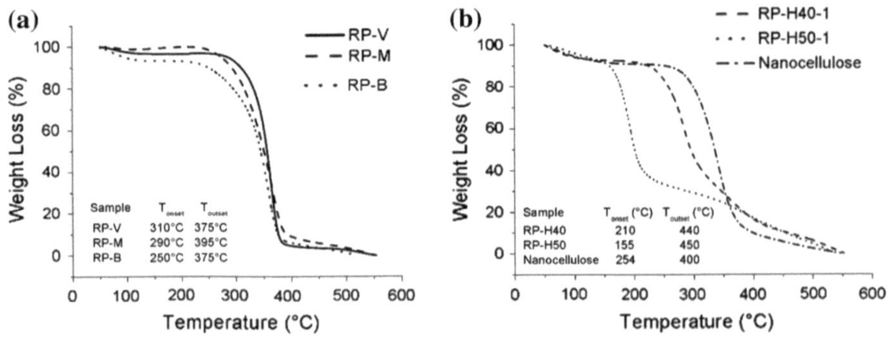

Fig. 2 **a** Pretreated samples, **b** hydrolyzed sample and commercial nanocellulose

Figure 2b shows the results for hydrolyzed samples and commercial nanocellulose, and the CNCs have lower thermal degradation than the pretreated samples. The lower T_{onset} might be due to sulfate groups on the surface of the CNCs, and the increase heat transfer rate due to the increased surface area of the nanoparticles [18]. Sulfuric acid hydrolysis leads to reduced thermal stability of the CNCs compared with the neat fibers, and increasing quantities of sulfate groups resulted in lower T_{onset}. This explains the higher instability and lower onset temperature of thermal decomposition for RP-H50. RP-H40 and RP-H50 started decomposing at lower temperatures than the commercial nanocellulose, which could indicate the presence of lignin that withstood the extraction procedures. However, the differences in T_{onset} could be due to a higher degree of order in the commercial nanocellulose in comparison to that obtained in this work (size distribution, molecular weight, etc.) [6].

3.4 *Scanning Electron Microscopy (SEM)*

The SEM images of RP-V and RP-M show that the mercerized fibers are more exposed with longitudinal cracks and less extractives on the surface (Fig. 3a, b, respectively). The alkali treatment removes the binding materials like hemicelluloses, lignin, pectin, and oils [19].

Figure 4 shows the RP-M and RP-B. Before bleaching, the lignin was partially removed, and the porosity increased. This promoted higher exposure of the fiber interior. This occurs due to attack by the combination of NaOH and H_2O_2, which cleave the binding structure of the lignin and hemicellulose through a saponification process. After this, carboxylic salt and alcohol can be formed, the fiber surface swells, and the interior structure (the cellulose microfibrils) is exposed (Fig. 4b) [13].

Figure 5 illustrates the architecture of the hydrolyzed paper waste samples. As expected, the samples have higher exposure of the crystalline region, which confirms the effectiveness of the acid in the dissolution of the amorphous regions.

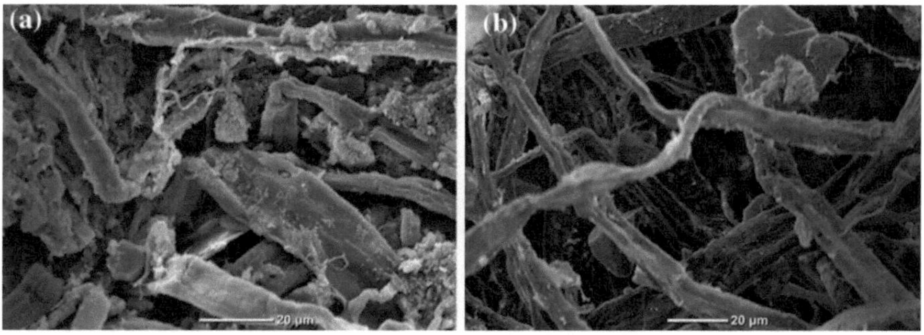

Fig. 3 SEM images of **a** RP-V, **b** RP-M

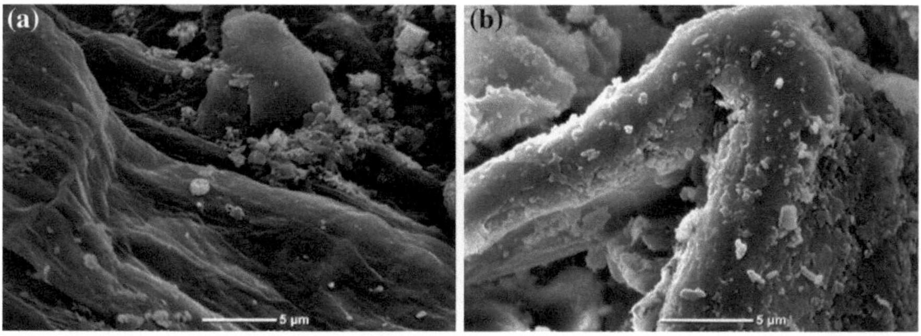

Fig. 4 SEM images of **a** RP-M, **b** RP-B

Fig. 5 SEM images of **a** RP-H40, **b** RP-H50

Furthermore, Fig. 5b shows greater orientation than Fig. 5a, indicating that the higher acid concentration was more effective.

Figure 6 shows the proposed CNCs isolation mechanism. The sludge pretreatment is meant to break the lignocellulosic complex, solubilize the non-cellulosic content, and preserve the nanofibers [13]. The acid hydrolysis process is related to

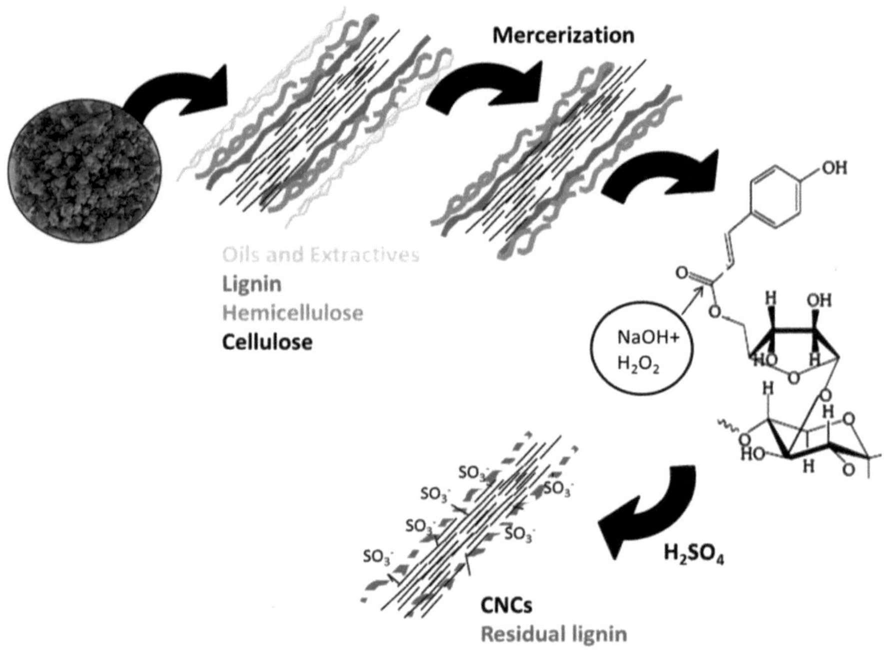

Fig. 6 Proposed CNC isolation mechanism

Fig. 7 Scattering intensity curves as a function of the hydrodynamic radius for hydrolyzed samples

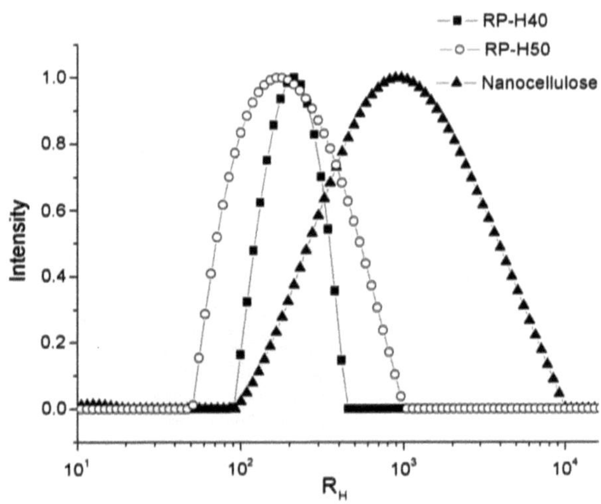

the breakage of $\beta - 1, 4$ glycosidic bonds in cellulose, which occurs because the low-density amorphous regions are more accessible to the acid and susceptible to the hydrolytic action than the crystalline domains [13]. The size of the CNCs is dependent on various factors, such as the cellulose sources, types of acid, and concentration of acid, as shown in this work (Fig. 7).

Table 4 Average, minimum, and maximum values of the hydrodynamic radius of the particles

Sample	R_H (nm)	$R_{H\ Min}$ (nm)	$R_{H\ Max}$ (nm)
RP-H40	195	80	350
RP-H50	165	48	862
Nanocellulose	946	115	7020

3.5 Dynamic Light Scattering (DLS)

DLS was used to determine the size of the particles in suspension. The advantage of this methodology is that it allows the particle measure down to 1 nm diameter [20]. In the case of non-spherical particles, the value calculated from the translational diffusion coefficient is called the hydrodynamic radius (R_H) [19]. The size distribution of the nanocrystals, presented in Fig. 7, is remarkably monodisperse, showing a peak in intensity between 80 and 150 nm for RP-H40 and between 50 and 800 nm for RP-H50, with average sizes of 195 and 165 nm, respectively. This occurs because RP-H50 has more residual sulfate groups on its surface, resulting in improved agglomeration of the nanoparticles, which explains the higher distribution of particle sizes. The acid concentration influences the particle size, which was confirmed by RP-H50 ($R_{H\ min}$ and average). Furthermore, the samples obtained are smaller than commercial nanocellulose, which indicates the successful isolation of CNCs by the proposed method. The diameters are shown in Table 4.

4 Conclusion

In this study, cellulose was successfully extracted using chemical treatment. The pretreatments decrease the lignin and extractives, present at the structure, which was confirmed by the composition analysis and FTIR results. TGA revealed a reduction in thermal stability, which was probably caused by sulfate groups on the CNC surface. In SEM, different morphologies were observed at each stage. The CNC's average sizes of the CNCs varied with the acid concentration (195 and 165 nm for 40 and 50% H_2SO_4, respectively). This confirms that the higher acid concentration produces smaller particle size. These fibers will be used in future works for the production of nanocomposites with enhanced properties.

Acknowledgements The authors thank the UFABC and CNPq. This work was supported by CNPq (grant numbers 306401/2013-4, 447180/2014-2, 163593/2015-9).

References

1. O'Sullivan, A.C.: The structure slowly unravels. Cellulose **4**, 173–207 (1997)
2. Abdul Khalil, H.P.S., Davoudpour, Y., Islam, M.D.N., Mustapha, A., Sudesh, K., Dungani, R., Jawaid, M.: Production and modification of nanofibrillated cellulose using various mechanical processes: a review. Carbohydr. Polym. **99**, 649–665 (2014)

3. Osong, S.H., Norgren, S., Engstrang, P.: Processing of wood-based microfibrillated cellulose and nanofibrillated cellulose, and applications relating to papermaking: a review. Cellulose **23**, 93–123 (2016)
4. Bettaieb, F., Khiari, R., Hassan, M.L., Belgacem, M.N., Bras, J., Dufresne, A., Mhenni, M.F.: Preparation and characterization of new cellulose nanocrystals from marine biomass Posidonia oceanica. Ind. Crops Prod. **72**, 175–182 (2015)
5. Abe, K., Yano, H.: Comparison of the characteristics of cellulose microfibril aggregates of wood, rice straw and potato tuber. Cellulose **16**, 1017–1023 (2009)
6. Morán, J.I., Alvarez, V.A., Cyras, V.P., Vázquez, A.: Extraction of cellulose and preparation of nanocellulose from sisal fibers. Cellulose **15**, 149–159 (2008)
7. Talebnia, F., Karakashev, D., Angelidaki, I.: Production of bioethanol from wheat straw: an overview on pretreatment, hydrolysis and fermentation. Bioresour. Technol. **101**, 4744–4753 (2010)
8. Li, X., Tabil, L.G., Panigrahia, S.: Chemical treatments of natural fiber for use in natural fiber-reinforced composites: a review. J. Polym. Environ. **15**, 25–33 (2007)
9. Correia, J.A.D.C., Júnior, J.E.M., Gonçalves, L.R.B., Rocha, M.V.P.: Alkaline hydrogen peroxide pretreatment of cashew apple bagasse for ethanol production: study of parameters. Bioresour. Technol. **139**, 249–256 (2013)
10. Fu, D., Mazza, G.: Aqueous ionic liquid pretreatment of straw. Bioresour. Technol. **102**, 7008–7011 (2011)
11. McIntosh, S., Vancov, T.: Enhanced enzyme saccharification of Sorghum bicolor straw using dilute alkali pretreatment. Bioresour. Technol. **101**, 6718–6727 (2010)
12. Mohan, D., Pittman, C.U., Steele, P.H.: Pryrolysis of wood/biomass for bio-oil: a critical review. Energy Fuel **20**, 848–889 (2006)
13. Lee, H.V., Hamid, S.B.A., Zain, S.K.: Review article: conversion of lignocellulosic biomass to nanocellulose: structure and chemical process. SCI World J. Article ID 631013, 20 (2014). doi:10.1155/2014/631013
14. Delgado-Aguilar, M., González, I., Tarrés, Q., Pèlach, M.A., Alcalà, M., Mutjé, P.: The key role of lignin in the preparation of low-cost lignocellulosic nanofibres for papermaking applications. Ind. Crops Prod. **80**, 295–300 (2016)
15. Moriana, R., Vilaplana, F., Ek, M.: Cellulose nanocrystals from forest residues as reinforcing agents for composites: A study from macro- to nano-dimensions. Carbohydr. Polym. **139**, 139–149 (2016)
16. Rojo, E., Peresin, M.S., Sampson, W.W., Hoeger, I.C., Vartiainen, J., Laine, J., Rojas, O.J.: Comprehensive elucidation of the effect of residual lignin on the physical, barrier, mechanical and surface properties of nanocellulose films. Green Chem. **17**, 1853–1866 (2015)
17. Rosa, M.F., Medeiros, E.S., Malmonge, J.A., Gregorski, K.S., Wood, D.F., Mattoso, L.H.C., Glenn, G., Orts, W.J., Imam, S.H.: Cellulose nanowhiskers from coconut husk fibers: Effect of preparation conditions on their thermal and morphological behavior. Carbohydr. Polym. **81**, 83–92 (2010)
18. Oun, A.A., Rhim, J-W.: Characterization of nanocelluloses isolated from Ushar (Calotropis procera) seed fiber: effect of isolation method. Mater. Lett. **168**, 146–150 (2016)
19. Chandra, J., George, N., Narayanankutty, S.K.: Isolation and characterization of cellulose nanofibrils from arecanut husk fibre. Carbohydr. Polym. **142**, 158–166 (2016)
20. Boluk, Y., Danumah, C.: Analysis of cellulose nanocrystal rod lengths by dynamic light scattering and electron microscopy. J. Nanopart. Res. **16**, 2174 (2014)
21. Tang, L., Huang, B., Lu, Q., Wang, S., Ou, W., Lin, W., Chen, X.: Ultrasonication-assisted manufacture of cellulose nanocrystals esterified with acetic acid. Bioresour. Technol. **127**, 100–105 (2013)
22. Beltramino, F., Roncero, M.B., Torres, A.L., Vidal, T., Valls, C.: Optimization of sulfuric acid hydrolysis conditions for preparation of nanocrystalline cellulose from enzymatically pretreated fibers. Cellulose **23**, 1777–1789 (2016)

23. Xu, F., Yu, J., Tesso, T., Dowell, F., Wang, D.: Qualitative and quantitative analysis of lignocellulosic biomass using infrared techniques: A mini-review. Appl. Energy **104**, 801–809 (2013)
24. Haafiz, M.K.H., Hassan, A., Zakaria, Z., Inuwa, I.M.: Isolation and characterization of cellulose nanowhiskers from oil palm biomass microcrystalline cellulose. Carbohydr. Polym. **103**, 119–125 (2014)
25. Chirayil, C.J., Joy, J., Mathew, L., Mozetic, M., Koetz, J., Thomas, S.: Isolation and characterization of cellulose nanofibrils from Helicteres isora plant. Ind. Crops Prod. **59**, 27–34 (2014)

Coffee Powder Reused as a Composite Material

Vasco Canavarro, Jorge L. Alves and Bárbara Rangel

1 Introduction

This article aims to present the different strategies that are currently being adopted in the coffee reuse industries. The relevance of this study is justified by the coffee sui generis characteristics, such as odor, visual appearance and texture, as well as the abundance of raw materials, most of which are normally discarded after serving their primary function. The results presented here in terms of tests will translate the different mixtures used in order to obtain the intended material.

1.1 World's Waste

In order to realize the enormity of the problem with waste materials, biological or artificial, we present the following data.

According to Eurostat (Statistical Office of the European Union), Portugal generated in 2012, 14.184.456 tons of waste [1]. Europe (28 Countries) ended the same year with 2.51422 billion tons and in the United States are daily generated 624.700 tons of solid waste and the forecast made in the document "What a Waste" [2] is that by 2025 will be 701,709 tons. As further demonstrated by Daniel

V. Canavarro (✉) · B. Rangel
University of Porto, Rua Dr. Roberto Frias, 4200-465 Porto, Portugal
e-mail: vasco.canavarro@gmail.com

B. Rangel
e-mail: brangel@fe.up.pt

J.L. Alves
INEGI, University of Porto, Rua Dr. Roberto Frias, 4200-465 Porto, Portugal
e-mail: falves@fe.up.pt

L.F.M. da Silva (ed.), *Materials Design and Applications*,
Advanced Structured Materials 65, DOI 10.1007/978-3-319-50784-2_9

Hoornweg, Perinaz Bhada-Tata and Chris Kennedy in their publication in the journal “Nature”, the amount of waste generated globally is set to increase by 2100, with the countries of sub-Saharan Africa at the forefront [3]. It can thus be inferred that the generation of waste from human beings, will hardly diminish in the near future and that a solution to its decline can be further reuse or recycling.

1.2 *Coffee Data and Statistics*

Given that coffee also turns out to be a waste at the end of its life cycle and to contextualize this study, it is necessary to know some numbers. Based on data provided by the International Coffee Organization, it is estimated that its overall consumption in 2014 was of 149.8 million bags of 60 kg [4]. In Portugal, in the same year, 823.000 bags were consumed and the per capita consumption was around 4.7 kg [5]. According to the same source, the annual coffee consumption from 2011 to 2014 increased by 2.4% (Fig. 1). It is further noted that this increase is gradual, having existed every year. It is therefore plausible that it will continue to recur in the near future.

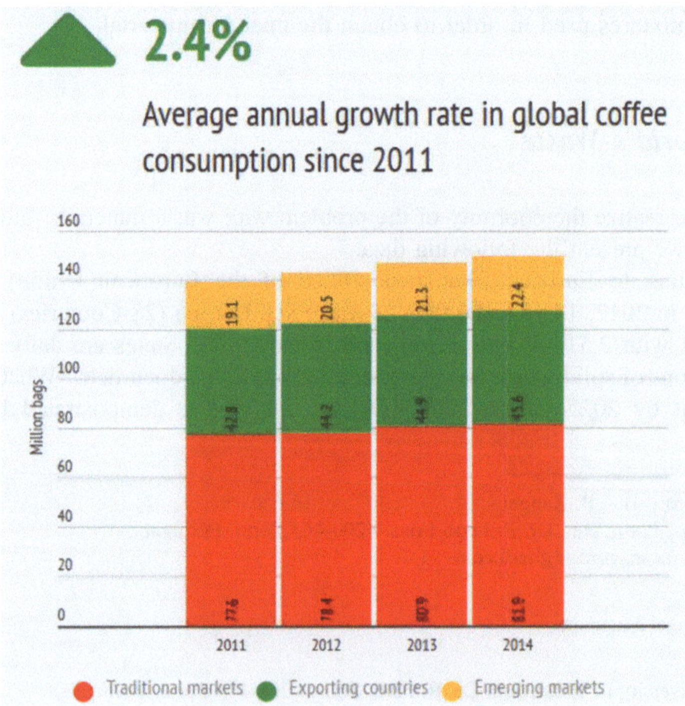

Fig. 1 Organization, international coffee. 2015 “The current state of the global coffee trade” [4]

2 Coffee Reuse Strategies

There are already reuse strategies in several areas, from agriculture to fuel. "The coffee grounds material has been consolidating itself as one of the most abundant biological resources of the world for use as green energy" [6]. However, in the Design area there are also ways of creating products with coffee grounds based materials beginning to be explored. The solutions that have been tested in this area are divided by the features presented by the final material, as a result of different mixtures between coffee grounds and other binding materials.

An example is Kaffeeform. "Kaffeeform is a recycled innovative material created with coffee grounds, vegetable fibres, cellulose and biopolymers. After five years of experimentation and research, an original formula was created to transform old coffee in new products" [7].

AgriDust is a design created by the Italian Product Designer Marina Ceccolini, which consists of using some food residue such as coffee grounds, peanut or orange peels mixed with a binder of potato starch base (Fig. 2). "AgriDust can be used to create packaging, plant pots and, in addition, using cold technology lends itself as a material for 3D printers, in which the classical extruder is replaced by a syringe" [8].

Another example of coffee use in creating raw material with 3D printing capabilities is the Hoop by Francesco Pacelli. "Francesco Pacelli, a member of + Lab at the Polytechnic University of Milan, frightened by the exponential amounts of food waste from the present to 2050, had an interest in ways to recycle waste in a sustainable manner, combining it with an ancient material, clay" [9]. The Hoop design by Francesco Pacelli has many similarities to the AgriDust by Marina Ceccolini, although the mixture added to coffee is completely different. In this particular case the mixture reached seeks a higher consistency. This material is also moulded and has the ability to be extruded with the aid of a syringe.

"Wound Up™ is a 3D printing filament created using coffee waste by-products" [10]. This material produced by the American company 3Dom uses coffee residues mixed with PLA (polylactic acid) to create a 3D printing material with visibly unique printing finishes. The filament produces a brownish colour and a natural noticeable grain (Fig. 3). This is the first of a line of 3Dom USA intriguing materials called composite c2renew. More bio-based products will be launched in the near future. One of the strongest features of this product is the ability to be printed on any machine capable of printing with PLA.

Fig. 2 AgriDust by Ceccolini [8]

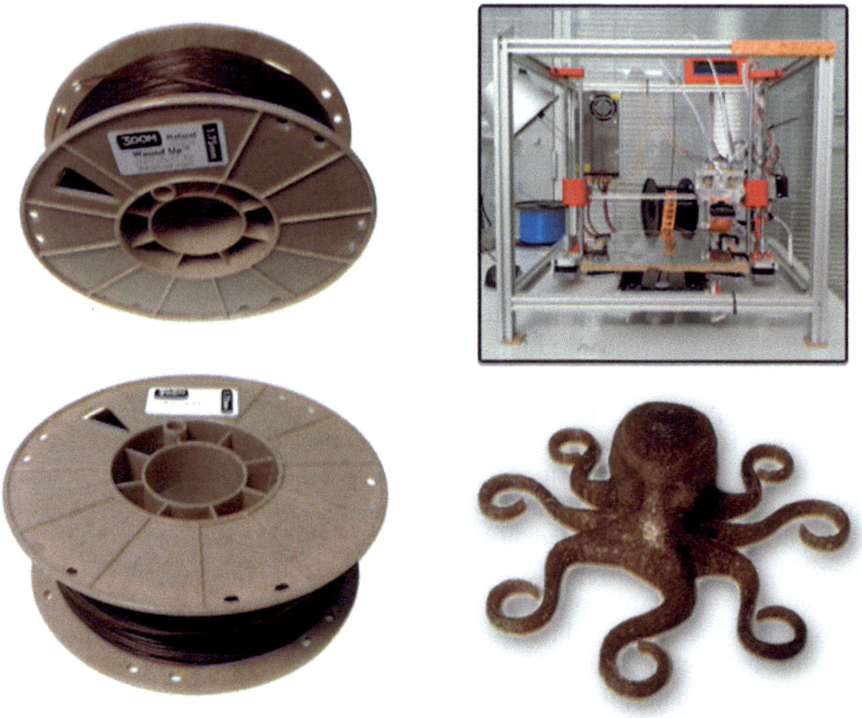

Fig. 3 Wound-up by 3Dom [10]

3 Experimental Work

3.1 Materials and Methods

The experimental work appears as a way to establish the main and most suitable material to add to coffee, in creating this material.

The first binder to be tested was Biresin 83, an epoxy resin with 30% catalyst and the ratios between this material and the coffee have varied, in an attempt to understand what types of results could be obtained.

Then starch (potato and corn) was tested, because it is a natural and biodegradable polymer. Several formulas of this material were tested and the cures were made at different temperatures and for various lengths of time.

Later, experiments were performed with pine resin, as it is a Portuguese material, and is readily available. Also in this case, the ratios varied and the cure was virtually instantaneous.

Because pine resin showed very weak results, it was decided to add other materials to the mix in an attempt to give greater strength to the final material. Therefore wax was selected, but the results were still unsatisfactory.

Finally, tests were done using Polylactide (PLA) as a binder and positive results began to appear. PLA, is a biodegradable thermoplastic derived from natural lactic acid from corn, maize or milk. Is a thermoplastic derived primarily from annually renewable resources (maize, corn or milk), and it is available in a number of grades, designed for ease of processing. In-line drying may be needed to reduce water content for extrusion and moulding. The recommended moulding temperature is 165–170 C.

It resembles clear polystyrene, provides good aesthetics (gloss and clarity), but it is stiff and brittle and needs modification using plasticizers for most practical applications. It can be processed like most thermoplastics into fibres, films, thermoformed or injection moulded.

PLA is transparent and has FDA approval for food packaging. PLA film and sheet can be printed and laminated. Biopolymers are, however, expensive, costing 2–6 times as much as commodity plastics like polypropylene.

Food packaging, plastic bags, plant pots, diapers, bottles, cold drink cups, sheet and film are some examples of the typical applications of this material (CES EDUPACK 2016, Granta Design, UK).

This information legitimates the use of this material in the project, having into account its features with regard to food contact.

3.2 Results and Discussion

In the first attempt for combinations between coffee and a binding material, the choice was an epoxy resin. Figure 4 presents the results obtained with this mixture, using different ratios coffee/resin.

The epoxy resin used was Biresin CR83 produced by the German company SIKA, with 30% catalyst and the curing time was, as usual, 24 h at room temperature. As a mould, simple plastic coffee cups available at any coffee vending machine were used.

Cups A and B represent tests performed to symbolise the existing coffee cups on the market. In object C the ratio used was a coffee part to four parts resin, while in object D the ratio was 1:2 and in E 1:1.

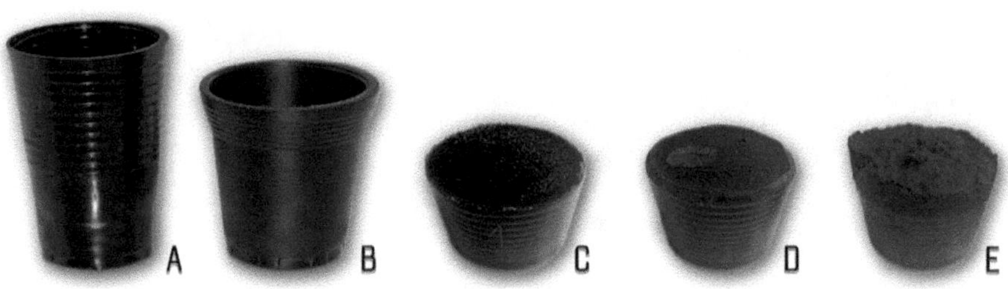

Fig. 4 Coffee cups produced with epoxy resin mixtures

These experiments were performed to demonstrate the possibility of mixture between coffee remains and a binder, but have never been seen as a possibility in this project, because the used resin is not safe for food contact.

The next step was the use of starch. Being a natural, biodegradable substance, both potato and corn starches were taken into consideration for this study as potential solution.

The way to materialize this experiment involves mixing the starch with cold water to dissolve it. Then there is the mixture with a certain percentage of coffee and the product obtained was placed in an oven, using different temperatures and time to reach the best properties (Fig. 5).

However, this experiment has proven to be unsuccessful, since all the results have shown, after some time, moisture signals itself through the appearance mould, as can be seen in Fig. 6.

At a later stage it was decided to test pine resin as a binder to ascertain whether its characteristics could give a reasonable material when mixed with coffee. It was quickly realized that this union would be fruitless, given the fragility of the resin itself. It can still be seen in Fig. 7 that most results have become brittle and not very rigid.

Fig. 5 Mixtures with starch. Bowls with potato starch and cold water mixture and addition of coffee powder (*left*). Industrial oven (*centre*). Results after heating (*right*)

Fig. 6 Degradation of the mixtures with starch. After one month, all results that used starch showed humidity problems with the presence of mold

Fig. 7 Coffee-pine resin mixtures. The samples aim to show the fragility of the mixture, since most of them are broken

Fig. 8 Coffee-pine resin and wax mixtures. The object on the *left* has a greater percentage of wax and the one on the *right* has a greater percentage of coffee powder

After a few tests with negative results, the option was to use the same pine resin, but with wax addition in an attempt to improve its physical characteristics. As shown in Fig. 8, very interesting visual results were produced, but the problem of resistance remained and the resulting material was still very brittle, thus this possibility was discarded.

Finally, and after several experiments, the material chosen for the mixture with coffee was PLA. There are several reasons for choosing this material, although the main reason is the perfect symbiosis with the coffee in the design of a perfectly mouldable and inert material in contact with food.

The figure below represents some of the first tests with PLA in conjunction with coffee residue (Fig. 9). The percentages were not taken into account, but it was quickly realized it would work for the fulfilment of the pre-defined goal.

After realizing that this would be the ideal material, an exemplificative 3D model was made in Solidworks and the file was subsequently saved in STL format. From there, a 3D printed piece was created and from that a mould was made out of silicon. For the creation of this mould, the silicone used was 950 VTX, provided by SLM and the catalyst proportion was 10%. The curing time was 24 h as usual with this type of resin. With the mould ready, various experiments took place with different percentages of both materials to understand what would be the most suitable mixture (Fig. 10).

Figure 11 presents 3 experiments carried out with this material, all with different percentages of coffee and PLA.

It is concluded, therefore, that the greater the proportion of coffee in the blend, the longer will have to be the heating of the two materials in order for them to merge. 40% is the threshold amount for this mixture. Although it is possible to increase this amount, the excess time and the thickness of the final material do not justify it.

Fig. 9 Samples of coffee-PLA mixtures. The samples presented here represent the experimental mixture between PLA and coffee powder

Fig. 10 3D printed part in PLA (*left*), Silicone mould (*centre*) and casted part (*right*)

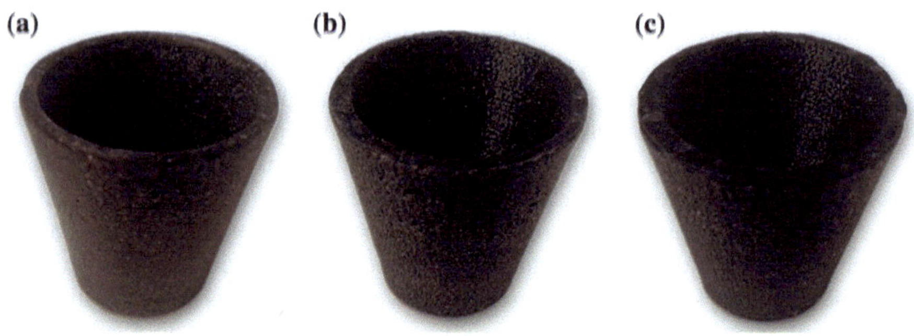

Fig. 11 Parts produced with different coffee-PLA ratios, **a** 20% coffee, **b** 30% coffee and **c** 40% coffee

The current state of this study can be seen in Fig. 12, of experiences with hot coffee in cups casted in the silicone mould. The parts behaves according to the goal, some insulation, withstands the contact with the hot coffee and mechanical resistance, fulfilling its function perfectly.

Fig. 12 Experimenting the material in contact with hot beverage (coffee). It is also noticeable that the heat from the hot coffee does not pass to the exterior

This mixture presents however a structural problem, since the part cannot be machine washed, due to the high temperatures to which it would be subjected. In this case, the material undergoes irreparable deformation and gets unusable. The piece has therefore to be washed by hand, preferably with cold water.

4 Design Proposal

From the point of view of design, the focus was to try to create an object from the developed formulation of coffee-PLA. In an attempt to generate a semantic cycle, it seems interesting that the primary object developed based on coffee waste is a cup of coffee (Fig. 13).

Fig. 13 Tray for coffee cups

Fig. 14 Containers for appetizers

Fig. 15 Plant pots

Taking into account the characteristics of the material created and that it is a mouldable material, the shape of the idealized objects is only limited by the imagination and of course the costs of the mould itself. Therefore, Figs. 14 and 15 are two examples of objects that could be crated through this technology.

The use for plant pots, since it is biodegradable and there are already a number of developments in this area using coffee as a fertilizer, presents interesting possibilities.

5 Conclusions

The developed material allows the coffee waste to be reused in the universe of Industrial and Product Design. There is still some way to go to improve this technology, but at this time one can already realize that coffee does not need to be just another piece of trash, it can be turned into a building material, very efficient and applicable to a large number of areas.

From an experimentation point of view, there have been several tests with different binding materials. It began with epoxy resin, followed by starch, in a later stage pine resin and wax, but the material that showed the best results, for the time being was PLA.

Nevertheless, the material has limitations, and cannot be used at high temperatures (washing machine). It can therefore be concluded that that there is still some way to improvement to create an ideal material, but the forecasts are quite positive.

Taking into account that it is a mouldable material, the creation of objects is only limited by the imagination.

Acknowledgements Authors gratefully acknowledge the funding of Project NORTE-01-0145-FEDER-000022—SciTech—Science and Technology for Competitive and Sustainable Industries, cofinanced by Programa Operacional Regional do Norte (NORTE2020), through Fundo Europeu de Desenvolvimento Regional (FEDER).

References

1. Eurostat: Generation of waste by economic activity (2015). http://ec.europa.eu/eurostat/tgm/table.do;jsessionid=ACitl4TZNMMwLpJeJFXMdIC5m7-JCB1VhQNenuSc-I-c_t9uViC8!-1505779883?tab=table&plugin=1&language=en&pcode=ten00106. Accessed 16 Dec 2015
2. Daniel, H., Perinaz, B.-T.: What a waste a global review of solid waste management. Urban Development Series Knowledge Papers No. 15 (2012)
3. Daniel, H., Perinaz, B.-T., Kennedy, C.: Waste production must peak this century. Nature **502** (2013)
4. Organization, International Coffee. The Current State of the Global Coffee Trade (2015). http://www.ico.org/monthly_coffee_trade_stats.asp. Accessed 10 Dec 2015
5. Organization, International Coffee. "Portugal—Latest facts and figures about the global coffee trade from the International Coffee Organization" (2015). https://infogr.am/_/S268nTQ9nsOy58h5VJZH. Accessed 10 Dec 2015
6. Gómez-de la Cruz, F.J., Cruz-Peragón, F., Casanova-Peláez, P.J., Palomar-Carnicero, J.M.: A vital stage in the large-scale production of biofuels from spent coffee grounds: the drying kinetics. Fuel Proc. Technol. **130**, 188–196 (2015). http://www.sciencedirect.com/science/article/pii/S0378382014004275
7. Kaffeeform: Kaffeeform (2015). http://www.kaffeeform.com/. Accessed 14 Dec 2015
8. Ceccolini, M.: AgriDust—Biodegradable material (2015). https://www.behance.net/gallery/24616719/AgriDust-Biodegradable-material. Accessed 16 Dec 2015
9. Grellier, C.: The future of 3D printing in coffee grounds (2015). http://www.makery.info/en/2015/01/19/lavenir-de-limpression-3d-dans-le-marc-de-cafe/. Accessed 14 Dec 2015
10. USA, 3Dom. Woun Up (2015). http://www.3domusa.com/shop/wound-coffee-filled-filament/. Accessed 16 Dec 2015

Comparison of Mechanical Properties of Polyester Composites Reinforced with Autochthonous Natural Fibres: Flax and Hemp

J. Rocha, J.E. Ribeiro and L. Queijo

1 Introduction

Nowadays, increasing pollution is causing climate change, and therefore the need to reduce carbon footprints has brought the growing use of autochthonous natural materials. On the other hand, there is new demand from the world population that requires the increased use of recyclable materials, and natural fibres are excellent candidates for this role. However, only in the last few years has this subject been studied for technical applications [1]. Natural fibre is a very interesting material, and in association with a matrix it forms a composite material. Natural fibre composites (NFCs) have some advantages compared with synthetic ones, with the most relevant being their lower density and higher specific strength and stiffness. Furthermore, they are a renewable resource whose production requires little energy and involves CO_2 absorption. Nevertheless, NFCs also have disadvantages, such as lower durability when compared to synthetic fibre composites, although this can be improved significantly with specific treatments; a greater variability of properties; and the fact that they suffer from higher moisture absorption, which results in swelling [2].

There are many factors that can influence the performance of natural fibre reinforced composites, of which the most important is the fibre selected [3]. The properties of natural fibre reinforced composites can also be influenced by the fibre dispersion or fibre volume fraction as well as the fibre orientation. In general, a high

J. Rocha · J.E. Ribeiro (✉) · L. Queijo
Polytechnic Institute of Bragança, C. Sta. Apolónia, 5301-857 Bragança, Portugal
e-mail: jribeiro@ipb.pt

J. Rocha
e-mail: jrocha@ipb.pt

L. Queijo
e-mail: lqueijo@ipb.pt

L.F.M. da Silva (ed.), *Materials Design and Applications*,
Advanced Structured Materials 65, DOI 10.1007/978-3-319-50784-2_10

fibre volume fraction is essential to achieve high performance of composites [4], while the fibre orientation yields composites with very different properties [5]. It is frequently observed that an increase in the fibre loading leads to the growth of tensile properties [6]. An additional factor that has an important influence on the mechanical properties of NFCs is the composites' interfacial strength, which can be changed using chemical treatments [7]. Hence, appropriate processing techniques and parameters must be cautiously selected in order to yield optimal composite materials [8]. Despite all these factors, matrix selection is also a very important factor that must be analysed to characterize the NFCs' mechanical properties [9]. In this work, the studied factors that affect the NFCs' mechanical performance are the fibre type and the interfacial strength, which are changed by chemical treatments.

Some of the most frequently used fibres in Portugal are hemp and flax. These fibres have their origins in plants that have been grown in the northeast of the country. In the territory that is now Portugal, flax growing and the manufacture of fabrics derived from it date from prehistory. There are traces of linen dated from 2500 BC in the Algarve region. More recently, in the nineteenth century, flax cultivation had great economic and social importance in the north of Portugal, having suffered a decline with the emergence of simpler and cheaper fibres, as was the case of cotton. In several historical periods, attempts were made to relaunch the linen industry [10]. In northeast Portugal, more specifically in the Vilariça valley, hemp was grown for manufacturing ropes in the National Cordage, which equipped caravels and other ships during the period of Portuguese discoveries throughout the fifteenth and sixteenth centuries. In the twentieth century, there were several hemp producers in the Vilariça valley. Industrial hemp growth is complex as it is a variant of cannabis (*Cannabis sativa*): these two variants differ only in terms of the tetrahydrocannabinol (THC) content, and thus the production of this plant is regulated by law [11].

The interfacial adhesion between fibres and matrix plays an important role in the mechanical properties of composites. As the stress is transferred from fibre to fibre across the matrix interface, a good interfacial adhesion is required to achieve a good reinforcement. Nevertheless, if the interface is too strong, fissures are able to propagate, which can reduce the toughness and strength. However, for fibre composites based on plants, the interaction between the fibres, which are usually hydrophilic, and matrices, which are generally hydrophobic, is very limited, leading to a low interfacial adhesion, which affects the mechanical properties. On the other hand, low humidity decreases the resistance of the properties over a long time period. To guarantee good adhesion, the matrix and fibres must be very close and, as is usual in any adhesion phenomenon, the property of wettability is fundamental to ensure the adhesion between the adhesive and adherent. In this particular case, insufficient wetting of the fibres originates interfacial flaws, which can act as stress concentrators [12] and affect the mechanical properties [13]. There are different types of fibre surface treatments, which can be classified as physical and chemical and can improve the fibres' wettability and therefore improve the interfacial strength [14–16].

Interfacial adhesion may take place by different mechanisms, namely chemical bonding, inter-diffusion bonding, mechanical interlocking, and electrostatic bonding [17]. To improve the interfacial adhesion in NFCs, the authors of this work used the chemical approach. This approach includes many different techniques using chemical products like zirconate, peroxide, benzyl, acryl, titanate, permanganate, acetyl, alkali, and silane, among others [18]. These products can be used singly or in combination [7]. The most commonly used products are alkali, acetyl, and silane [19] and for this reason alkali–silane treatment [20] was selected for implementation in this work. Alkaline treatment consists in immersing the fibres in an alkaline solution, normally NaOH, for a period of time. This treatment removes fibre constituents including lignin, hemicellulose, pectin, and wax, which exposes cellulose and increases the surface roughness per unit area, providing an improved interfacial adhesion [21]. Silane treatment generally involves moistening the fibres in a solution of diluted silane in a water/alcohol mixture that will be broken down into silanol and alcohol in the presence of water. Silanol reacts with the cellulose OH groups in the natural fibres, forming stable covalent bonds in the cell walls [22]. Silane treatment improves the amount of reticulation in the interface region and increases the fibre surface area, causing a stronger adhesion between the matrix and the fibre [23].

To evaluate the influence of fibre type and chemical treatment, it is necessary to develop experimental work in which combinations of different factors are used, and for that reason it is important to organize experiments systematically using design of experiments techniques. These techniques make it possible to conveniently organize the experiments and then process the results statistically. The first published works that used design of experiments were done by Fisher [24], who initially used these techniques in the field of agriculture. His technique was based on factorial design, in which he created an orthogonal array with multi-factors and different levels for each factor.

Factorial design has different approaches depending on the number of experiments chosen by the researcher, and it can be general or fractional. In general factorial design, all possible combinations must be implemented, so this approach is only viable if the number of factors and levels is small. However, if the number of factors and levels is high, it is impractical to implement all the experiments because of economic and time costs. There are some methods that use the fractional approach [25], but the most popular method was developed by Genichi Taguchi (Tokamachi, Japan, 1924–2012) [26]. In the present work, only two factors were used: types of fibres and surface treatments, and considering this small number of factors, it is appropriate to implement a design of experiment model based on the general factorial approach.

2 Experimental Procedure

2.1 Materials and Properties

NFCs are constituted by the matrix and fibres. In this case, the matrix used was polyester with 1% catalyst to activate the polymerization. The natural fibres used were flax and hemp produced in northeast Portugal. The typical mechanical properties of these materials are listed in Table 1.

For the alkali-silane treatment, different chemical solutions were prepared according to the specificities of each treatment. Thus, solutions of sodium hydroxide (NaOH) with concentrations of 2 and 5% by mass were used for the alkali treatment. On the other hand, the silane treatment was performed using 5% by weight of 3-aminopropyltriethoxysilane, which was diluted in a 50% aqueous solution of methanol.

2.2 Plan of Experiments

The experiments were conducted in accordance with a standard orthogonal array, where each column defines the factor control. So, the first column corresponds to the type of fibre (A) and the second one is used for the fibre surface alkali–silane treatment (B), as can be seen in Table 2.

As mentioned before, in this work, two types of natural fibres were used: hemp and flax, some of which were used untreated while others were subjected to alkali–silane treatment of the fibre surface with NaOH concentrations of 2 and 5%.

Table 1 Mechanical properties of matrix and natural fibres

		Density (g/cm^3)	Tensile strength (MPa)	Young's modulus (GPa)
Matrix	Polyester [29]	1.2	70–103	2.1–4.4
Natural fibre	Flax [1]	1.5	345–1830	27–80
	Hemp [1]	1.5	550–1100	58–70

Table 2 Orthogonal array of experiments

Experiment	A	B
1	1	1
2	2	1
3	1	2
4	2	2
5	1	3
6	2	3

Table 3 Factors and respective levels

		Levels		
		1	2	3
Factors	A—Type of fibre	Hemp	Flax	–
	B—Surface alkali–silane treatment	Untreated	2% NaOH	5% NaOH

The reason for the choice of these concentrations was based on the paper published by Asumani et al. [20]. According to their work, the maximum improvement in tensile strength occurred for the 5% NaOH concentration, while and 2% concentration corresponds to an intermediate value between the untreated case and the case with an NaOH concentration of 5%. Table 3 lists the levels for each factor.

2.3 Specimen Manufacture and Tensile Tests

The specimens used in the tensile tests to determine the tensile strength of each combination of factors and levels were cut from a composite plate (200 × 200 1 mm^3) manufactured previously. The fibres of these composite plates were aligned in one direction corresponding to the direction of tensile load application. Firstly, to prepare the composite plate, the untreated and treated fibres were wrapped around a steel plate with a thickness of 2 mm using a lathe. To perform this operation, the metallic plate was first greased with release agent and then fixed in the chuck. Next, the spindle was set in very slow rotation, wrapping the fibres around the plate. During the process it was ensured that the fibres were very close and aligned in one direction. Subsequently to this operation, the metallic plate with the fibres was immersed in polyester resin (99% resin with 1% catalyst) during approximately 30 s, and while unpolymerized, placed into a two-part mould. This mould was placed on a press table and a pressure of 1 MPa [27] was applied on the composite plate. After 12 h of polymerization in the press, the composite was removed from the mould and the metallic plate was taken away from the middle of the composite. By cutting the composite set in the same plane as the direction of the fibres, it was possible to obtain two composite plates, as seen in Fig. 1.

The specimens were cut from the composite plates using a laser system (X252 from GCC). The dimensions and geometry of specimens were chosen according to the ASTM D 3039 M standard [28]. In Fig. 2, it is possible to observe some examples of the specimens used for the tensile tests. For each composite plate, corresponding to a combination of the different factors and levels, three specimens were cut, to give a total of 18.

The alkali–silane treatment of fibres was done in two steps. The first step was the alkali treatment using sodium hydroxide (NaOH) solutions with concentrations of 2 and 5% by mass. Hemp and flax fibre mats were immersed in the NaOH solution for 24 h at a temperature of 45 °C. After this period, the fibres were washed with tap water and submerged in distilled water that contained 1% acetic acid to neutralize

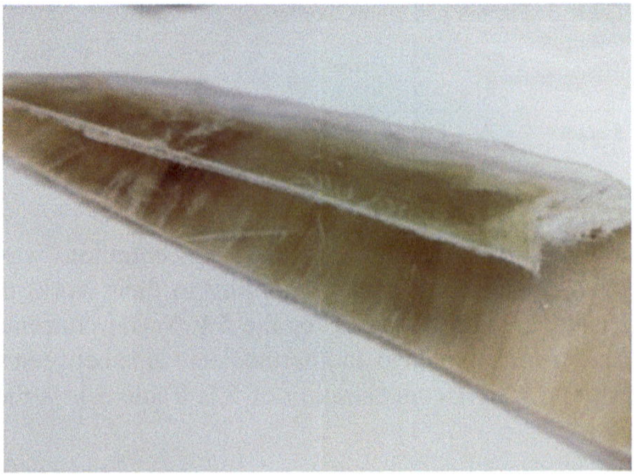

Fig. 1 Two composite plates obtained after the manufacturing process

Fig. 2 Composite specimens for tensile tests

the residual NaOH. These fibres were then dried in an oven at 45 ºC during 12 h. The second step was the silane treatment. For this treatment, a solution of 5% 3-aminopropyltriethoxysilane by weight (weight of silane relative to the weight of hemp or flax mat) diluted in a 50% aqueous solution of methanol was used. The pH of the solution was kept between 4 and 5 using acetic acid. The mats were submerged in the solution for 4 h at 28 °C. After that period, they were washed with distilled water and finally dried in an oven at 45 °C for a period of 12 h.

After manufacturing the specimens, the tensile tests were performed using a universal test machine (Instron 4485). The specimens were fixed in the machine grips and the tensile test was performed at a speed of 1 mm/s. For each experiment (Table 1), three tensile tests were executed, corresponding to a total of 18 trials.

3 Results and Discussion

In Table 4 and Fig. 3, it is possible to observe the average of the results obtained. We can see that Experiments 1, 5, and 6 give higher values of tensile strength, especially Experiment 6, in which a flax composite was tested using the alkali–silane surface treatment with 5% NaOH concentration.

In Table 4, is also possible to verify that Experiment 5 has the highest value of standard deviation (25.4 MPa), which means that the data are widely spread (less reliable), while Experiments 2, 4, and 6 have relatively low values of standard deviation for the NFCs: 4.6, 4.7, and 7.8 MPa, respectively; in other words, the experimental data are closely clustered around the mean (more reliable).

Using the data presented in Table 4, it is possible to obtain an analysis of variance (ANOVA). The ANOVA is a group of statistical models used to analyse the changes among group averages and their associated procedures, developed by Ronald Fisher [24]. Table 5 presents the ANOVA for the tensile tests performed.

Table 4 Experimental results: average tensile strength and standard deviation

Experiment	A	B	Average tensile strength (MPa)	Standard deviation (MPa)
1	1	1	112.8	11.8
2	2	1	49.7	4.6
3	1	2	70.8	12.7
4	2	2	32.2	4.7
5	1	3	102.0	25.4
6	2	3	113.4	7.8

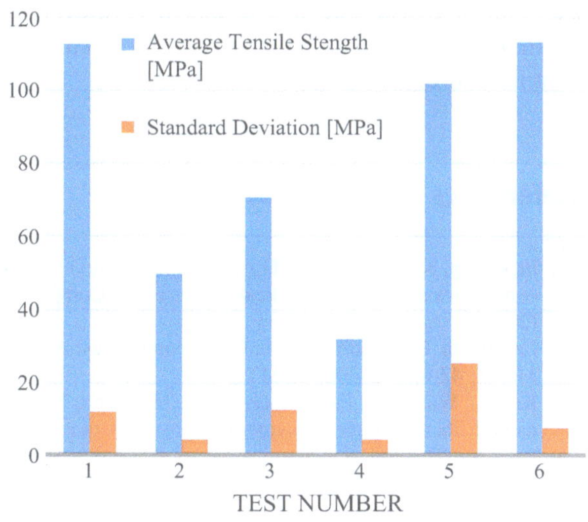

Fig. 3 The average tensile strength and standard deviation for each tensile test

Table 5 ANOVA results

Source	DF	Adj SS	Adj MS	F-Value	Percentage contribution
A—Type of fibre	1	1358	1358.1	1.89	22.8
B—Surface treatment	2	3157	1578.6	2.19	53.0
Residual	2	1441	720.3		24.2
Total	5	5956			

In this table, DF refers to the degrees of freedom, Adj. SS is the sum of squares, and Adj. MS is the mean squares. The F test is a statistical tool to determine which parameters affect the quality more significantly. Observing Table 5, it turns out that the chemical surface treatment of the fibre has the largest influence on the tensile strength, with a contribution of 53.0%.

4 Conclusions

The influence of two types of fibres and alkali–silane surface treatment on the tensile strength of NFCs was studied in this work. An array of experiments was implemented using three levels for the surface treatment and two levels for the types of fibres.

To determine the tensile strength of NFCs, a total of 18 tensile tests were performed. From these tests it was verified that the combination that gives the highest value of tensile strength is the flax composite associated with the alkali–silane surface treatment with an NaOH concentration of 5%, and for this case the average tensile strength was 113.4 MPa. Based on the experimental results and using the ANOVA approach, it was determined that the most influential factor for maximizing the tensile strength was the alkali–silane surface treatment, with a contribution of 53.0%.

References

1. Pickering, K.L., Efendy, M.G.A., Le, T.M.: A review of recent developments in natural fibre composites and their mechanical performance. Compos. Part A Appl. Sci. Manuf. **83**, 98–112 (2016)
2. La Mantia, F.P., Morreale, M.: Green composites: A brief review. Compos. Part A Appl. Sci. Manuf. **42**(6), 579–588 (2011)
3. Ku, H., Wang, H., Pattarachaiyakoop, N., Trada, M.: A review on the tensile properties of natural fiber reinforced polymer composites. Compos. Part B Eng. **42**(4), 856–873 (2011)
4. Pan, N.: Theoretical determination of the optimal fiber volume fraction and fiber-matrix property compatibility of short fiber composites. Polym. Compos. **14**(2), 85–93 (1993)

5. Yahaya, R., Sapuan, S.M., Jawaid, M., Leman, Z., Zainudin, E.S.: Effect of fibre orientations on the mechanical properties of kenaf–aramid hybrid composites for spall-liner application. Def. Technol. **12**(1), 52–58 (2015)
6. Ahmad, I., Baharum, A., Abdullah, I.: Effect of extrusion rate and fiber loading on mechanical properties of Twaron fiber-thermoplastic natural rubber (TPNR) composites. J. Reinf. Plast. Compos. **25**(9), 957–967 (2006)
7. Rong, M.Z., Zhang, M.Q., Liu, Y., Yang, G.C., Zeng, H.M.: The effect of fiber treatment on the mechanical properties of unidirectional sisal-reinforced epoxy composites. Compos. Sci. Technol. **61**(10), 1437–1447 (2001)
8. Ho, M., Wang, H., Lee, J., Ho, C., Lau, K., Leng, J., Hui, D.: Critical factors on manufacturing processes of natural fibre composites. Compos. Part B **8**(8), 3549–3562 (2012)
9. Wollerdorfera, M., Bader, H.: Influence of natural fibres on the mechanical properties of biodegradable polymers. Ind. Crops Prod. **48**(3), 1887–1899 (2012)
10. Alves, J.F.: A Indústria Têxtil do Vale do Ave. chapter of book Património e Industria no Vale Ave, Edited by Câmara Municipal de Vila Nova de Famalicão, Vila nova de Famalicão, pp. 372–389 (2002)
11. Santos, M.O.: O Cânhamo como material de construção: Viabilidade e Oportunidade. Universidade Fernando Pessoa (2013)
12. Chen, P., Lu, C., Yu, Q., Gao, Y., Li, J., Li, X.: Influence of fiber wettability on the interfacial adhesion of continuous fiber-reinforced PPESK composite. J. Appl. Polym. Sci. **102**(3), 2544–2551 (2006)
13. Wu, X.F., Dzenis, Y.A.: Droplet on a fiber: geometrical shape and contact angle. Acta Mech. **185**(3–4), 215–225 (2006)
14. Bénard, Q., Fois, M., Grisel, M.: Roughness and fibre reinforcement effect onto wettability of composite surfaces. Appl. Surf. Sci. **253**(10), 4753–4758 (2007)
15. Sinha, E., Panigrahi, S.: Effect of plasma treatment on structure, wettability of jute fiber and flexural strength of its composite. J. Compos. Mater. **43**(17), 1791–1802 (2009)
16. Liu, Z., Sun, C., Liu, Z.-W., Lu, J.: Adjustable wettability of methyl methacrylate modified ramie fiber. J. Appl. Polym. Sci. **109**(5), 2888–2894 (2008)
17. Matthews, F.L., Rawlings, R.D.: Composite Materials: Engineering and Science. Woodhead Publishing, New York (1999)
18. Kabir, M.M., Wang, H., Lau, K.T., Cardona, F.: Chemical treatments on plant-based natural fibre reinforced polymer composites: An overview. Compos. Part B Eng. **43**(7), 2883–2892 (2012)
19. Ismail, H., Shuhelmy, S., Edyham, M.R.: The effects of a silane coupling agent on curing characteristics and mechanical properties of bamboo fibre filled natural rubber composites. Eur. Polym. J. **38**(1), 39–47 (2002)
20. Asumani, O.M.L., Reid, R.G., Paskaramoorthy, R.: The effects of alkali-silane treatment on the tensile and flexural properties of short fibre non-woven kenaf reinforced polypropylene composites. Compos. Part A Appl. Sci. Manuf. **43**(9), 1431–1440 (2012)
21. Li, X., Tabil, L.G., Panigrahi, S.: Chemical treatments of natural fiber for use in natural fiber-reinforced composites: a review. J. Polym. Environ. **15**(1), 25–33 (2007)
22. Agrawal, R., Saxena, N., Sharma, K., Thomas, S., Sreekala, M.: Activation energy and crystallization kinetics of untreated and treated oil palm fibre reinforced phenol formaldehyde composites. Mater. Sci. Eng. A **277**(1–2), 77–82 (2000)
23. Huda, M.S., Drzal, L.T., Mohanty, A.K., Misra, M.: Effect of fiber surface-treatments on the properties of laminated biocomposites from polylactic acid (PLA) and kenaf fibers. Compos. Sci. Technol. **68**(2), 424–432 (2008)
24. Fisher, R.A. In: Crew, F.A.E., Ward Cutler, D. (eds.) Statistical Methods for Research Workers, 5th edn. London (1934)
25. Montgomery, D.: Design and Analysis of Experiments, 8th edn. Wiley, Danvers (2013)
26. Wu, Y., Wu, A.: Taguchi Methods for Robust Design. ASME Press, New York (2000)
27. Batch, G.L., Cumiskey, S., Macosko, C.W.: Compaction of fiber reinforcements. Polym. Compos. **23**(3), 307–318 (2002)

28. ASTM International: ASTM D 3039 M—standard test method for tensile properties of polymer matrix composite materials. In: Annual Book of ASTM Standards, vol. 15, no. 03, pp. 1–13 (2000)
29. Abdul-hamead, A.A., Kasim, T., Mohammed, A.A.: Mechanical properties for polyester resin reinforce with Fe weave wire. Int. J. Appl. Innov. Eng. Manage. **3**(7), 8–12 (2014)

Advanced Epoxy-Based Anticorrosion Coatings Containing Graphite Oxide

Monica Periolatto, Elisabetta Di Francia, Marco Sangermano, Sabrina Grassini and Pasquale Russo Spena

1 Introduction

The use of polymeric coatings on metals substrates, such as steel or aluminum, is a smart and very well known strategy to protect metals against corrosion [1]. While the polymeric matrix plays its role as a binder which should show uniformity and well adherence, different fillers could be added to the coating with the aim to inhibit corrosion processes or to act as a barrier for corrosive species, such as water and oxygen [2]. Coating acts, therefore, its protection towards the metal surface by providing a barrier between the metal and the aggressive agents present in the environment. The corrosion protection properties of an organic coating is exploited by the good barrier properties of the organic materials as well as the good maintenance of adhesion to the metal substrate under the environmental conditions [3].

M. Periolatto (✉) · P. Russo Spena
Faculty of Science and Technology, Free University of Bozen-Bolzano,
P.zza Università 1, 39100 Bolzano, Italy
e-mail: monica.periolatto@polito.it; monica.periolatto@unibz.it

P. Russo Spena
e-mail: pasquale.russospena@unibz.it

E. Di Francia · M. Sangermano · S. Grassini
Dipartimento di Scienza Applicata e Tecnologia, Politecnico di Torino,
C.so Duca degli Abruzzi 24, 10129 Torino, Italy
e-mail: elisabetta.difrancia@polito.it

M. Sangermano
e-mail: marco.sangermano@polito.it

S. Grassini
e-mail: sabrina.grassini@polito.it

L.F.M. da Silva (ed.), *Materials Design and Applications*,
Advanced Structured Materials 65, DOI 10.1007/978-3-319-50784-2_11

It is well known that one of the most effective corrosion protection systems for metals contains chromates [4], which act as a corrosion inhibitor due to the high oxidation properties of Cr^{6+} ions. Notwithstanding the very good corrosion protection, chromates dispersion into polymeric coatings is environmental unfriendly and during last years alternative strategies have been investigate and proposed. The alternative strategies have been developed taking into account that the polymeric coatings can exploit a good protective effectiveness thanks to their good barrier properties against gases and vapors, which could prevent oxygen and water diffusion.

With this in mind, hybrid organic-inorganic materials have been proposed as corrosion protective coatings, prepared by sol-gel process or using nanobuilding blocks such as inorganic nanoclusters, inorganic nanoparticles or nanolayered materials [5].

In particular, polymeric nanocomposite coatings containing nanolayered fillers, such as clays, exhibit enhanced gas barrier properties together with higher thermal stability and good mechanical properties [6]. In fact there are many reports in literature about the corrosion resistance of nanocomposite coatings based on polymers, such as polyimide [7], polyurethane [8] and epoxy [9] containing layered silicates. The well dispersed layered filler acts on diffusion properties by increasing the length of the diffusion pathways for reactive agents, such as oxygen and water, leading to a significant enhancement of the corrosion protection properties, with respect to the neat polymeric coating.

More recently, graphene has been proposed as a filler to confer enhanced barrier properties to the polymeric matrix [10–12]. The graphene-lamellar structure is an interesting alternative to the use of organoclays fillers to enhance the gas barrier properties of polymeric matrices. In previous works a significant increase of the water vapor barrier properties of UV-cured acrylic [13] and epoxy [14] coatings is observed. The idea to use the UV-curing technology is due to its high efficiency, environmental protection and energy saving [15]. Most often, a UV curable formulation consists of oligomers and multifunctional monomers that polymerize to form a highly crosslinked polymer network in the presence of UV radiation and a photoinitiator.

The present study focuses on the preparation of epoxy-based photocrosslinked polymers containing graphite oxide as high barrier fillers and their characterization as corrosion protective coatings.

Materials and methods are explained in detail in a first experimental part: processes for sample preparation and characterization techniques are reported, highlighting the main process parameters considered and monitored in the study. Obtained results are reported and deeply discussed in a following session, focusing on kinetic of polymerization, transparency, hydrorepellency, impedance measurements and morphological analysis. Finally, relevant results are summarized in the conclusions.

2 Experimental

2.1 Materials

The UV-curable epoxy resin was the bis-cycloaliphatic diepoxy resin 3,4-epoxycyclohexylmethyl-3',4'-epoxycyclohexyl carboxylate (CE) and the triphenylsulfoniumhexafluoroantimonate ($Ph_3S^+SbF_6^-$) was used as photoinitiator. Both products were purchased from Sigma-Aldrich and used as received without any further purification.

Graphite oxide (GO) was obtained using natural graphite flakes (purum powder ≤ 0.1 mm, Sigma-Aldrich) following the Brödie method. Graphene produced through this method led to the formation of single graphene layers or stacks of up to 7 sheets with hydroxyl, carbonyl and epoxy groups on their surface [16].

2.2 Sample Preparation

Graphite oxide was directly dispersed, in the range between 0.05 and 1wt%, into the UV-curable epoxy resin using an IKA Ultraturrax at 30000 rpm for 5 min. The mixtures were then placed in an ultrasonic bath, at room temperature, for 60 min. The photoinitiator was added just before UV-irradiation in a proportion of 2 wt% with respect to the resin. The formulations were spread on stainless steel samples and UV-irradiated for 1 min with a light intensity of 60 mW/cm^2. The average coating thicknesses were determined with a micrometer by performing three measurements on each sample, at least.

2.3 Sample Characterization

The progress of the photopolymerization process was followed by FT-IR in a Thermo-Nicolet 5700 spectrophotometer. The spectra were obtained co-adding 7 individual scans at 8 cm^{-1} of resolution. 25 μm-thickness films were deposited on silicon wafers and simultaneously exposed to a UV-beam (Hamamatsu LC8 portable UV-lamp), with an intensity of 60 mW/cm^2 for 3 min to induce polymerization, and to the IR-beam, to make an in situ evaluation of the reaction extent. The conversion of the epoxy group was followed by monitoring the decrease in the absorbance of the epoxy groups, centered at 750 cm^{-1}, and normalized by the carbonyl peak, centered at around 1700 cm^{-1}. The conversion curves as a function of irradiation time is the average result of 3 runs for each sample. Gel content values of the cured coatings were obtained by measuring the weight loss after 24 h in chloroform at room temperature (ASTM D2765-84).

Static contact angle measurements were carried out with water ($\gamma = 72.1$ mN m^{-1}) by means of a Kruss G10 instrument, equipped with a video camera and a DSA 10 image analyzer, at room temperature; the drop profile was evaluated by sessile drop technique.

Film samples were equilibrated with distilled water at room temperature for water uptake measurements. The increase of weight, due to water uptake, was monitored at different times, till equilibrium was reached. The wet samples were surface dried with filter paper and immediately weighed. Starting dry weights were determined after sample conditioning.

Water-vapor permeability specimens were prepared as coating films using a PET substrate, taken as reference the uncoated form. Afterwards, the obtained samples were mounted on an Extrasolution MultiPerm instrument with a surface reducing frame, in order to realize an exposed surface of 2.27 cm^{-2}. Analyses were carried out under a pressure of 1 atm, a temperature of 23 °C, and a humidity level of 100%.

The film transparency was evaluated in terms of transmittance % at 510 cm^{-1} wavelength by Unicam UV2 (ATI Unicam, Cambridge, UK) spectrophotometer.

Finally, the corrosion-protective properties of the coatings deposited onto the stainless steel substrate were assessed by means of electrochemical impedance spectroscopy (EIS). Impedance measurements were carried out as a function of the immersion in an aerated 0.1 M NaCl solution, every 24 h, for a total exposure of 168 h. The impedance spectra were recorded at the open-circuit potential (E_{OCP}) at room temperature, by applying a sinusoidal signal, with amplitude of 10 mV, in the $10^5 \div 10^{-2}$ Hz frequency range with five points per decade. A three-electrode electrochemical configuration was used with a clamp on electrolyte cell attached by a rubber O-ring; an Ag–AgCl reference electrode and a platinum counter-electrode were used. For comparison purpose the impedance spectra were normalized to an exposed area of 1 cm^2 and scaled to a nominal coating thickness of 100 μm.

After the EIS measurements the coating morphology was investigated by field emission scanning electron microcopy (Supra 40 Zeizz FESEM microscope), for observing the coating homogenity, possible cracks and defects in the structure and failure mechanisms.

3 Results and Discussion

UV-cured cycloaliphatic epoxy nanocomposite coatings were prepared in the presence of graphene oxide nanoplatelets. Graphene oxide was chosen as filler for the crosslinked system because of its lamellar morphology which could generate a tortuous diffusion path able to hinder the diffusion of reactive gas such as water vapor. Furthermore, the graphite oxide shows a good compatibility with the epoxy resin due to the presence of the polar functional groups on its surface, and at the same time should enhance the hydrophobicity of the coating decreasing the water uptake. As mentioned above, from previous studies [13, 14] it is observed that when GO is added in such a low amount there is not a significant influence on photocuring kinetic and final epoxy group conversion. These data were confirmed by FT-IR analysis, which showed that epoxy group conversion was around 89% for the pristine epoxy resin and decreased to about 79% for the formulation containing 0.5 wt% of GO, after one minute of irradiation.

Always high gel content values (above 90%) were measured for all the UV-cured films. This indicates the formation of a tight crosslinked network and the absence of extractable monomers or oligomers in the cured system.

Despite the dark gray/black color of GO, the film transparency, for all the considered GO add-ons, is evident at sight and confirmed by UV-VIS measurements. The presence of 0.05% wt of GO does not affect the appearance of the neat epoxy film. A slight decrease in transmittance is related to the higher amounts of dispersed GO but with transmittance values always higher than the required limit of 85%: 96.5% for 0.5% GO and 92.2% for 1% GO. This result confirms the efficiency of the coupled Ultraturrax and sonication mixing methods to obtain a good dispersion of GO in the polymer matrix.

The surface properties of the UV-cured films were tested by contact angle measurements with distilled water. The obtained CA values are listed in Table 1. The presence of GO does not affect the surface properties of epoxy coating up to a content of 1 wt%, where the contact angle reaches a value of 90°. Therefore, GO can confer surface hydrorepellency while, at lower amounts, graphene is confined inside the bulk of the film so that the water drop get in touch mainly with the hydrophilic polymer.

Water uptake of the crosslinked composite films was measured and reported in terms of % weight increase as a function of time. The water uptake curves are reported in Fig. 1 and they show that the extent of water uptake decreases in presence of GO, indicative of the moisture barrier provided by the filler. This is a very important results, since it shows that the presence of the carbonious filler limit somehow the possibility of water uptake, which could be attributed both to the hydrophobicity of the filler as well as to its effect on water diffusion decrease, which is a good effect for corrosion protection.

From previous investigations, it was also observed that the addition of GO induced a relevant effect on the barrier behavior of the crosslinked coating. The permeability values decreased of about 40% for the crosslinked films containing 0.5 wt% of GO, and of about 60% for 1% GO weighted samples. This is another important effect to enhance the corrosion protective properties of the graphene-epoxy coating. Besides, it is very well known that UV-cured epoxy films are characterized by high adhesion properties on stainless still substrates.

Table 1 Static contact angles measured with distilled water ($\gamma = 72.1$ mN m^{-1})

GO (wt%)	Static water CA (°)
PET substrate	91
0	78
0.05	80
0.5	75
1.0	90

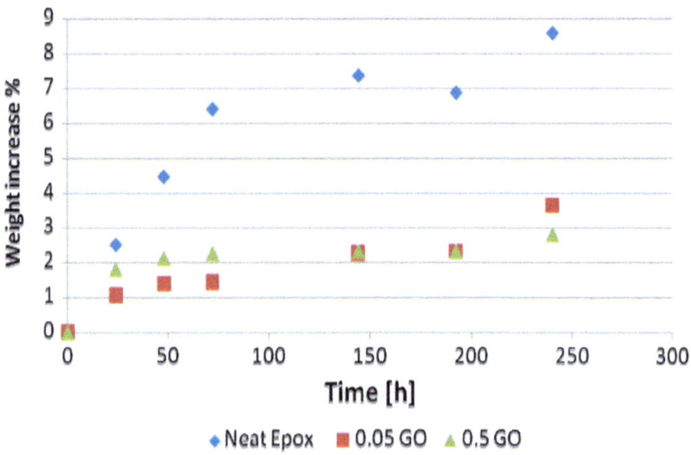

Fig. 1 Water uptake measurements

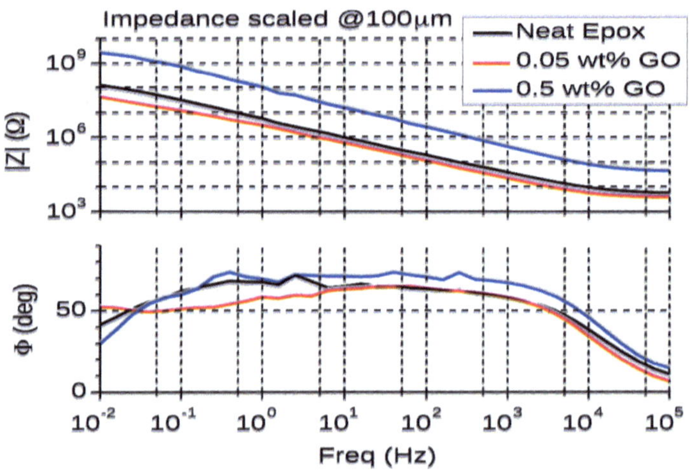

Fig. 2 Impedance spectra recorded after 48 h of immersion in the 0.1 M NaCl solution. The impedance spectra were normalized to an exposed area of 1 cm^2 and scaled to a coating thickness of 100 μm

Figure 2 shows the impedance spectra collected after 48 h of immersion in the aerated NaCl solution.

The impedance data put in evidence the good protective capability of the UV-cured epoxy coating. The impedance modulus, |Z|, increases up to 10^9 Ω cm^2 by increasing the GO wt% highlighting the higher protective properties of the 0.5 wt % GO coating. The impedance phase of about 80° accounts for a quasi-capacitive behavior, proving the high insulating properties of all the coatings. In addition Fig. 3 shows the morphology of the 0.05 wt% and 0.5 wt% GO-UV cured epoxy

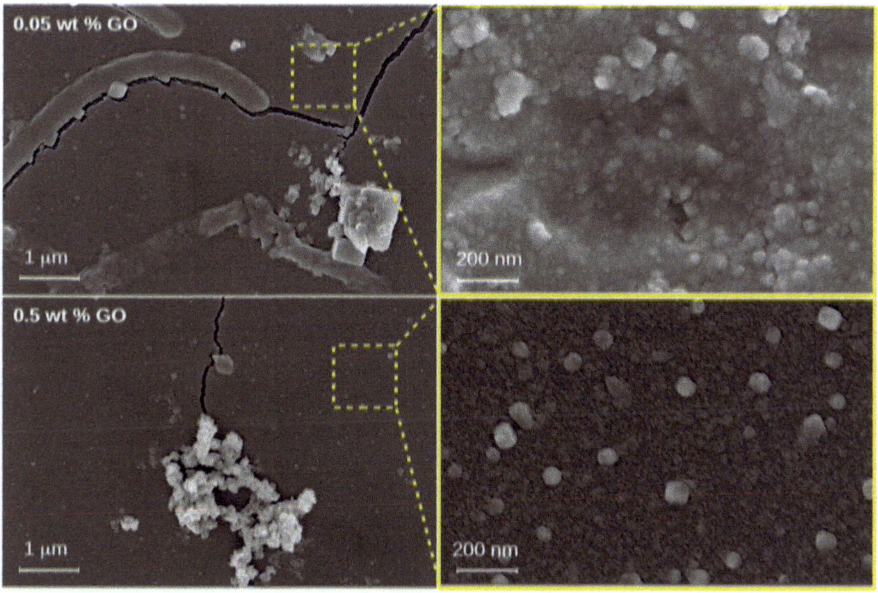

Fig. 3 FESEM images of the UV-cured graphene-epoxy coating deposited on stainless steel at the end of the impedance measurements, after 168 h of immersion in the 0.1 M NaCl solution

coatings, respectively, at the end of the EIS measurements, after 168 h of immersion in the aggressive solution. As it is possible to observe, the 0.5% GO coating is characterized by a more homogeneous morphology and a lower presence of defects, which turn out in a slight higher impedance phase and in a higher impedance modulus value. This higher impedance might be also related to a lower dielectric constant of the coating, even thought other measurements have to be performed to investigate this aspect. However, the protective capability of the 0.5% GO coating are quite promising as confirmed by the impedance measurements carried out as a function of the immersion time. As a matter of fact, no significant variation of the impedance spectra is observed; the impedance modulus and phase values remain almost constant with an amplitude change of less than 10% after 168 h of immersion in the aggressive solution.

4 Conclusions

An effective protective coating was obtained by dispersion of graphene oxide in a cycloaliphatic epoxy photocured polymer matrix.

Better results were obtained with a graphene oxide amount of 0.5% on resin weight. When GO is added in such a low amount there is not a significant influence on photocuring kinetic and final epoxy group conversion, as confirmed by real time

FTIR characterization. High gel content values (above 90%) were measured for all the UV-cured films and the film transparency was kept unvaried.

The extent of water uptake decreases in presence of GO, indicative of the moisture barrier provided by the filler. Moreover, the permeability values decreased of about 40% for the crosslinked films containing 0.5 wt% of GO, and of about 60% for 1% GO weighted samples.

The impedance data put in evidence the good protective capability of the UV-cured epoxy coating, while the impedance phase of about 80° accounts for a quasi-capacitive behavior, proving the high insulating properties of all the coatings. The 0.5% GO coating is characterized by a more homogeneous morphology and a lower presence of defects, which turn out in a slight higher impedance phase and in a higher impedance modulus value. All these properties where stable in time.

In conclusion, the epoxy composite film with 0.5 wt% GO filler showed promising properties in terms of water vapor barrier and corrosion protection, even though using very low amounts of GO dispersed in the polymeric matrix. It suggests many potential applications in anticorrosive coatings, electronic devices, pharmaceuticals and food packaging.

References

1. Wicks, Z.W., Jones, F.N., Pappas, S.P.: Organic Coatings: Science and Technology. Wiley, New York (1999)
2. Leidheiser, H.: Corrosion Control by Organic Coatings. Science Press, Princeton, NJ (1979)
3. Grundmeier, G., Schmidt, W., Stratmann, M.: Corrosion protection by organic coatings: electrochemical mechanism and novel methods of investigation. Electrochim. Acta **45**, 2515–2533 (2000)
4. Osborne, J.H.: Observations on chromate conversion coatings from a sol-gel perspective. Prog. Org. Coat. **41**, 280–286 (2001)
5. Sanchez, C., Soler-Ilia, C., Ribot, F., Myer, C.R., Cabuil, V., Lalot, T.: Designed hybrid organic-inorganic nanocomposites from functional nanobuilding blocks. Chem. Mater. **13** (10), 3061–3083 (2001)
6. Alexandre, M., Dubois, P.: Polymer layered silicate nanocomposites: preparation, properties and uses of a new class of materials. Mater. Sci. Eng. R **28**, 1–63 (2000)
7. Yu, H., Yeh, J.M., Lion, S.J., Chang, Y.P.: Acta Mater. **54**, 475 (2004)
8. Solarski, S., Benali, S., Rochery, M., Devaux, E., Alexandre, M., Monteverde, F., Dubois, P.: Synthesis of a polyurethane/clay nanocomposite used as coating: interactions between the counterions of clay and the isocyanate and incidence on the nanocomposite structure. J. Appl. Polym. Sci. **95**, 238–244 (2004)
9. Becker, O., Varley, R.J., Simon, G.P.: Thermal stability and water uptake of high performance epoxy layered silicate nanocomposites. Eur. Polym. J. **40**, 187–195 (2004)
10. Zhang, W.L., Liu, Y.D., Choi, H.J.: Fabrication of semiconducting graphene oxide/polyaniline composite particles and their electrorheological response under an applied electric field. Carbon **50**, 290–296 (2012)
11. Zhang, W.L., Park, B.J., Choi, H.J.: Colloidal graphene oxide/polyaniline nanocomposite and its electrorheology. Chem. Commun. **46**, 5596–5598 (2010)
12. Zhang, W.L., Liu, Y.D., Choi, H.J.: J. Mater. Chem. **21**, 6919 (2011)

13. Rizza, G., Melilli, G., Sangermano, M.: UV curing of perfluoropolyether oligomers containing graphene nanosheets to enhance water-vapor barrier properties. Macromol. Chem. Phys. **215**, 1588–1592 (2014)
14. Sangermano, M., Periolatto, M., Signore, V., RussoSpena, P.: Improvement of the water-vapor barrier properties of an UV-cured epoxy coating containing graphite oxide nanoplatelets. Prog. Org. Coat. (2016)
15. Fouassier, J.P., Rabek, J.F.: Radiation Curing in Polymer Science and Technology, vol. 1–4. Elsevier, London (1993)
16. Sinha Ray, S., Okamoto, M.: Polymer/layered silicate nanocomposites: a review from preparation to processing. Prog. Polym. Sci. **28**, 1539–1641 (2003)

Part IV
Design

A Numerical Study of Fenestral Otosclerosis

B. Areias, M.P.L. Parente, F. Gentil and R.M. Natal Jorge

1 Introduction

Otosclerosis is a focal disease of localized bone remodeling. The first hearing loss and related symptoms usually arise between the ages 15 and 45 in 90% of the cases, although earlier and somewhat later occurrences have been reported. It is bilateral in 90% of the females and 80% of the males [1–3]. The otosclerosis foci has been described throughout the labyrinthine capsule, but the oval window is the principal localization in approximately 80–90% of cases. Figure 1 shows the types of fenestral otosclerosis: (A) anterior focus, the fissula ante fenestram, which lies in front of the oval window, is the principal place where otosclerosis appears; (B) posterior focus, where otosclerosis may appear behind the oval window; (C) circumferential, where the disease process spreads around the margin of the stapes footplate; (D) biscuit type, where the disease process involves the footplate but the annular ligament is free; (E) obliterative type, where the disease process completely obliterates the oval window niche [4, 5].

B. Areias (✉) · M.P.L. Parente · R.M. Natal Jorge
INEGI, Institute of Science and Innovation in Mechanical and Industrial Engineering, Porto, Portugal
e-mail: bareias@inegi.up.pt

M.P.L. Parente
e-mail: mparente@fe.up.pt

R.M. Natal Jorge
e-mail: rnatal@fe.up.pt

M.P.L. Parente · R.M. Natal Jorge
FEUP, Faculty of Engineering, University of Porto, Porto, Portugal

F. Gentil
Clínica ORL – Dr. Eurico de Almeida,
Escola Superior de Tecnologia da Saúde do Porto, Porto, Portugal
e-mail: fernanda.fgnanda@gmail.com

L.F.M. da Silva (ed.), *Materials Design and Applications*,
Advanced Structured Materials 65, DOI 10.1007/978-3-319-50784-2_12

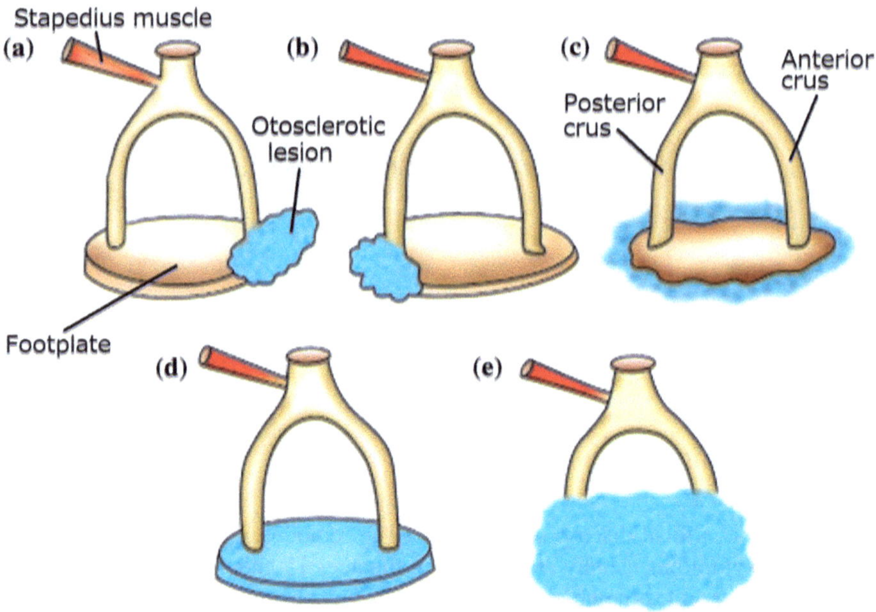

Fig. 1 Types of fenestral otosclerotic lesions that cause the fixation of the stapes footplate, figure adapted from [4]. **a** Anterior focus; **b** Posterior focus; **c** Circumferential; **d** Biscuit type lesion; **e** Obliterative lesion

Histologically, otosclerosis is characterized by three phases: the first phase shows bone absorption, the tissue is highly vascularized and macrophages and active osteoclasts are present; in the second phase a new bone formation occurs, starting around the vessels, leading to the characteristic blue mantles of Manasse; in the last phase, the otosclerotic foci appears as a scar with rare cells and calcification [6, 7].

There is no known nonsurgical treatment for this disease, although some physicians believe the use of sodium fluoride can reduce the enzyme activity derived from otosclerosis, aside from diminishing the rate of osteoblast-induced bone remodelling [8, 9]. This treatment is being used in the presence of cochlear otosclerosis. Amplification with a hearing aid may also help.

When there is a fixation of the stapes footplate, a surgical procedure is performed [4]. Stapedectomy is an operation in which the stapes suprastructures and footplate are removed and replaced by a prosthesis. On the other hand, in stapedotomy, a small hole is performed in the fixed footplate under microscope magnification and a piston prosthesis is inserted in the hole and connected to the long process of the incus from the other side. Husban [10] showed clearly that stapedotomy gives better hearing results and fewer complications than stapedectomy. Partial fixation of the malleus has been reported in literature in about 37.5% of the patients with otosclerosis.

Otosclerosis leads to a conductive hearing loss of 15–30 dB. Isolated congenital fixation of the malleus is rare, and may result from bony fusion of its head to the roof of the tympanic cavity or persistence of a bony bar attaching it to the posterior wall [11]. The partial fixation is caused by calcification of the anterior mallear ligament and the anterior mallear process [12].

Malleostapedotomy involves connecting the malleus directly to the vestibule through the oval window, instead of connecting the incus and the stapes [13]. Fisch et al. [14] analyzed 80 cases of revision surgery for otosclerosis, and found partial malleus fixation, total malleus fixation and incus fixation in 37.5%, 8.7% and 13.7% of the cases, respectively. Based on the work of Fisch et al. [14], malleostapedotomy yields better functional hearing results than incus stapedotomy in revision surgery for otosclerosis.

To help understanding the influence and the alterations caused by this disease and thus contribute to the development of more effective treatment techniques and prosthesis, a numerical study is proposed. A 3D model of the human ear was developed and the finite element method was employed to obtain the umbo and stapes footplate displacements. The results were validated by comparison with the previously published data of Huber et al. [12, 15]. The software *Abaqus Standard* [16] was used for the various analyses. A direct-solution steady-state dynamic analysis was performed to determine the frequency response. This type of analysis provides the steady-state amplitude and phase angle of the system response due to a harmonic excitation at a given frequency. The analysis is done by applying the loading at a series of different frequencies [16].

2 Materials and Methods

2.1 Geometrical Model

The project "The Visible Ear" [17] was the basis for the construction of the finite element mesh. In this project, a set of high quality images of a frozen temporal bone were obtained through a cryosectioning procedure. The slice thickness was 25 μm and high resolution images were captured every 50 μm. A total of 605 images with 24 bits RGB and a resolution of 50 μm/pixel were obtained. A total of 26 different structures of the human ear were identified through manual segmentation of the entire set of images. Using the segmented structures, the finite element mesh was constructed, including the tympanic membrane (TM), ossicular bones (malleus;

incus; stapes), cochlea, temporal bone and the air in tympanic cavity. The model also included two tendons (stapedius tendon and tensor tympani tendon), six ligaments (superior mallear ligament; lateral mallear ligament; anterior mallear ligament; superior incudal ligament; posterior incudal ligament; stapedius annular ligament) and the incudomalleolar and incudostapedial joints.

2.2 Material Properties

The mechanical properties attributed to the structural components of the model, ligaments, tendons and the acoustic medium are shown in Table 1. The mechanical properties of the structural components, ligaments and tendons were based on the work of Sun et al. [18] and Gentil et al. [19]. Linear elastic properties were incorporated in all parts of the 3D finite element model. The Poisson's ratio was assumed to be 0.3 for all components of the ear [18, 20, 21]. Raleigh's proportional damping with the coefficients $\alpha = 0\,s^{-1}$ and $\beta = 0.0001$ s were introduced in the ear components and in the ligaments and tendons [22, 23]. The TM was modelled with one layer [24], which was divided into *pars tensa* with orthotropic material properties and *pars flaccida* with isotropic material properties. The fluid inside the cochlea was modelled with acoustic elements; a density of 1000 kg/m^3 and a bulk modulus of 220 GPa was used in this medium [25].

2.3 Boundary Conditions

The TM was excited by a sinusoidal sound pressure level (SPL) of 80 dB. A frequency range of 100 Hz to 10 kHz was used. The nodes of the temporal bone on the exterior sectioned surface were considered fixed in all degrees of freedom. The magnitude of the displacements in the umbo and the stapes footplate for the normal ear and for an ear with otosclerosis were obtained.

The stapedius annular ligament was discretized with 641 linear tetrahedral (C3D4) elements (Fig. 2). To simulate the fenestral otosclerosis, the Young's modulus of the annular ligament was changed (increased) until the sound attenuation at low frequencies was approximately 30 dB, as proposed by Huber et al. [12]. The whole model was discretized with 140 441 nodes and 703 303 elements, of which 578 774 are linear tetrahedral (C3D4) elements, 124 522 are linear tetrahedral acoustic (AC3D4) elements and 7 are linear truss (T3D2) elements. The tendons and ligaments were simulated using truss T3D2 elements. The fluid inside the cochlea and the air in the tympanic cavity were meshed with acoustic AC3D4 elements and the remaining parts of the ear with tetrahedral C3D4 elements.

Table 1 Material properties assignment to finite element model

Human ear components	Density, ρ $\left[\frac{kg}{m^3}\right]$	Young's modulus, E [Pa]	Damping
Tympanic membrane	1.2×10^3	*Pars tensa* 3.2×10^7(*radial*) 2.0×10^7(*circumferential*) *Pars flaccida* 1.0×10^7	$\alpha = 0\,s^{-1}$ $\beta = 0.0001\,s$
Malleus	2.55×10^3 (*Head*) 4.53×10^3 (*Neck*) 3.70×10^3 (*Handle*)	1.41×10^{10}	
Incus	2.36×10^3 (*Body*) 2.26×10^3 (*Short process*) 5.08×10^3 (*Long process*)	1.41×10^{10}	
Stapes	2.2×10^3	1.41×10^{10}	
Incudomalleolar joint	3.2×10^3	1.41×10^{10}	
Incudostapedial joint	1.2×10^3	6.0×10^5	
Temporal bone	2.0×10^3	1.41×10^{10}	
Round window	1.2×10^3	1.0×10^7	
Ligaments and tendons			
Superior mallear ligament	2.5×10^3	4.9×10^4	
Lateral mallear ligament	2.5×10^3	6.7×10^4	
Anterior mallear ligament	2.5×10^3	2.1×10^6	
Superior incudal ligament	2.5×10^3	4.9×10^4	
Posterior incudal ligament	2.5×10^3	6.5×10^5	
Tensor tympani tendon	2.5×10^3	2.6×10^6	
Stapedius tendon	2.5×10^3	5.2×10^5	
Tympanic annulus	1.2×10^3	6.0×10^4	
Stapedius annular ligament	2.5×10^3	2.0×10^4	
Acoustic medium	**Density**, ρ $\left[\frac{kg}{m^3}\right]$	**Bulk modulus**, B [Pa]	
Cochlear fluid	1000	220×10^9	
Air in the tympanic cavity	1.164	1.01×10^5	

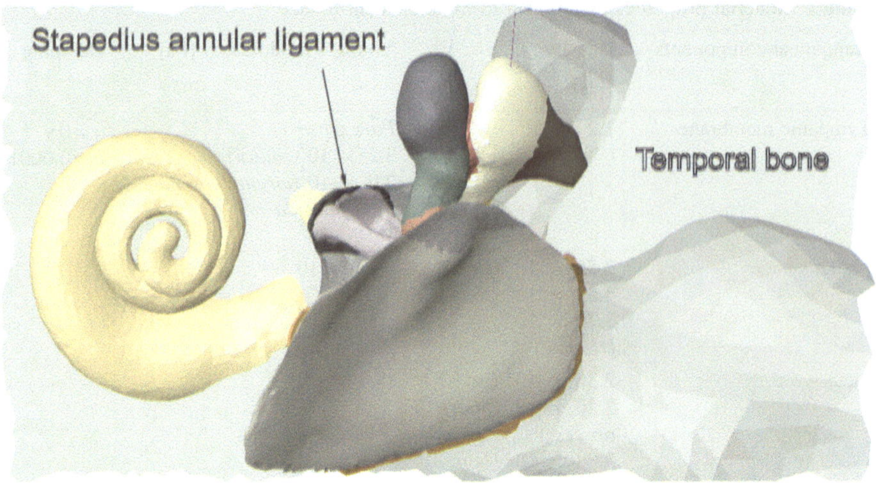

Fig. 2 Representation of the finite element model with the stapedius annular ligament

The connection between the acoustic elements and the structural elements of the ear was performed using the *TIE* command available in *Abaqus Standard* [16]. This option allows imposing acoustic-mechanical interactions between pairs of surfaces.

3 Results

The influence of the fenestral otosclerosis in the normal ear was studied. Figure 3 shows the magnitude of the displacement in the umbo, obtained through the numerical model developed in this work, and also the results achieved by Huber et al. [12]. The solid lines show the data for the normal ear, the dashed lines represent the ear with otosclerosis, and the dotted lines the displacement reduction caused by the otosclerosis. Figure 3 also shows the range of experimental values archived by Huber et al. [15], based on 90 experimental measurements in individuals with normal hearing. The data was obtained using a Doppler interferometry laser (LDI). Huber et al. [12] used the finite element method to study the presence of otosclerosis in the annular ligament, increasing the Young's modulus of the annular ligament until a sound conduction attenuation of approximately 30 dB in the stapes footplate, mainly at low frequencies, was achieved (see Fig. 4). In our work the Young's modulus of the annular ligament was increased from 0.02 Pa ($E_{normalear}$) to 6 Pa ($E_{otosclerosis}$) or otherwise 300 times higher. The numerical results obtained for the umbo displacement are close to the experimental lower limit obtained by Huber et al. [15]. In the model with otosclerosis there was a maximum reduction in the magnitude of the umbo displacement of 5 dB, located at 700 Hz. The umbo displacement response at low frequencies was almost constant. However,

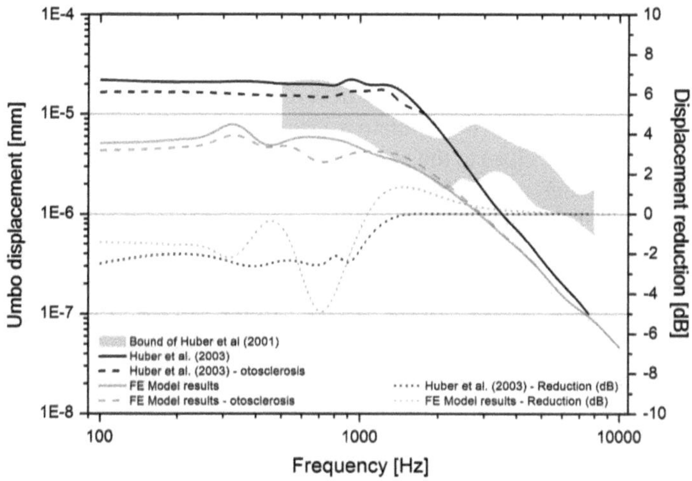

Fig. 3 Umbo displacements (normal and otosclerosis); Reduction of umbo displacement

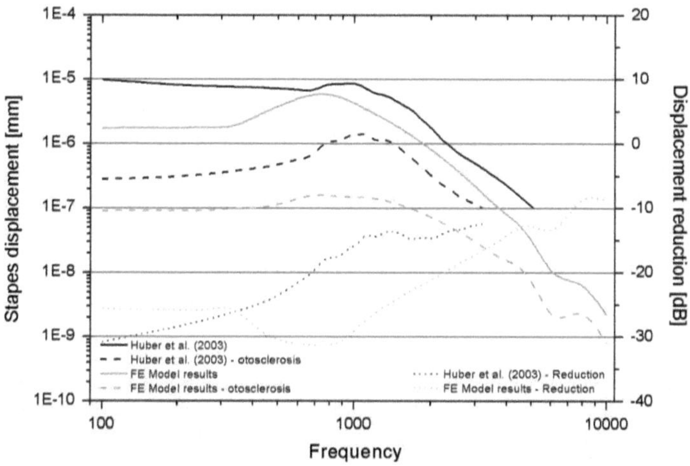

Fig. 4 Stapes footplate displacements (normal and otosclerosis); Reduction of stapes footplate displacements

at high frequencies, the umbo displacement decreases with the increase of the frequency.

Figure 4 shows the magnitude of the stapes displacement. A similarity in the results with those published by Huber et al. [12] was obtained. Otosclerosis results in a reduction of approximately 30 dB at low frequencies, which tends to decrease with the increase of the frequency. Comparing the magnitude of the umbo displacement with the stapes displacement, it can be concluded that the loss was more pronounced in the stapes footplate and at low frequencies.

4 Discussion

The Young's modulus of the annular ligament increased from 0.02 Pa (normal ear) to 6 Pa (ear with otosclerosis), to simulate a hearing loss of about 30 dB. The umbo displacement results are close to the experimental lower limit obtained by Huber et al. [15], the displacement was slightly affected by the alteration introduced in the Young's modulus of the annular ligament. A maximum umbo displacement reduction of 5 dB was obtained at low frequencies. Otosclerosis results in a reduction of stapes displacement at low frequencies of approximately 30 dB. The reduction tends to decrease with the increase of the frequency. Overall the hearing loss was more pronounced in the stapes footplate and at low frequencies. The numerical model reported here highlights the consequences of the fenestral otosclerosis present in the human ear.

5 Conclusions

Fenestral otosclerosis involves the stapes and results in formation of new, abnormal spongy bone, especially around the oval window, which results in the fixation of the stapes. The efficient transmission of sound is prevented because the stapes footplate cannot vibrate normally.

The study reported here allowed to obtain the impact of the fenestral otosclerosis on the hearing loss, by computing the displacements in the umbo and in the stapes footplate. The displacement reduction in the umbo and in the stapes footplate were also calculated. The finite element method provided by the software Abaqus Standard was used. The numerical results showed good agreement with the experimental and numerical results of Huber et al. [12, 15]. Overall the fenestral otosclerosis affects the motion of the stapes footplate, without change significantly the umbo response.

Acknowledgements The authors gratefully acknowledge the support from the Portuguese Foundation of Science under Grant IF/00159/2014, and the funding of Project NORTE-01-0145-FEDER-000022—SciTech—Science and Technology for Competitive and Sustainable Industries, co-financed by Programa Operacional Regional do Norte (NORTE2020), through Fundo Europeu de Desenvolvimento Regional (FEDER).

References

1. Miller, M.H., Schein, J.D.: Selected complex auditory disorders. J. Rehabil. Res. Dev. **42**(4), 1–8 (2005). doi:10.1682/Jrrd.2004.10.0136
2. Purohit, B., Hermans, R.: Imaging in otosclerosis: a pictorial review. Insights Imaging **5**(2), 245–252 (2014)
3. Gentil, F., Parente, M., Martins, P., Santos, C., Almeida, E., Ferreira, A., Natal, R.: Numerical study of Hough technique in surgery of otosclerosis, using the finite element method. Acta Bioeng Biomech. **17**(4), 149–153 (2015). doi:10.5277/Abb-00289-2015-03
4. Bansal, M.: Diseases of Ear, Nose and Throat. Jaypee Brothers, Medical Publishers Pvt. Limited (2012)

5. Sataloff, J., Sataloff, R.T., Sataloff, R.T.: Occupational Hearing Loss, 3rd edn. CRC Taylor & Francis, Boca Raton, FL (2006)
6. Mendonça, J.A., Ribeiro, R.: Ostoclerose. Revista de Ciências Médicas **14**(5) (2012)
7. Arnold, W., Häusler, R.: Otosclerosis and Stapes Surgery. Karger (2007)
8. Oliveira Vicente, A., Yamashita, H.K., Cruz, O.L.M., Suzuki, F.B., de Oliveira, Penido N.: The effectiveness of audiometric evaluation in drug treatment for otospongiosis. Braz. J. Otorhinolaryngol. **78**(2), 73–79 (2012)
9. Brunner, L.S., Suddarth, D.S., Smeltzer, S.C.O.C., Bare, B.G.: Brunner and Suddarth's Textbook of Medical-Surgical Nursing. Lippincott Williams & Wilkins (2004)
10. Husban, H.A.: Outcome of management of otosclerosis by stapedotomy compared to stapedectomy in a Jordanian population. Oman Med. J. **28**(1), 36–38 (2013). doi:10.5001/omj.2013.08
11. Kurosaki, Y., Tanaka, Y.O., Itai, Y.: Malleus bar as a rare cause of congenital malleus fixation: CT demonstration. Am. J. Neuroradiol. **19**(7), 1229–1230 (1998)
12. Huber, A., Koike, T., Wada, H., Nandapalan, V., Fisch, U.: Fixation of the anterior mallear ligament: diagnosis and consequences for hearing results in stapes surgery. Ann. Otol. Rhinol. Laryn. **112**(4), 348–355 (2003)
13. Ahn, S.H., da Kim, H., Choi, J.Y., Kim, B.G.: Two cases of malleostapedotomy in congenital oval window atresia. Korean J Audiol. **17**(3), 152–155 (2013). doi:10.7874/kja.2013.17.3.152
14. Fisch, U., Acar, G.O., Huber, A.M.: Malleostapedotomy in revision surgery for otosclerosis. Otol. Neurotol. **22**(6), 776–785 (2001)
15. Huber, A.M., Schwab, C., Linder, T., Stoeckli, S.J., Ferrazzini, M., Dillier, N., Fisch, U.: Evaluation of eardrum laser Doppler interferometry as a diagnostic tool. Laryngoscope **111** (3), 501–507 (2001). doi:10.1097/00005537-200103000-00022
16. Hibbit, D.K.B., Sorenson, P.: ABAQUS Analysis User's Manual Version 6.14-1, vol. 2014. Dassault Systèmes (2014)
17. Wang, H., Merchant, S.N., Sorensen, M.S.: A downloadable three-dimensional virtual model of the visible ear. ORL; J. Oto-Rhino-Laryngol. Relat. Spec. **69**(2), 63–67 (2007). doi:10.1159/000097369
18. Sun, Q., Gan, R.Z., Chang, K.H., Dormer, K.J.: Computer-integrated finite element modeling of human middle ear. Biomech. Model Mechanobiol. **1**(2), 109–122 (2002). doi:10.1007/s10237-002-0014-z
19. Gentil, F., Parente, M., Martins, P., Garbe, C., Jorge, R.N., Ferreira, A., Tavares, J.M.R.S.: The influence of the mechanical behaviour of the middle ear ligaments: a finite element analysis. Proc. Inst. Mech. Eng. [H] **225**(1), 68–76 (2011). doi:10.1243/09544119jeim783
20. Areias, B.A.F.: Simulação biomecânica do ouvido humano, incluindo patologias do ouvido médio. Faculdade de Engenharia da Universidade do Porto (2014)
21. Gentil, F.: Estudo biomecânico do ouvido médio. Universidade do Porto (2008)
22. Lee, C.F., Chen, P.R., Lee, W.J., Chen, J.H., Liu, T.C.: Three-dimensional reconstruction and modeling of middle ear biomechanics by high-resolution computed tomography and finite element analysis. The Laryngoscope **116**(5), 711–716 (2006)
23. Liu, Y., Li, S., Sun, X.: Numerical analysis of ossicular chain lesion of human ear. Acta. Mech. Sin. **25**(2), 241–247 (2009)
24. Ferris, P., Prendergast, P.J.: Middle-ear dynamics before and after ossicular replacement. J. Biomech. **33**(5), 581–590 (2000). doi:10.1016/S0021-9290(99)00213-4
25. Gan, R., Reeves, B., Wang, X.: Modeling of sound transmission from ear canal to cochlea. Ann. Biomed. Eng. **35**(12), 2180–2195 (2007). doi:10.1007/s10439-007-9366-y

Development and Validation of a Numerical Model for the Optimization of a Brace for Lower Limb

G. Bellavita, M. Cocconcelli, D. Castagnetti and R. Rubini

1 Introduction

AFO Orthosis is an external aid, characterized by a geometry adhering to the foot sole and to the patience's posterior tibial tract, that never exceeds the knee; it has the purpose to act as a corrective device for diseases that cause, mainly, plantar and back flexion of the ankle problems [1]. To meet the wide variety of diseases and patients, to which an orthosis must cope, its design is influenced by two factors: a purely geometrical, which must ensure the comfort of the user, and a second related to the mechanical response that the orthosis should provide to act as a corrective element. The corrective aspect and the consequent allocation process of the orthosis starts from the comparison between the patient's gait under observation and the "normal" gait, that is produced by statistical profiles reflecting the endemic variability in patients without specific diseases [2].

Two are the main phases of the step cycle: the support and the oscillation, each divided into the following sub-phases [1]:

G. Bellavita · M. Cocconcelli (✉) · D. Castagnetti · R. Rubini
Department of Sciences and Methods for Engineering, University of Modena and Reggio Emilia, Via G. Amendola, 2 – Pad. Morselli – Campus S. Lazzaro, Modena, Italy
e-mail: marco.cocconcelli@unimore.it

G. Bellavita
e-mail: giuseppe.bellavita@gmail.com

D. Castagnetti
e-mail: davide.castagnetti@unimore.it

R. Rubini
e-mail: riccardo.rubini@unimore.it

L.F.M. da Silva (ed.), *Materials Design and Applications*,
Advanced Structured Materials 65, DOI 10.1007/978-3-319-50784-2_13

Support: this phase affects 60% of the entire period, and it is divided into:

(a) **Initial Contact (10% of the cycle)**. The heel exchanges energy with the ground in order to support the action of body weight and the inertia given by the dynamics of the action. In this situation, both foots are in simultaneous contact with the ground, defining a short region of double support. After the initial support, the ankle has a slight rotation in response to the load of the body weight that causes a vertical displacement of the center of gravity.
(b) **Reaction to the load (10–30% of the cycle)**. The entire sole of the foot is in contact with the ground. The pressure distribution evolves from being localized on the heel to distributed over the whole sole of the foot. Simultaneously, the foot not interested in the exchange of energy with the soil, performs a forward swing.
(c) **Median step (30–50% of the cycle)**. In the moment in which the heel is lifted from the ground, the even distribution of pressure on the sole of the foot increases in the metatarsal region and on the toe tips.
(d) **Foot deadlift (50–60% of the cycle)**. The last step consists in the detachment of the toes of the foot from the floor, thus starting the swing step for this foot.

Oscillation: this step deals with the remaining 40% of the entire length of the step, and is organized in three steps.

(a) **Initial pendulum (60–73%)**. The lower limb moves forward immediately after the detachment due to the rotation of the hip, knee and back of the foot.
(b) **Median pendulum (73–85%)**. The limb moves forward, at the same time the ankle rotates and retrieves the extension of the foot due to the thrust with the ground.
(c) **Final pendulum (85–100%)**. To complete the gait cycle, the thigh decelerates and the foot prepares to the contact between the heel and the ground.

The step of the cycle phases is made possible by the complex kinematics and dynamics of the lower limbs, including the bones, the muscular system and the joints of hip, knee and ankle. The direct and inverse dynamic analysis of the lower limbs is made possible by acquiring the position and the spatial orientation of the bones during the gait and by measuring the reactions between soil and foot. These measurements, repeated on a statistically significant sample, have allowed defining —in literature—a series of "normality" curves which describe the angular excursions of the joints with reference to the single cycle of step. Significant variations in measurements taken on a patient compared to normal curves are an important indication to the physician regarding the patient's pathology [3] and the subsequent actions to correct the posture, for example through the use of orthoses. In fact, the orthoses increase the bending stiffness of the ankle [4], restoring muscle action that many patients are no longer able to impart.

In particular, the use of composite materials [5, 6] has enabled a considerable reduction of the weight of the orthosis, at the same time increasing the stiffness. The patient in need of a carbon AFO undergoes a visit from a specialist orthopedic center: the carbon orthosis is fabricated manually and then optimized according to the tests on the patient. These tests allow to calibrate the stiffness of the orthosis to the correct value [7], according to the medical evidence. Usually, the stiffness of the orthoses needs to be reduced by introducing an appropriate notch between the plantar and the leg. The calibration of the stiffness is still an operation that is performed manually by most orthopedic centers, thus being susceptible to errors and subsequent delays in the realization [2].

The purpose of this work is to achieve and experimentally validate a numerical model to evaluate the structural response of the orthosis during the step cycle and in particular to provide the stiffness as a function of its geometry [8], in order to use this model improving the redesign of the orthosis in the calibration step [9]. In literature there are similar studies—especially on plastic AFO—starting from the work of Chu et al. [10, 11], who developed a 3D FEM model to determine the maximum peak stress occuring in a polypropylene AFO in the gait. Papi et al. [12] focused on polypropylene AFO, measuring strain on the AFO during walking by the use of strain gauges. Copilusi et al. [13] developed a modular knee orthosis with flexible elements starting from a locomotion system kinematic analysis and a FEM analysis. Lee et al. [14] developed a design guide to find out the structural characteristics of polypropylene of AFO used for hemiplegics patients. Zou et al. [15] developed a finite element analysis (FEA) models to predict the mechanical behavior of carbon fiber reinforced AFOs, in order to speed-up the production and reduce the costs. Also Stier et al. [16] presented a three dimensional multi-scale structural analysis methodology to speed up the design process of a composite AFO, and an advanced FEM model [17] of a carbon fiber reinforced plastics orthosis, taking into account the possible delamination effects that constitutes one of the major failure modes.

2 Computational Model

The work deals with an ad hoc carbon AFO made from Centro Ortopedico Emiliano (Ottobock Italia). Due to confidential restrictions from the supplier, the number and the arrangement of the carbon fibers were not known. Reverse engineering was performed on the orthoses, by scanning its surface. Subsequently, the three dimensional model registered by the scanner was imported into the commercial software Rhinoceros, for processing surface nurbs. The reconstruction process consists in connecting the points acquired through two-dimensional surfaces [18]. This procedure leads to geometrical discontinuities, substantially due to incorrect arrangement of the points measured from the scanner, thus needing a post processing that allows their correction (Fig. 1). Through direct measurements on the orthoses we defined its exact geometry. A peculiar feature of the orthoses is its discontinuous and decreasing thickness in the direction of the metatarsal part; this appears to be consistent with the presence of areas with a different number of layers. This comes from the need for lowering the stiffness of the distal region.

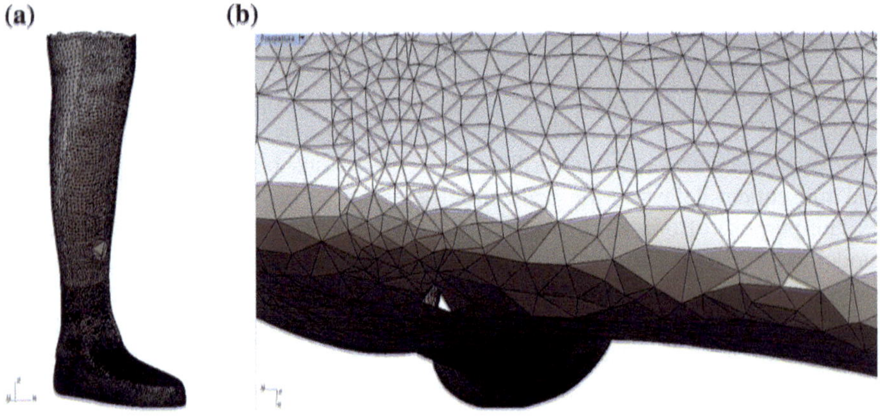

Fig. 1 Three-dimensional scan of orthosis (**a**) details of the gaps in the model reconstruction (**b**)

Subsequently, the model built on two-dimensional surfaces is imported into Solidworks for a further refinement of the geometric continuity and the creation of a solid element, experimentally considering the size and curvature on real orthosis. Figure 2 shows the final three-dimensional graphic model.

The finite element (FE) analysis aims to describe the stresses between the soil and orthoses during the propulsive action of the foot in the gait cycle [19–21]. The FE model (Fig. 3a) describes the orthosis through three dimensional shell elements with an average side length equal to 4 mm (Fig. 3b). The mechanical properties of the composite laminate was defined with regard to the single layer as reported in Table 1. The composite layup was defined for each different region of the orthosis: it ranges from a minimum of three to a maximum of six layers, from the tip of the insole to the bootleg. Figure 4 shows the layup of the laminate in the different regions of the orthosis. To reproduce the experimental test condition, the bar applying the load to the sole of the orthoses is modelled through a cylindrical rigid analytical surface with the same diameter of the rod employed. The contact between the rod and the plantar was described without friction, using a "hard" contact law in the normal direction. The orthosis model is fully constrained at ankle level (Fig. 3a) and is loaded by the rigid rod that apply a deflection equal to 10 mm in the vertical direction acting on the metatarsal bone part, so as to reproduce the experimental test condition.

The model describes the non-linearities of the system in terms of large displacements (plantar flexion), non-linearity of the material (carbon fiber), and the boundary conditions (contact between soil and orthosis). The model was implemented on the ABAQUS commercial software [22–25], and involves 93575° of freedom.

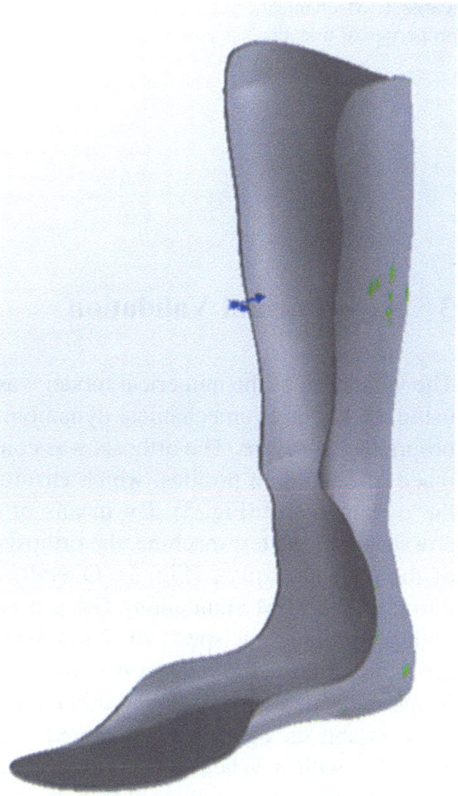

Fig. 2 Three-dimensional model of the orthosis

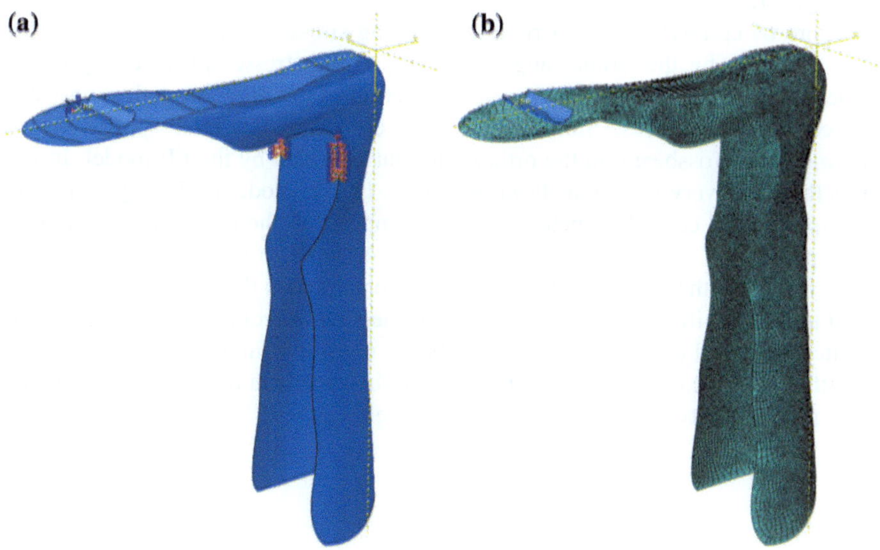

Fig. 3 Finite element model of the orthosis (**a**) and mesh (**b**)

Table 1 Mechanical properties of a layer

Mechanical properties	Value
E_{11}	113 (GPa)
E_{22}	7 (GPa)
G_{12}	4.5 (GPa)
G_{13}	2 (GPa)
G_{23}	2 (GPa)
ν_{12}	0.3

3 Experimental Validation

The validation of the numerical model was carried out through an experimental test using a 5 kN electromechanical dynamometer (Galdabini Sun 500), controlled by a proprietary software. The orthosis was constrained at the ankle by means of a frame made of aluminum profiles, which ensures the correct position and orientation on the dynamometer (Fig. 5). By means of a steel rod applied to the upper mobile crossbeam of the test machine, the orthosis was loaded in bending on the lower face of the metatarsal plate (Fig. 5). Overall, it is therefore reproduced the same condition of the FEM simulation. The test was carried out in quasi-static conditions, with a compression speed of 2 mm/min. Table 2 shows the details of the test procedure. In order to register the strain and the corresponding stress of the orthosis in the metatarsal area, a unidirectional strain gauge was bonded to the sole of the orthosis and its signal was registered via a National Instruments acquisition card NI-9237, with a Wheatstone bridge connection. At the end of each step of the experimental test procedure (Table 2), the strain data provided by the strain gauge was registered.

Figure 6a compares the experimental bending stresses registered on the surface of the orthosis by the strain gauge (in blue), to the stresses provided by the FE model (in red) in the same position, as a function of the different deflection values applied to the orthoses. Similarly, Fig. 6b compares the load applied by the dynamometer crossbeam on the orthosis to that provided by the FE model. In both cases there is a very good match: in particular, the FE model is slightly stiffer than the experimental case. This behavior can be imputed to the not ideal constraints of the experimental case. The only deviation is observed to the maximum displacement, in which the forecast of the stress underestimates the experimentally measured trend. Significant is the linear trend of the stress according to the load, which justifies the initial assumption of being able to consider - in a first approximation - the stiffness of the orthosis as constant during the cycle. On the whole, these results show the reliability of the developed FE model.

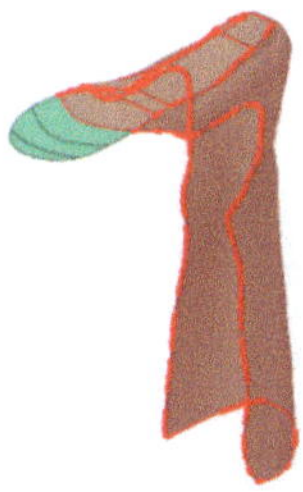

Layer	Thickness	Orientation angle
1	0.3	45
2	0.3	-45
3	0.3	0
4	0.3	0
5	0.3	-45
6	0.3	45

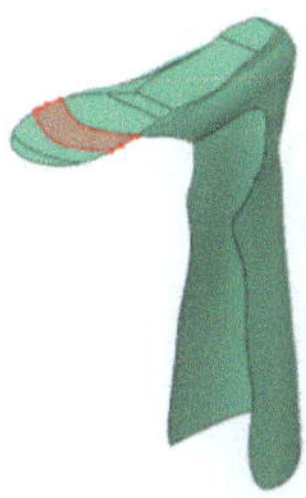

Layer	Thickness	Orientation angle
1	0.3	45
2	0.3	-45
3	0.3	0
4	0.3	-45
5	0.3	45

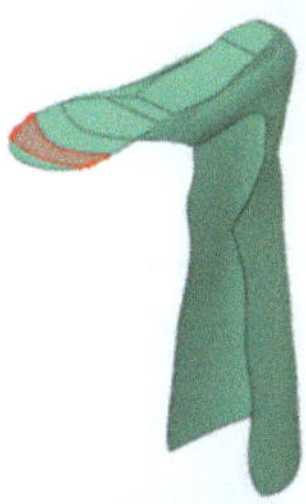

Layer	Thickness	Orientation angle
1	0.3	45
2	0.3	-45

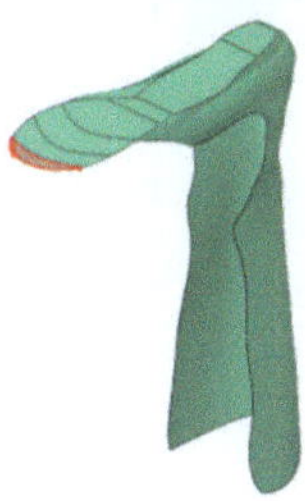

Layer	Thickness	Orientation angle
1	0.3	45
2	0.3	0
3	0.3	-45

Fig. 4 Layup of the composite laminate in different regions of the orthosis

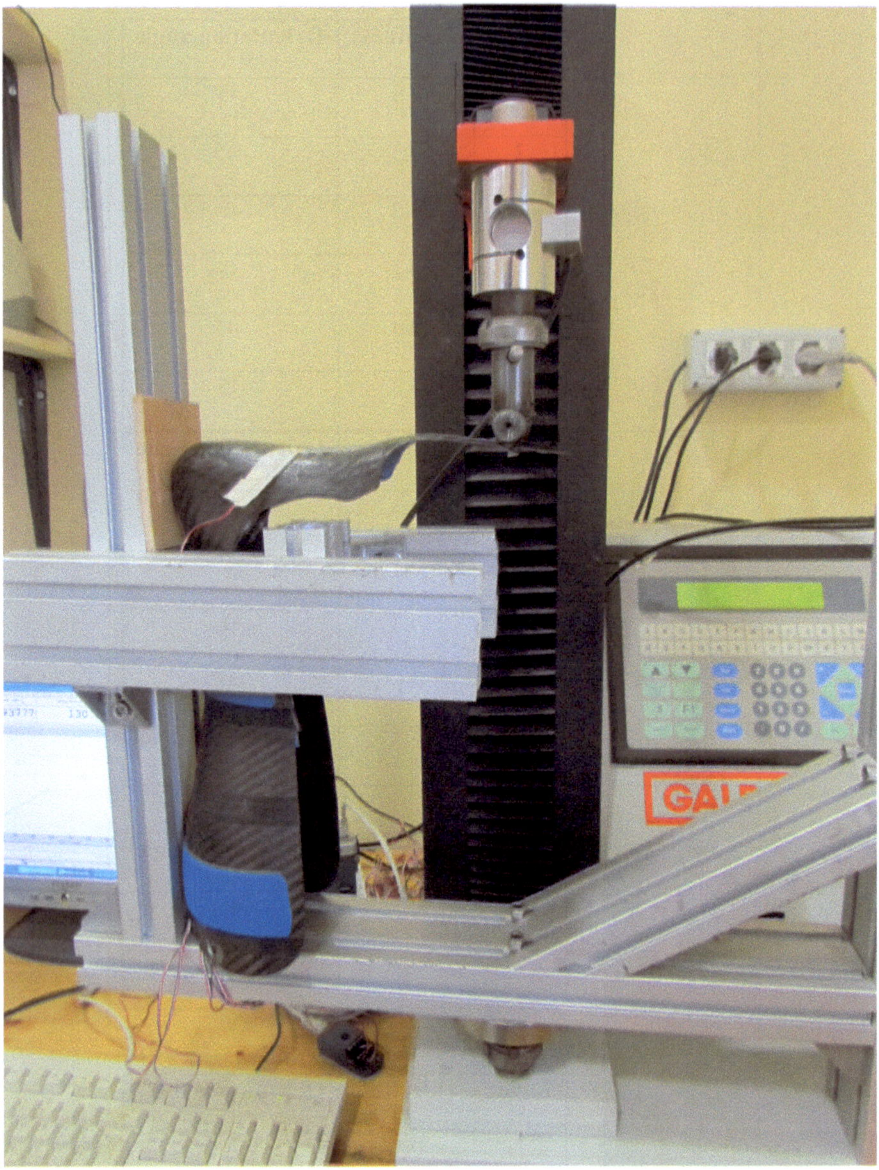

Fig. 5 Picture of the experimental test set-up

Table 2 Steps of the experimental test procedure

Step	Procedure
1	Deflection up to 2 mm (2 mm/min rate)
2	Stop (10 s): strain measurement with the extensometer
3	Deflection up to 4 mm (2 mm/min rate)
4	Stop (10 s): strain measurement with the extensometer
5	Deflection up to 6 mm (2 mm/min rate)
6	Stop (10 s): strain measurement with the extensometer
7	Deflection up to 8 mm (2 mm/min rate)
8	Stop (10 s): strain measurement with the extensometer
9	Deflection up to 10 mm (2 mm/min rate)
10	Stop (10 s): strain measurement with the extensometer

Fig. 6 Comparison between the computational prediction and the experimental results relative to the

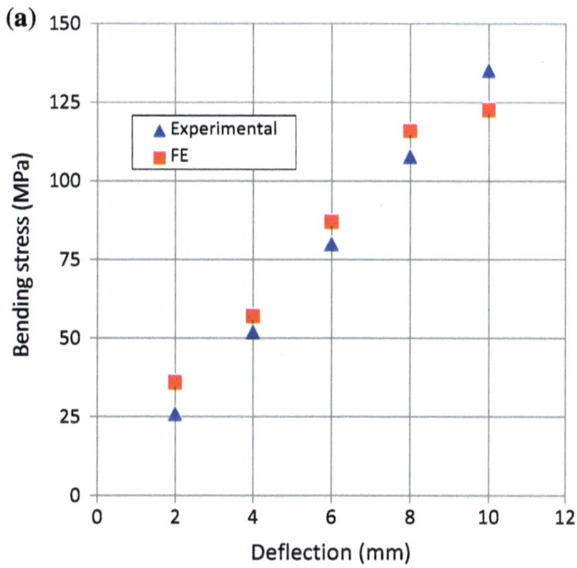

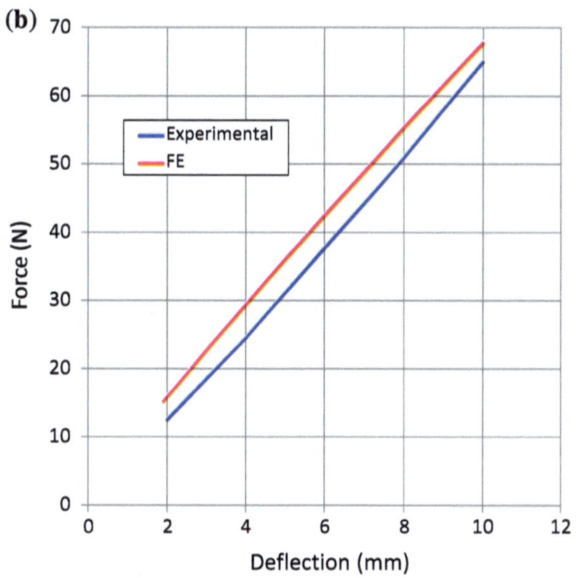

4 Structural Modifications

The proposed FE model allows to correlate the stiffness of the orthosis to its geometry. The stiffness of the orthosis is a remarkably important parameter since the first aim of the orthosis is to modify the ankle measured stiffness, making it a corrective device that allows overcoming the patient's deficiencies. Gait analysis monitors the degree of plantar and dorsal rotation of the foot, by modelling the ankle as a torsion bar, with reference to the malleolus. Consequently, it is possible to evaluate how the plantar and dorsal rotation of the foot deviate from the normal curve (Fig. 7).

Secondly, it is possible that the orthosis—being a rehabilitative tool—needs to be replaced or modified over time, as a function of the patient's improvements. By assuming a regression of the disease, it is possible to operate by removing material to obtain a reduction of the stiffness of the orthosis. The proposed method is the same adopted by orthopedic technicians, and the goal is to use the validated FE model to predict how the notches on the orthosis changes its stiffness, thus choosing the most appropriate operation, saving time and material.

The analysis uses a local reference system with X-axis orthogonal to the sagittal plane, Y-axis orthogonal to the frontal plane, Z-axis perpendicular to the transverse plane and the center above the malleolus. From the center of the reference system nine concentric notches with increasing radius were defined, to simulate the possible removal of material carried by the orthopedic laboratory. For each notch geometry the FE simulation was performed similar to that described in the previous section and taking into account the same load condition and constraints. Figure 8 shows the equivalent Von Mises stress maps for some of the examined geometries. The prediction of the force as a function of applied deflection has allowed, finally, to calculate the stiffness coefficient for each of the studied configurations.

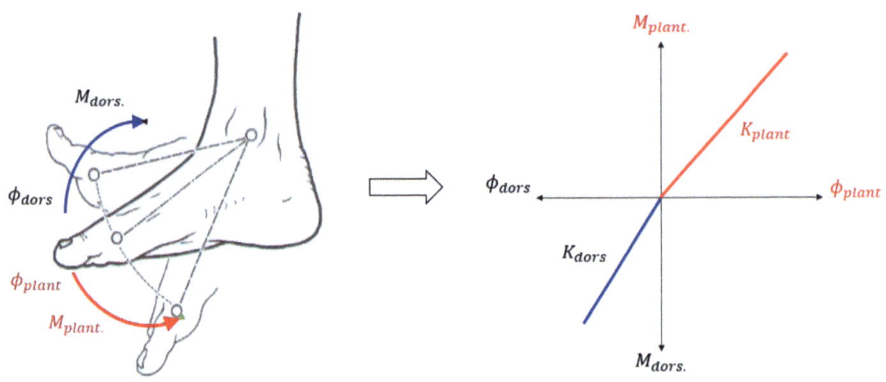

Fig. 7 Assessment of plantar and dorsal flexion (*on the left*) and the corresponding stiffness of an equivalent flexional spring (*on the right*)

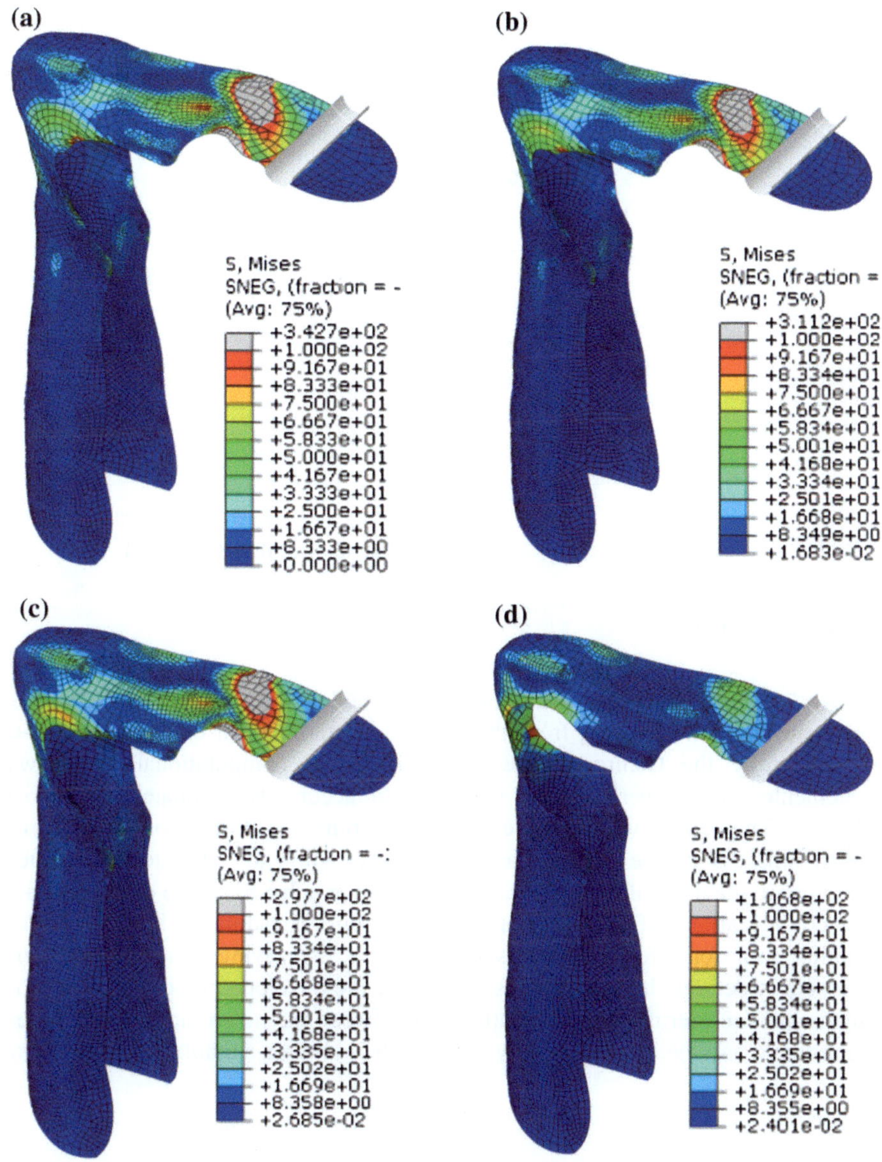

Fig. 8 Von Mises stress prediction for configuration 2 (**a**), 4 (**b**), 6 (**c**), 9 (**d**)

Figure 9 shows the value of the notch radius that must be realized to provide a given bending stiffness of the orthosis. The relation is highly non-linear: in particular, the stiffness is quite stable for a notch radius included between 10 and 30 mm and then collapses for more extended notches, due to the low residual resistant section.

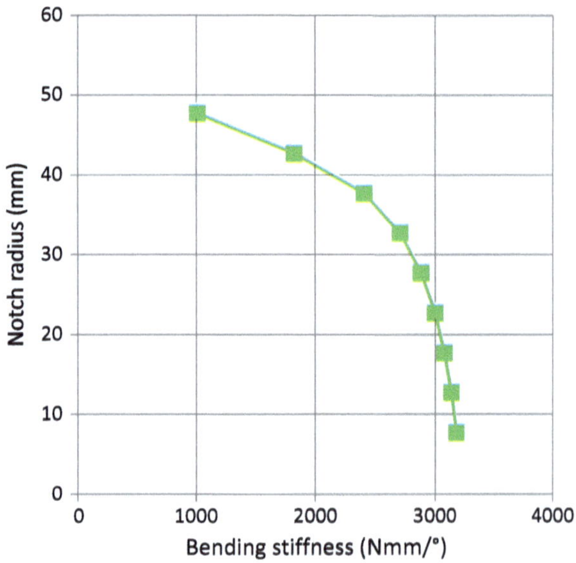

Fig. 9 Trend of the notch radius as a function of the bending stiffness of the orthosis

5 Conclusions

The work developed a numerical finite element model of an Ankle-Foot-Orthosis in composite material. Starting from the three-dimensional scan of the orthosis—ad hoc made by the Centro Ortopedico Emiliano—a computational model was implemented and experimentally validated to accurately simulate the orthosis behavior. The model was, then, used to determine the correlation between the overall stiffness of the orthosis and the notch radius to calibrate the stiffness according to the need of the patient. This correlation is based on the examination of nine configurations of circular notches of increasing radius: it is an important planning tool for effective actions on the orthosis, saving time and avoiding excessive structural changes. The usefulness of the developed model is even more remarkable considering that the relationship between the notch radius and stiffness is non-linear and therefore not easily predictable without a suitable calculation tool.

References

1. Perry, J.: Gait Analysis. Elsevier, Amsterdam, The Netherlands (2005)
2. Ferri, S.: Classificazione di pazienti affetti da diplegia infantile mediante cluster analysis. B.S. Thesis, University of Modena and Reggio Emilia, Reggio Emilia, Italy (2014). (In Italian)
3. Anderson, F.C., Pandy, M.G.: Individual muscle contribution to support in normal walking. Gait & Posture **17**, 159–169 (2003)
4. Harper, N.G., Esposito, E.R., Wilken, J.M., Neptune, R.R.: The influence of ankle-foot orthosis stiffness on walking performance in individuals with lower-limb impairments. Clin. Biomech. **29**(8), 877–884 (2014)

5. Rettig, O., Wolf, S., Doederlein, L.: Kinetics of a carbon spring AFO-orthosis and its influence on the kinetic of gait. Gait & Posture **18**(Suppl. 2), 94 (2003)
6. Bartonek, A., Eriksson, M., Gutierrez-Farewik, E.M.: A new carbon fibre spring orthosis for children with plantarflexor weakness. Gait & Posture. **25**, 652–656 (2007)
7. Bregman, D.J.J., van der Krogt, M.M., de Groot, V., Harlaar, J., Wisse, M., Collins, S.H.: The effect of ankle foot orthosis stiffness on the energy cost of walking: a simulation study. Clin. Biomech. **26**(9), 955–961 (2011)
8. Syngellakis, S., Arnold, M.A.: Modelling considerations in finite element analyses of ankle foot orthoses. WIT Trans. Ecol. Environ. **160**, 183–194 (2012)
9. Ginestra, P.S., Ceretti, E., Fiorentino, A.: Potential of modeling and simulations of bioengineered devices: Endoprostheses, prostheses and orthoses. Proc. Inst. Mech. Eng. **230** (7), 607–638 (2016)
10. Chu, T.-M., Reddy, N.P.: Stress distribution in the ankle-foot orthosis used to correct pathological gait. J. Rehabil. Res. Dev. **32**(4), 349–360 (1995)
11. Chu, T.-M., Reddy, N.P., Padovan, J.: Three-dimensional finite element stress analysis of the polypropylene, ankle-foot orthosis: static analysis. Med. Eng. Phys. **17**(5), 372–379 (1995)
12. Papi, E., Maclean, J., Bowers, R.J., Solomonidis, S.E.: Determination of loads carried by polypropylene ankle-foot orthoses: a preliminary study. Proc. Inst. Mech. Eng. **229**(1), 40–51 (2015)
13. Copilusi, C., Dumitru, N., Margine, A.: Modular knee orthosis fem analysis from kinematic considerations. Mech. Mach. Sci. **7**, 431–439 (2013)
14. Lee, Y.-S., Choi, Y.-J., Kim, H.-S., Lee, H.-S., Cho, K.-H.: A study on the structural stress analysis of plastic ankle foot orthosis (AFO) under dorsiflexion and plantarflextion conditions. Int. J. Mod. Phys. B **20**(25–27), 4559–4564 (2006)
15. Zou, D., He, T., Dailey, M., Smith, K.E., Silva, M.J., Sinacore, D.R., Mueller, M.J., Hastings, M.K.: Experimental and computational analysis of composite ankle-foot orthosis. J. Rehabil. Res. Dev. **51**(10), 1525–1536 (2014)
16. Stier, B., Simon, J.-W., Reese, S.: Numerical and experimental investigation of the structural behavior of a carbon fiber reinforced ankle-foot orthosis. Med. Eng. Phys. **37**(5), 505–511 (2015)
17. Stier, B., Simon, J.-W., Reese, S.: Finite element analysis of layered fiber composite structures accounting for the material's microstructure and delamination. Appl. Compos. Mater. **22**(2), 171–187 (2015)
18. Rhinoceros NURBS modeling for Windows, Robert McNeel & Associates, Seattle, WA, United States
19. Chung, D.D.L.: Carbon Fiber Composites. Butterworth-Heinemann, Elsevier, Oxford, United Kingdom (1994)
20. Gay, D., Hoa, S.V.: Composite Materials: Design and Applications, 2nd edn. CRC Press, Taylor & Francis Group, Cleveland, Ohio (2007)
21. Cherouat, A., Borouchaki, H.: Present state of art of composite fabric forming: geometrical and mechanical approaches. Materials **2**, 1835–1857 (2009)
22. Abaqus V.6.11 User manual. Providence, Rhode Islands, United States
23. Berge, J., Lerneryd, E., Heintz, P.: Spring orthosis analysis. Finite element modeling and optimization of a composite material. In: SIMULIA Customer Conference. Dassault Systemes, Vélizy-Villacoublay Cedex, France, (2011)
24. Svärd, L.: Composites failure modeling and optimization of a spring orthosis. M.S. Thesis, Chalmers University of Technology, Gothenburg, Sweden (2012)
25. Miconi, A.: Caratterizzazione meccanica e chimico-fisica del peek caricato con fibre di carbonio per la realizzazione di protesi acetabolari dell'anca. M.S. Thesis, University of Padova, Padova, Italy (2010). (in Italian)

Part V
Power Generation

Electrical and Geometrical Optimization for a 2DoF Non-linear Energy Harvester

E. Boco, R. Frizzell and J. Punch

1 Introduction

Energy harvesting is regarded as one of the most promising strategies to replace fossil fuels in the future. In fact, energy is present in the ambient in many forms, such as solar radiation, thermal gradients, wind and vibrations. Several different techniques to convert them to usable electrical energy have been studied, usually via new materials such as photothermal and photochemical devices, nano-engineered p-n junctions and piezoelectrics, or the creation of new structures like windmill blades, hydroelectric turbines and so on. This paper deals with vibrational energy harvesting since vibrations are one of the most widespread energy source in the ambient: every object is vibrating at a certain frequency and amplitude. The conversion method present in this paper is electromagnetic induction: having magnets and coils in relative motion, current is induced in the coils depending on the relative velocity between the components, following the Faraday law of induction.

A few years ago, vibrational energy harvesters were mainly based on spring-mass resonator structures, which allowed them to respond to a single frequency excitation with high efficiency, but which perform relatively poorly at non-resonant frequencies. This is an important limitation, considering that real vibrations are often characterized by wide and variable spectra [6, 8, 12, 18]. Recently, the focus has shifted to non-linear systems, which are less powerful for converting single frequencies, but

E. Boco (✉) · J. Punch
Stokes Laboratories, Engineering Research Building,
University of Limerick, Limerick, Ireland
e-mail: eisabetta.boco@ul.ie

J. Punch
e-mail: jeff.punch@ul.ie

R. Frizzell
Efficient Energy Transfer (ηET) Department, Bell Labs, Nokia, Blanchardstown Business and Technology Park, Snugborough Road, Dublin 15, Blanchardstown, Ireland
e-mail: ronan.frizzell@nokia-bell-labs.com

L.F.M. da Silva (ed.), *Materials Design and Applications*,
Advanced Structured Materials 65, DOI 10.1007/978-3-319-50784-2_14

which can have a wider bandwidth: it was thus demonstrated that non-linear systems are more effective than linear systems if tested under noise excitation [1, 9–11, 14, 21, 23]. However, one of the more important energy harvesting limitations at the moment is the low power that every technique is able to achieve, so the development of strategies to optimize the power output from each device is now becoming fundamental for any practical development [2, 5]. This is particularly complex for non-linear devices, since power output depends on a large number of the system's parameters and its value and spectral shape can vary with the input conditions [4].

In this paper, a model is presented to optimize both the electrical and the height of a non-linear energy harvester which features cubic stiffening effect due to the presence of magnetic springs. In Sect. 2 the device is described in detail. In Sect. 3 the finite element models for the Halbach stack's magnetic field and for the magnetic springs are discussed. Finally, the estimated fields are then used in Sect. 4 to build a numerical equation-based model in Matlab to optimise the electrical and mechanical configurations. Electrical optimisation is achieved considering the number of turns per coil under harmonic excitation, while the cap height is varied under coloured noise excitation, when optimising the mechanical response.

2 The Device

The harvester consists of two masses oscillating one inside the other between two sets of magnetic springs which enable collisions in order to transfer momentum from the heavier to the lighter mass, as shown in Fig. 1 [3, 7]. Moreover, the use of magnetic springs allows the device to respond at a lower resonance frequency if compared to

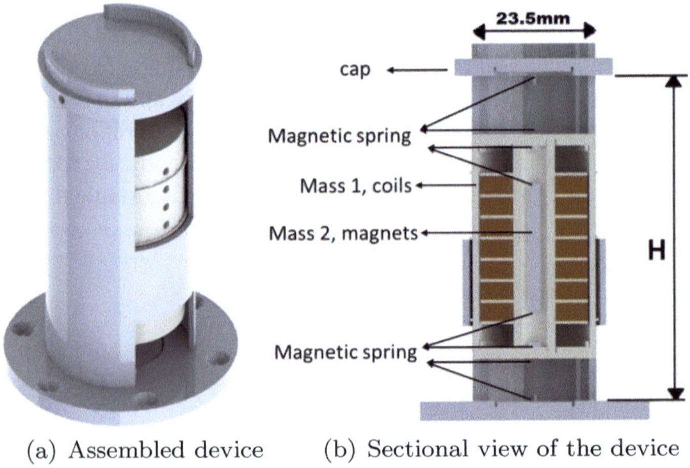

(a) Assembled device (b) Sectional view of the device

Fig. 1 The energy harvester consisting of seven coils and a stack of magnets relatively oscillating mechanical springs and with a higher restitution coefficient, which is valuable for a device which can be operated by low frequency vibrations, such as human motion [20]. The height of the cap can vary between $H = 50.10$ mm and $H = 57.45$ mm in order to change the shape of the harvester's spectral response. The outer (heavier) mass is made of seven coils and the inner (lighter) mass consists on a stack of five magnets arranged in a Halbach configuration: alternating diametrically and axially polarized magnets, as discussed in details in Sect. 3.1, in order to obtain a magnetic flux which closes on the stack itself, together with a higher gradient, and an associated higher induced current [22].

In the electrical optimization model, the harvester was excited using a harmonic signal and three sets of coils were simulated, having $N = 1500$, $N = 2800$ and $N = 4800$ turns per coil, in order to compare the results with experimental data. In the geometrical optimization, the height of the cap changed over the range $H = 50.10$ mm to $H = 57.45$ mm, while the coils have $N = 2800$ turns each. In both cases each coil was connected in such a way to minimise the cancelling currents. The voltage and power output were computed across an optimal load resistance, which was found experimentally.

3 Finite Element Model

In this Section, the Comsol finite element models for both the Halbach stack and the magnetic springs are presented. A distinction is made between the internal and the external springs: the length of the first is fixed, while for the latter analytic expressions are found for different values of the cap height.

3.1 Halbach Stack

The Halbach configuration for a stack of magnets consists of alternating diametrically and axially polarized magnets, in order to have the magnetic field lines closing on the stack itself, as shown in Fig. 2. The produced magnetic field is one-sided, and its shape is as follows:

$$H(x, y) = H_0 e^{ikx} e^{iky} \tag{1}$$

where H_0 is the field at the surface of the array, and k is the spatial wave-number. On the other side, the flux ideally cancels by interference.

The simulated stack was made of five NdFeB magnets having equal radius and thickness, $r_m = t_m = 3.175$ mm. The surrounding element was air as defined in the built-in materials, having relative permeability $\mu_r = 1$ and electrical conductivity $\sigma = 0$.

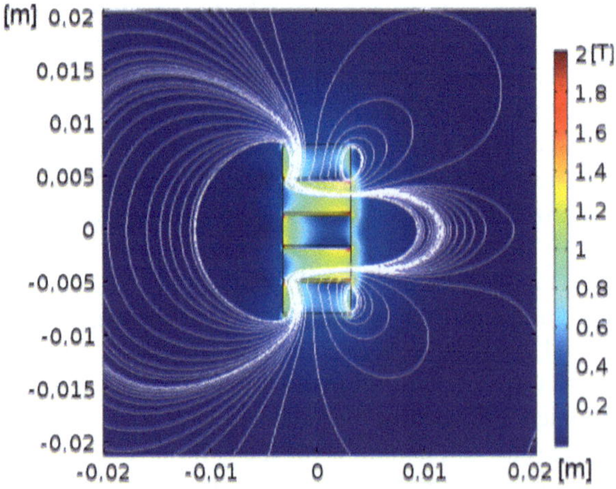

Fig. 2 Halbach configuration for the stack. The *magnetic flux lines* are in *white* and the *colour scale* refers to the magnetic flux density norm in T

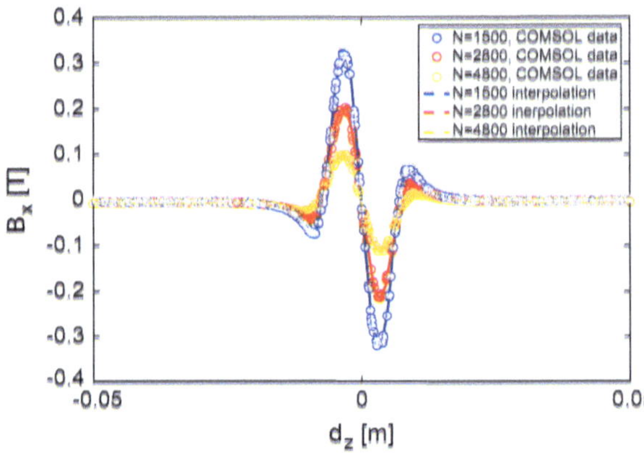

Fig. 3 Radial magnetic field in the central point of the thickness for each set of coils along the *vertical axis* of figure (**a**)

The radial component of the magnetic flux was thus computed at $r = 5.175$ mm, $r = 5.9375$ mm and $r = 7.2$ mm distance from the centre of the stack, which are the medium points of the thickness for the $N = 1500$, $N = 2800$ and $N = 4800$ turns set of coils respectively. The radial component of the magnetic flux density variation as a function of the position along the stack length is shown in Fig. 3.

3.2 Magnetic Springs

In this Section, the finite element simulations for the external and internal springs are shown, and the Matlab fitting is discussed for each case.

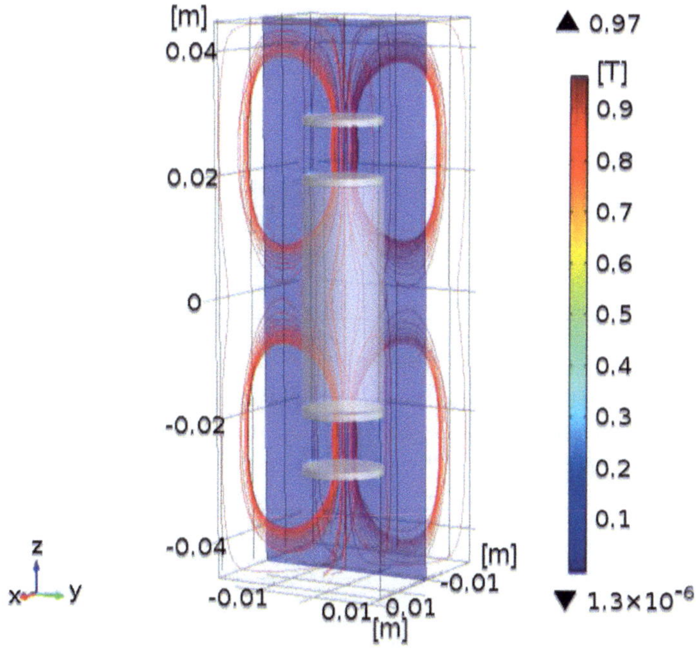

Fig. 4 3D model for the external springs. The *colour scale* refers to the magnetic field norm, and the *red lines* are the field lines

3.2.1 External Springs

A magnetic spring consists of two magnets configured to repel each other: the repulsive force is larger when the two magnets are closer, and weaker when the magnets are distant, behaving similarly to compressed and uncompressed mechanical springs. For the external springs, the simulated NdFeB magnets had a radius of $r = 6.35$ mm and a thickness of $t = 1.6$ mm, and were surrounded by air. The external mass was defined as an air cylinder, in order to not induce perturbations to the springs field, as shown in Fig. 4. A sweep of the mass position was then performed, and for each case the total force on the centre of the mass was computed, in order to obtain a force behaviour similar to a mechanical spring with a cubic stiffness. The same simulation was performed lowering the cap in the range $H = 50$ mm to $H = 57$ mm, and the complete set of results is shown in Fig. 5a, together with the interpolated function, using a $f = -k_1y^3 - k_3y$ shape. The cubic stiffening effect is the most commonly used in literature for magnetic springs, and comes from the coupled effect of the upper and lower repulsive forces [15, 17].

The variation in the linear and non-linear fitting coefficients for the range of cap heights was then studied: in Fig. 5b, c, the logarithmic values for k_1 and k_3 for each cap height are shown, together with a cubic polynomial fit. The polynomial order

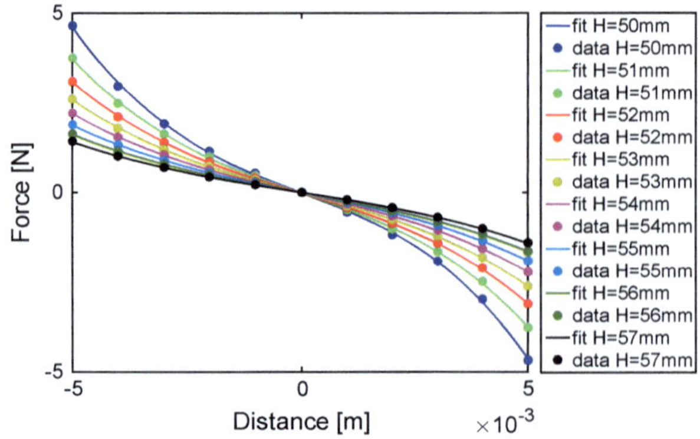

(a) Magnetic spring force as a function of cap height, Matlab interpolation for the Comsol data points with cubic fitting function.

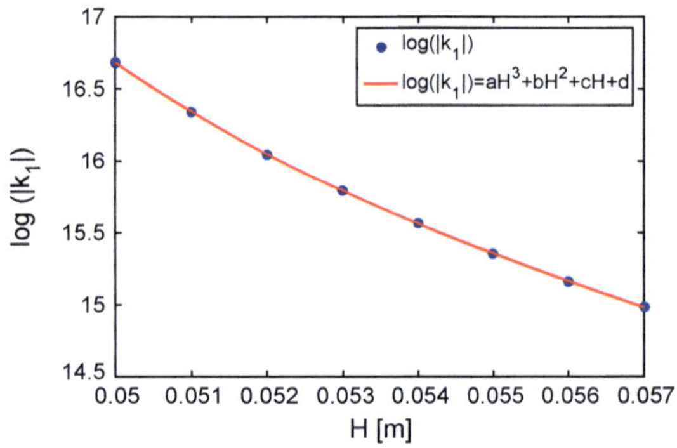

(b) Cubic polynomial fitting for k_1 as a function of cap height

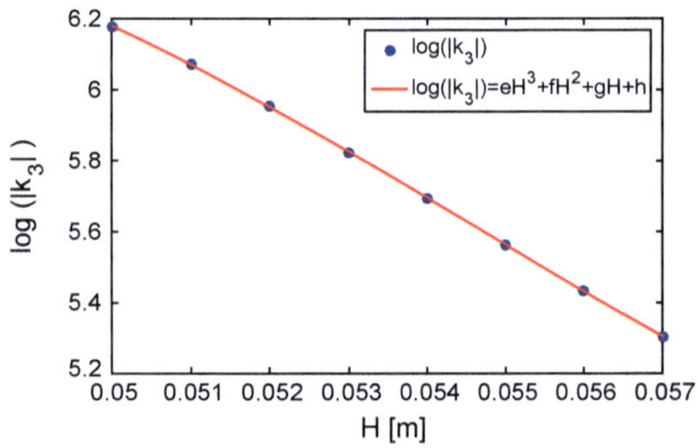

(c) Cubic polynomial fitting for k_3 as a function of cap height

Fig. 5 Logarithmic fit for the variation in the springs coefficients as a function of cap height

Table 1 Fit parameters for the spring force as a function of cap height

	a	b	c	d
$log(\|k_1\|)$	−1.304e6	2.225e5	−1.281e4	264.6
$log(\|k_3\|)$	4.387e5	−7.189e4	3795	−58.68

Fig. 6 3D numerical model and fitting for the internal springs

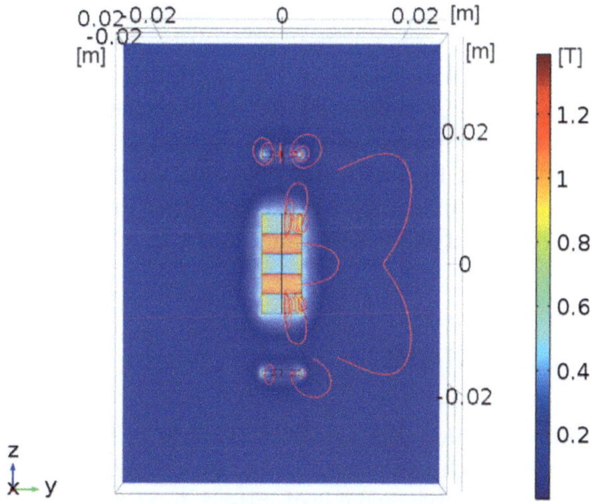

(a) 3D model of the Halbach stack. The colour scale refers to the magnetic flux density norm and the red lines to the magnetic field lines

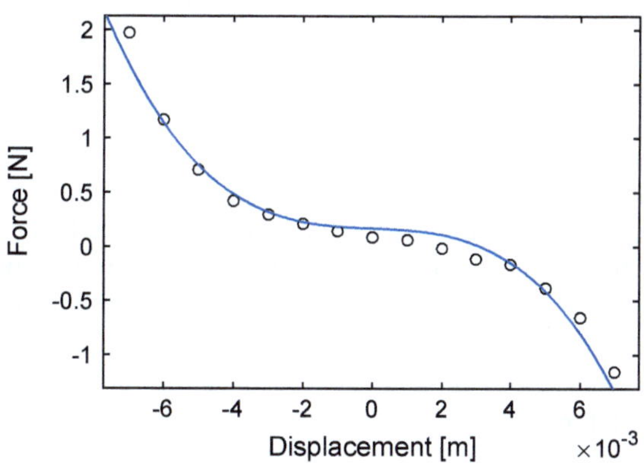

(b) Cubic polynomial fitting for the internal springs.

was chosen as the smallest one which was still capable of reproducing the behaviour of the data.

The obtained values for k_1 and k_3 with 95% confidence bounds are listed in Table 1.

The resulting values were used, as described in Sect. 4, to describe the magnetic spring force while lowering the cap, without any need to simulate each case separately.

3.2.2 Internal Springs

A similar procedure to the one described in Sect. 3.2.1 was followed to simulate the internal springs, shown in Fig. 6a. In this case, the total force on the lighter mass was calculated subtracting the effect of the repulsive forces on the bottom and the top magnets instead of being computed on the centre of mass. This avoided contribution from the interactions between the magnets forming the stack. The resulting data points for the force and cubic fitting are shown in Fig. 6b.

The fitted parameters values were: $k_1 = 3 \cdot 10^6$ N/m^3 and $k_3 = 36.7$ N/m.

4 Full Model

In this Section, the numerical models in Matlab are implemented. In Sect. 4.1, a model which can reproduce the experimental optimization condition is presented, and a predictive model is discussed. In Sect. 4.2, the geometrical optimization based on varying cap height in the presence of noise excitation is discussed, in order to predict the existence of an optimal stiffness under real excitation conditions. The predictions are then compared with experimental data.

4.1 Electrical Optimization

A 2 Degree of Freedom electromagnetic energy harvester can be modelled using two mechanical equations (one for each mechanical DoF) and one equation related to the electrical conversion which is coupled to the other two. The conversion effectively behaves as a damping, appearing in the mechanical part multiplied by the velocity of the mass. For a 1 Degree of Freedom linear energy harvester, the power optimization law is well known, and it is based on the total power dissipated by a simple linear oscillator:

$$P_{tot}(\omega/\omega_N) = \frac{m\xi_{tot}A^2(\omega/\omega_N)^3\omega^3}{[1-(\omega/\omega_N)^2]^2 + [2\xi_{tot}(\omega/\omega_N)]^2} \tag{2}$$

where m is the oscillator mass, ξ_{tot} is the total damping which can be split into mechanical and electrical components as $\xi_{tot} = \xi_{mech} + \xi_{em}$, A is the amplitude of the displacement given to the device, and ω and ω_N respectively are the excitation and the natural angular frequencies. Splitting the damping into mechanical and the electrical parts and solving the equation for $\omega = \omega_N$, the power dissipated through the electrical conversion can be computed:

$$P(\omega_N)_{em} = \frac{mA^2\omega_N^3\xi_{em}}{4(\xi_{em} + \xi_{mech})^2} \tag{3}$$

It is easy to differentiate this expression in order to reveal that the maximum power is dissipated electrically when $\xi_{em} = \xi_{mech}$ [2].

However, considering the non-linearities in the system and additional DoFs, the derivation of a simple expression to describe the optimization condition depending on few parameters become impossible. Moreover, approximation techniques to obtain the analytical solutions must be used, which are not valid for strong non-linear behaviour and which diverge easily from the real solution. For this reason, a numerical model was built to describe the system based on the three electro-mechanical equations shown in Eq. 4, and integrated numerically in Matlab [19]:

$$\begin{cases} m_1\ddot{z}_1 = -(d_1 + d_2)\dot{z}_1 - k_a z_1^3 - k_b z_1 + k_c(z_2 - z_1)^3 + k_d(z_2 - z_1) + d_2\dot{z}_2 + \\ \quad - d_{em}(\dot{z}_1 - \dot{z}_2) - F_{ext} - (m_1 + m_2)g \\ m_2\ddot{z}_2 = -d_2\dot{z}_2 - k_c(z_2 - z_1) - k_d(z_2 - z_1)^3 + d_2\dot{z}_1 - m_2 g - d_{em}(\dot{z}_2 - \dot{z}_1) \\ \dot{V}_L = -V_L(R_L + R_C)/L + V_{ind}RL/L \end{cases} \tag{4}$$

where the subscripts 1 and 2 refer respectively to the external and the internal mass, subscripts a and b are used to identify respectively k_1 and k_3 for the external mass, while c and d refer to coefficients for the internal mass. The displacement z is relative to the centre of the harvester. For the damping, $\xi = d/2m\omega_N$. To compute the open circuit inducted electromotive force, V_{ind}, each coil was divided into 10 smaller coils, in which the magnetic field shown in Fig. 3 is considered to have a constant value equal to its amplitude at the centre of each coil considered. This value was determined for each coil integration step as [13]:

$$V_{ind,step} = B_c l(\dot{z}_2 - \dot{z}_1) \tag{5}$$

where l is the length of the wire for the coil integration step. The electromagnetic damping was similarly computed for each coil as:

$$\gamma_{em} = \frac{B_c^{2l^2}}{R_L + R_C} \tag{6}$$

where $R_{C,coil}$ is the coil resistance for each coil.

Subsequently, the voltage for each integration step was summed in series, while the voltage induced in each whole coil was summed to the next in order to have the

Fig. 7 Experimental and simulated electrical optimization for the harvester at a = 0.4 g

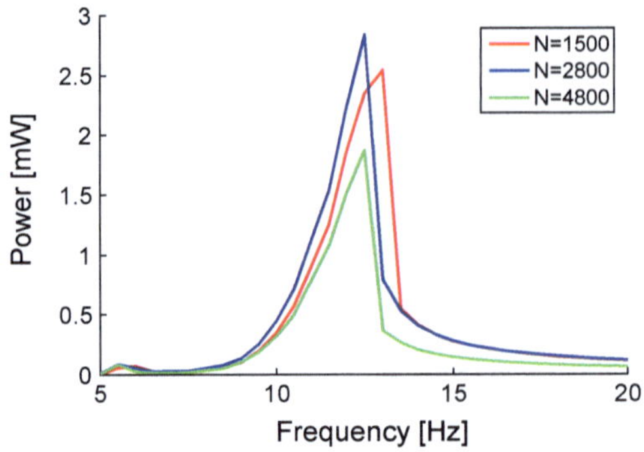

(a) Simulated power output for three sets of coils having $N = 1500$, $N = 2800$ and $N = 4800$ turns for each coil

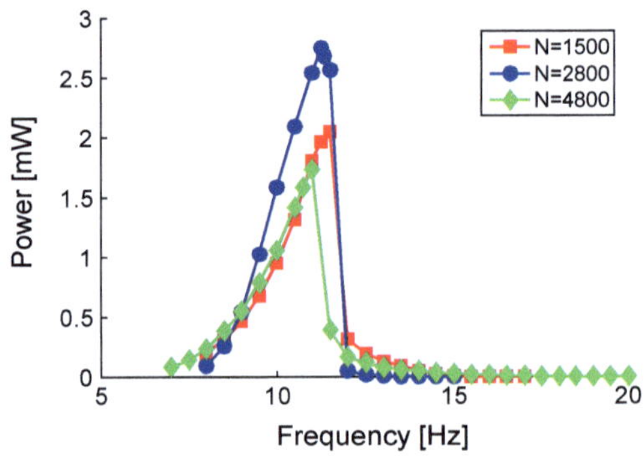

(b) Experimental power output as a function of frequency for $N = 1500$, $N = 2800$ and $N = 4800$

smallest possible cancelling currents in the seven coils. For each considered N for the set of coils, the magnetic field produced by the stack was computed in the center of the coil thickness, as already explained in Sect. 3.

Figure 7a shows the results of the simulation for three coils of varying N ($N = 1500$, $N = 2800$ and $N = 4800$). Figure 7b shows the experimental verification of the model, which demonstrates that it is capable of reproducing the experimental optimal value of $N = 2800$.

The tests were performed at $a = 0.4$ g acceleration with a cap height $H = 57.45$ mm. The model is able to predict the optimization condition as well as the power output within 91% confidence on the best coil, enabling predictive simulations to be run as guidelines for the construction of the device.

4.2 Geometrical Optimization

The same motion and conversion equations were used to model the power output as a function of the cap height. In this case, coloured noise was used as excitation. In order to generate the coloured noise, the Euler-Maruyama integration of stochastic processes was used [16]. A stochastic process x can be defined as:

$$x = x_0 + \int_{t0}^{t} G(x)\xi ds \tag{7}$$

where ξ is defined to satisfy $\int_{t0}^{t} \xi ds = W$. W is called a Wiener process and it has the following properties:

- W(0) = 0;
- s and t are two time instants so that $0 \leq s \leq t \leq T$ then $W_t - W_s$ follows a normal distribution with zero mean and $\sigma^2 = t - s$;
- u and v two time instants so that $0 \leq s \leq t \leq u \leq v \leq T$ then $W_t - W_s$ and $W_v - W_u$ are independent;

Thus Eq. (7) can be rewritten as $\int_0^{W_t} G(x_t)dW_s$, and it can be integrated with Euler-Maruyama obtaining:

$$x(T) = x(t_0) + \sum_{j=0}^{N-1} G(x(t_j), t_j)dW_j \tag{8}$$

In the numerical simulation, the Wiener increment is defined as $dW = \sqrt{(dt)}g(t)$, where g(t) is a Gaussian stochastic white process with zero mean, and it can be obtained using the random number generator in Matlab. Then the coloured noise is generated by integrating with Euler-Maruyama Eq. (9):

$$\frac{d\chi}{dt} = -\frac{1}{\tau}\chi + \frac{\sigma}{\tau}g(t) \tag{9}$$

as described for a general random process x(t). The resulting signal is exponentially correlated, which means that the correlation between two time instants is:

$$< \chi(t)\chi(t') > = < \chi >^2 e^{-|t-t'|/\tau} \tag{10}$$

This is clearly not a completely random signal, for which the correlation between two instants would be zero. In fact, a coloured noise is defined as a white noise with a frequency cut-off: its spectrum is limited by a time τ after which the spectral amplitude is the half of its initial value, and this brings it to the non-zero correlation. The coloured noise was used as the excitation F_{ext} in the system defined in Eq. (4) The results for cap heights varying between $H = 50$ mm to $H = 57$ mm is shown in Fig. 8, together with experimental data.

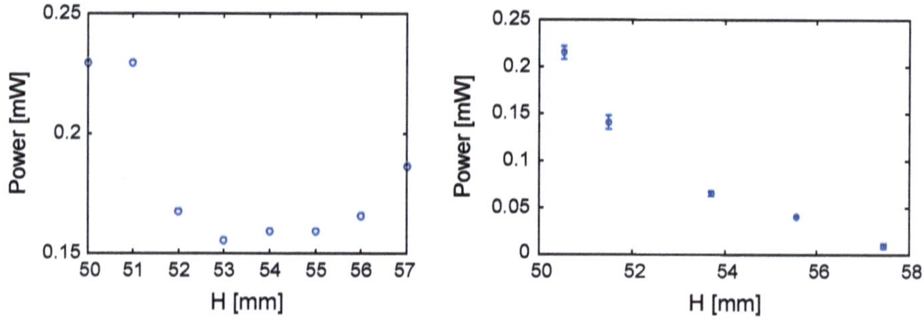

(a) Simulated power output as a function of the cap height H

(b) Experimental power out put as a function of the cap height H

Fig. 8 Simulated and experimental results changing H for the optimization of the power output

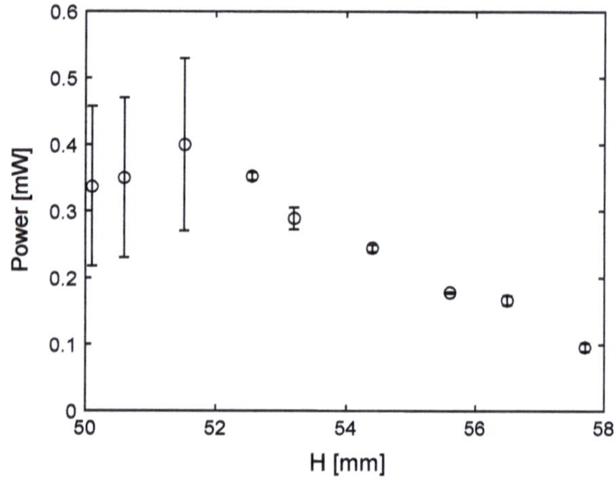

Fig. 9 Experimental power output with 1.6 mm thickness magnets

The model was able to predict the power output shape, and the simulation result for the maximum power is compatible with the data points, if considering the simulated uncertainty 0.06 mW (not plotted for better visibility of the data points). The decrease in power output for higher cap heights is smaller than the experimental behaviour, and this is due to the broader spectrum of the simulated noise, which is not affected by the mechanical filtering effect of the shaker at low frequencies.

Experimental analysis with stronger magnets (double the thickness) and the same excitation condition as described above suggests that a maximum in the power output is actually reached with stiffer magnetic springs, as shown in Fig. 9. The larger uncertainty at low values of the cap is due to the fact that in that condition the harvester is less stationary, due to the higher non-linear effect of the springs. This offers further possibilities for improvements of the harvester and will be investigated numerically and experimentally in future work.

5 Conclusions

In this paper, a numerical model integrated with finite element simulations for reproducing and predicting the electrical and geometrical optimization of a non-linear energy harvester was presented and validated.

Finite element analysis was firstly used to predict the magnetic field produced by a Halbach stack of magnets, which is the internal mass of the device. Secondly, the same strategy was used to model the force implemented by four magnets in repulsive configuration forming two magnetic springs, both for the external and for the internal mass. This analysis was repeated for different distances between the top and the bottom spring, in order to simulate the lowering of the harvester's cap.

Subsequently, the fitted data obtained by the finite element studies were integrated in the numerical equation-based simulation. Initially, a harmonic excitation was used as the input in order to reproduce the experimental data from the electrical optimization. Then, coloured noise was generated as an input, and the lowering of the cap was simulated. The Matlab results were compared with experimental data, in order to show the ability of the model to represent the real device. A further set of data was shown, which demonstrated that stiffer magnetic springs resulted in increased power output for a particular cap height, allowing the harvester to respond better at higher frequencies.

Both models were able to reproduce the experimental optimal condition, allowing their use as a guideline for the harvester design.

Acknowledgements The authors acknowledge the financial support of Science Foundation Ireland under Grant No.10/CE/I1853 and the Irish Research Council (IRC) for funding under their Enterprise Partnership Scheme (EPS). Bell Labs Ireland would also like to thank the Industrial Development Agency (IDA) Ireland for their continued support.

References

1. Ashraf, K., Md Khir, M. Dennis, J., Baharudin, Z.: A wideband, frequency up-converting bounded vibration energy harvester for a low-frequency environment. Smart Mater. Struct. **22**(2), 025, 018 (2013)
2. Beeby, S.P., Torah, R.N., Tudor, M.J., Glynne-Jones, P., O'Donnell, T., Saha, C.R., Roy, S.: A micro electromagnetic generator for vibration energy harvesting. J. Micromechanics Microengineering **17**(7), 1257 (2007)
3. Boco, E., Nico, V., O'Donoghue, D., Frizzell, R., Kelly, G., Punch, J.: A 2DOF vibrational energy harvester expoliting velocity amplification: modeling and testing. In: CCIS, vol. 579, pp. 165–183. Springer (2015)
4. Cammarano, A., Neild, S., Burrow, S., Wagg, D., Inman, D.: Optimum resistive loads for vibration-based electromagnetic energy harvesters with a stiffening nonlinearity. J. Intell. Mater. Syst. Struct. **25**(14), 1757–1770 (2014)
5. Challa, V.R., Prasad, M.G., Fisher, F.T.: A coupled piezoelectric–electromagnetic energy harvesting technique for achieving increased power output through damping matching. Smart Mater. Struct. **18**(9) (2009)

6. Choi, W.J., Jeon, Y., Jeong, J.H., Sood, R., Kim, S.G.: Energy harvesting mems device based on thin film piezoelectric cantilevers. J. Electroceramics **17**(2–4), 543–548 (2006)
7. Cottone, F., Basset, P., Vocca, H., Gammaitoni, L., Bourouina, T.: Bistable electromagnetic generator based on buckled beams for vibration energy harvesting. J. Intell. Mater. Syst. Struct. **12**, 1–12 (2013)
8. El-hami, M., Glynne-Jones, P., White, N., Hill, M., Beeby, S., James, E., Brown, A., Ross, J.: Design and fabrication of a new vibration-based electromechanical power generator. Sens. Actuators A Phys. **92**(13), 335–342 (2001)
9. Cottone, F., Vocca, H., Gammaitoni, L.: Nonlinear energy harvesting. Phys. Rev, Lett (2009)
10. Galchev, T., Kim, H., Najafi, K.: A parametric frequency increased power generator for scavenging low frequency ambient vibrations. Proc. Chem. **1**(1), 1439–1442 (2009)
11. Galchev, T., McCullagh, J., Peterson, R., Najafi, K.: Harvesting traffic-induced vibrations for structural health monitoring of bridges. J. Micromechanics and Microengineering, **21**(10), 104, 005 (2011)
12. Kulkarni, S., Roy, S., O'Donnell, T., Beeby, S., Tudor, J.: Vibration based electromagnetic micropower generator on silicon. J. Appl. Phys. **99**(8), 1–4 (2006)
13. Kwon, S.D., Park, J., Law, K.: Electromagnetic energy harvester with repulsively stacked multilayer magnets for low frequency vibrations. Smart Mater. Struct. **22**, 1–12 (2013)
14. Leadenham, S., Erturk, A.: M-shaped asymmetric nonlinear oscillator for broadband vibration energy harvesting: harmonic balance analysis and experimental validation. J. Sound Vib. **333**(23), 6209–6223 (2014)
15. Liu, H., Gudla, S., Hassani, F.A., Heng, C.H., Lian, Y., Lee, C.: Investigation of the nonlinear electromagnetic energy harvesters from hand shaking. IEEE Sens. J. **15**(4), 2356–2364 (2015)
16. Manella, R.: Integration of stochastic differential equations on a computer. Int. J. Mod. Phys. C **13**(09), 1177–1194 (2002)
17. Mann, B.P., Sims, N.D.: Energy harvesting from the nonlinear oscillations of magnetic levitation. J. Sound Vib. **319**, 515–530 (2009)
18. Mizuno, M., Chetwynd, D.G.: Investigation of a resonance microgenerator. J. Micromechanics and Microengineering, **13**, 209-216 (2003). (Institute of Physics Publishing)
19. Nico, V., O'Donoghue, D., Frizzell, R., Kelly, G., Punch, J.: A multiple degree-of-freedom velocity-amplified vibrational energy harvester: Part b modelling. In: Proceedings of ASME 2014 Conference on Smart Materials, Adaptive Structures and Intelligent Systems 2 (2014)
20. Nico, V., Boco, E., Frizzell, R., Punch, J.: A high figure of merit vibrational energy harvester for low frequency applications. Appl. Phys. Lett. **108**(1), 013, 902 (2016)
21. Soliman, M.S., Abdel-Rahman, E.M., El-Saadany, E.F., Mansour, R.R.: A design procedure for wideband micropower generators. J. Microelectromechanical Syst. **18**(6), 1288–1299 (2009)
22. Tang, X., Lin, T., Zuo, L.: Design and optimization of a small-scale linear electromagnetic energy harvester. IEEE/ASME Trans. Mechatron. **19**, 615–622 (2014)
23. Yang, B., Lee, C.: Non-resonant electromagnetic wideband energy harvesting mechanism for low frequency vibrations. Microsyst. Technol. **16**, 961–966 (2010)

Nonlinear Analysis of a Two-Degree-of-Freedom Energy Harvester

Valeria Nico, Ronan Frizzell and Jeff Punch

1 Introduction

The presence of wireless sensor networks is widespread in many aspects of everyday life [13], in part associated with the concept of "smart cities", which are urban environments in which sensors monitor both environmental conditions and other activities within the city. In the past few years, great improvements in handling data in smart cities has been achieved, however the problem of powering the sensors has not been satisfactorily solved. Sensors are usually battery powered to ensure the independence of each sensor from the mains supply, but batteries have a limited lifetime and must be replaced and disposed of. Moreover, their replacement in a large network can be difficult since some of the sensors could be embedded in structures and therefore difficult to reach [7].

A possible solution for powering a large network comes from the conversion of the ambient energy present in the environment. Among all the possible sources, kinetic energy available from ambient vibrations is one of the most common forms, and it can be converted by using variable capacitors, piezoelectric or magnetostrictive materials and through electromagnetic induction. Conventional vibrational energy harvesters (VEHs) are generally based on a linear mass-spring resonator configuration for which the resonant frequency of the device must be tuned to the main

V. Nico (✉) · J. Punch
CONNECT, Stockes Laboratories, Engineering and Research Building,
University of Limerick, Limerick, Ireland
e-mail: valeria.nico@ul.ie

J. Punch
e-mail: jeff.punch@ul.ie

R. Frizzell
Efficient Energy Transfer (ηET) Department, Bell Labs, Nokia,
Blanchardstown Business and Technology Park, Snugborough Road,
Blanchardstown, Dublin 15, Dublin, Ireland
e-mail: ronan.frizzell@nokia-bell-labs.com

L.F.M. da Silva (ed.), *Materials Design and Applications*,
Advanced Structured Materials 65, DOI 10.1007/978-3-319-50784-2_15

frequency of the ambient vibrations. To overcome the problems of narrow bandwidth and the high resonant frequencies associated with small-scale devices, a 2-Degree-of-Freedom (2*DoF*) electromagnetic velocity-amplified VEH [11] has been developed. The device comprises two masses oscillating between two sets of nonlinear magnetic springs. Impacts between the two masses can occur, which allows momentum transfer from the heavier to the lighter mass, providing velocity amplification. The dynamic behaviour of the system, and hence the output voltage, is highly influenced by the use of magnetic springs that introduce a hardening nonlinearity in the system.

Conventional power spectra show the frequency distribution of the output signal of the harvester, without any phase information that is relevant for a nonlinear system [6]. For this reason, the nonlinear response of the device is analysed in this paper using bispectral techniques such as bicoherence, to show the increasing nonlinearity when the external acceleration is increased. Bispectral analysis was first introduced in 1963 by Hasselman for studying ocean waves [5], and since then it has been used to study fluid mechanics [14] and mechanical systems [3], quantum mechanics systems [8] and more recently in the study of electroencephalographic signals [1]. To the best of the authors' knowledge, the method only appears once in the open literature on kinetic energy harvesters [9], where it was applied to the analysis of a piezoelectric membrane.

In this paper, bispectral analysis is performed on experimental data to show the presence of quadratic phase coupling (QPC) in the Fourier modes of the output voltage of the harvester, and to reveal how the increasing excitation leads to a period-doubling. The analysis reveals evidence of the excitation of modes not present in the external acceleration, which enlarge the bandwidth of the device; these modes are associated with the magnetic springs.

Section 2 will present a brief introduction to bispectral analysis, Sect. 3 will describe the prototype and the experimental set-up and, finally, experimental results will be provided in Sect. 4 to show the effectiveness of bispectral analysis techniques in detecting nonlinearities and providing a greater understanding of the complex dynamics of velocity-amplified VEHs.

2 Bispectral Analysis Theory

The general aspects of higher order analysis techniques, in particular bispectral techniques, are presented in this section. A discrete-time real valued stationary random process $x(t)$ can be expressed by means of its Fourier transform [6]:

$$x(t) = \sum_{k=-\infty}^{\infty} X_k e^{-i\omega_k t} \tag{1}$$

where $X_{-k} = X_k^*$ is the Fourier coefficient at radial frequency $\omega_k = 2\pi k/T$, the subscript k is a frequency index, T is the record length of $x(t)$, and the asterisk indicates the complex conjugate.

By means of the expectation value, or average, operator E [·], it is possible to define the power spectrum of the process $x(t)$ [6]:

$$P_{xx}(k) = E[X_k X_k^*] \tag{2}$$

where $P_{xx}(k)$ is real valued and nonnegative ($P_{xx}(k) \geq 0$), symmetric ($P_{xx}(k) = P_{xx}(-k)$) and shows the frequency distribution of the modes forming the signal. However, the power spectrum has no phase information so it cannot be used to study nonlinear interactions. For this purpose, the concept of higher order spectral analysis was developed [6] and, in this way, it is possible to isolate phase couplings between interacting Fourier components, which is an effective method for developing a detailed understanding of the complex dynamics of nonlinear systems.

One of the most commonly used detectors for couplings is the bispectrum, defined as [6]:

$$B(k, l) = E[X_k X_l X_{k+l}^*] \tag{3}$$

The bispectrum at the frequency pair (k, l) is proportional to the spectrum of the signal at the frequencies k, l and $k + l$; for this reason, it is common to define its normalised magnitude, the bicoherence, as:

$$b^2(k, l) = \frac{| B(k, l) |^2}{E[| X_k X_l |^2] E[| X_{k+l} |^2]}. \tag{4}$$

Considering three modes with radial frequencies ω_k, ω_l and $\omega_{k+l} = \omega_k + \omega_l$, b^2 represents the fraction of power at the radial frequency sum $\omega_k + \omega_l$ that has quadratic phase coupling (QPC) between ω_k and ω_l [12]. If the three components are spontaneously excited modes, it is expected that each component has a random phase; when the statistical averaging in Eq. 3 is carried out, the bispectrum, and hence the bicoherence, will vanish. On the contrary, if the three modes are quadratically phase coupled to each other ($\theta_{k+l} = \theta_k + \theta_l$), the averaging will not lead to a zero value of the bispectrum, and the bicoherence will have a value close to unity.

As an example, a discrete test signal is defined as $x(t) = y(t) + z(t) + \xi(t)$ where:

$$y(t) = cos(\omega_1 t + \theta_1) + cos(\omega_2 t + \theta_2) \tag{5}$$

$$z(t) = cos[(\omega_1 + \omega_2)t + \theta_1 + \theta_2] \tag{6}$$

in which θ_1 and θ_2 are uniformly distributed random phases in $[-\pi; \pi]$, $\omega_1 = 2\pi f_1$, $\omega_2 = 2\pi f_2$, $f_1 = 30$ Hz and $f_2 = 20$ Hz, and $\xi(t)$ is a Gaussian noise term [6]. Figure 1 shows the bicoherence plot of the test signal. Since $X_{-k} = X_k^*$, the bicoherence satisfies symmetry relations and, in particular, for a discrete time series with Nyquist

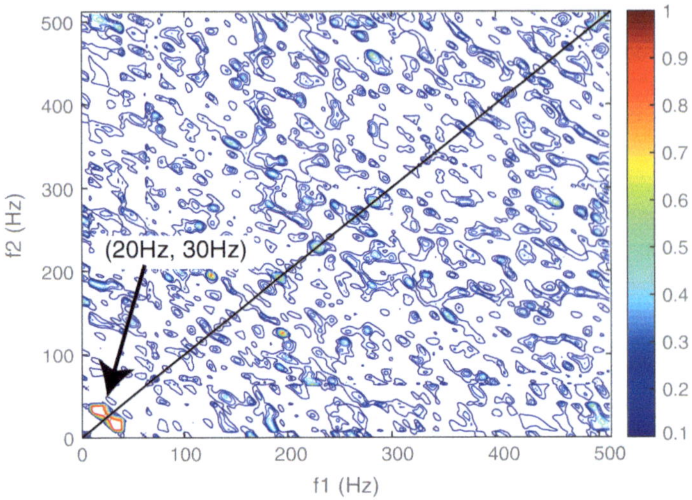

Fig. 1 Bicoherence of a test signal formed by three coupled waves

frequency f_N, the bicoherence is completely defined by values in a triangle of vertices at ($f_1 = 0, f_2 = 0$), ($f_1 = f_{N/2}, f_2 = f_{N/2}$) and ($f_1 = f_N, f_2 = 0$). Moreover it is symmetric to the bisector of the $f_1 - f_2$ plane [12], as shown in Fig. 1. For this reason in Fig. 1, only the lower part of the plot is relevant for the analysis. The presence of a peak at the pair (30 Hz, 20 Hz) of amplitude $b^2 = 1$ is evident, due to strong QPC between the two frequencies. In the next section bicoherence plots of the experimental data will be presented to identify the presence of QPC in the harvester response.

3 Harvester Description

The harvester under examination is shown in Fig. 2. It is based on a concept developed by the authors [4], and fully characterised in a paper by the same group [11]. It consists of two masses, an external larger mass (Mass 1) and an internal smaller one (Mass 2), relativity oscillating one inside the other between two sets of magnetic springs. Stoppers between the masses and between Mass 1 and the housing are present to prevent damage to the magnets. The use of magnetic springs reduces the mechanical energy losses compared to the case with mechanical springs, and it introduces a hardening nonlinearity in the system that enhances the bandwidth. The effect of the nonlinear springs under increasing levels of external excitation is increased, will be studied in detail in Sect. 4 using bicoherence plots.

Impacts between the two masses can occur that allow momentum transfer from Mass 1 to Mass 2, providing velocity amplification. Since according to Faraday's Law, the induced voltage is proportional to the relative velocity between a magnet and a coil, electromagnetic induction is chosen as the transduction mechanism

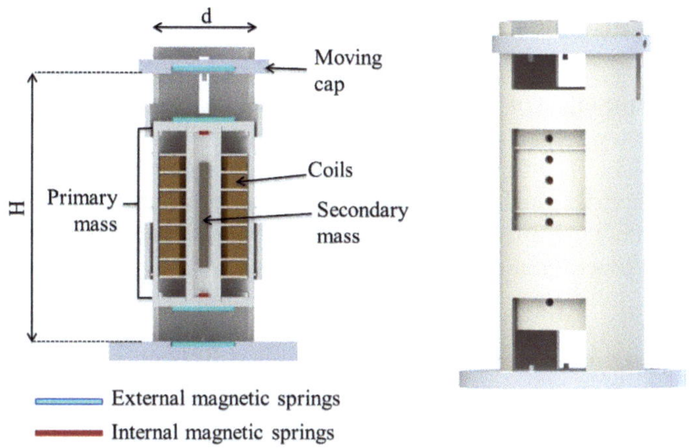

(a) Section of the prototype. (b) Assembly of the prototype.

Fig. 2 Schematic of the harvester

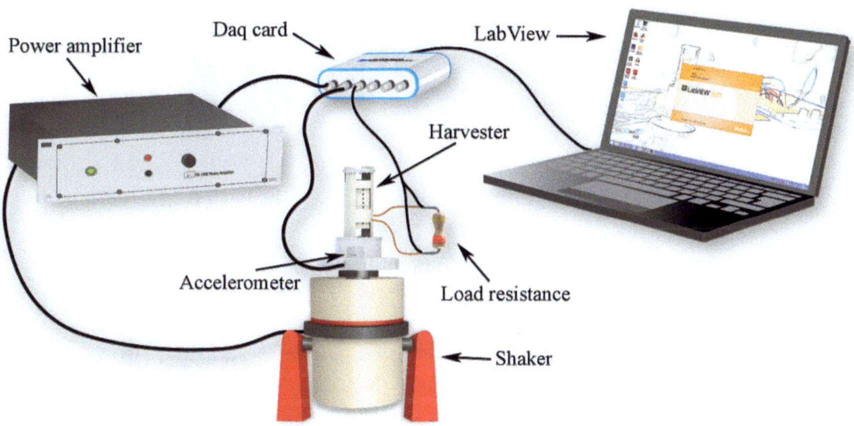

Fig. 3 Schematic of the experimental setup

because it is easily implemented in a device that exploits velocity amplification: the larger mass is made up of seven coils, while the smaller mass is made up of 5 magnets in a Halbach configuration. A detailed description of the device can be found in [2, 10, 11].

A schematic of the experimental setup is presented in Fig. 3.

The harvester was mounted on an LSD V406 electromagnetic shaker. A PCB Piezotronic accelerometer was placed on the head of the shaker, below the harvester, to provide a feedback which was used to ensure that the shaker was given the desired level of excitation. Labview was used to supply a voltage signal to the shaker through an LDS PA100E amplifier with a variable gain control, and to acquire the signals from the accelerometer and the harvester.

4 Experimental Results

In order to study the effect of acceleration on the output voltage, the system was tested at a fixed frequency $f_{ext} = 21.5$ Hz for two different acceleration (rms) levels: $a_1 = 0.4\,g$ and $a_2 = 0.8\,g$. The frequency was chosen far from the resonant frequency of the device ($f_{res} = 16$ Hz) so that any possible nonlinearities were due to the increasing acceleration and not to impacts with the stoppers (which were obseved to happen at f_{res} even at very low excitation levels). The signals for the two different excitations

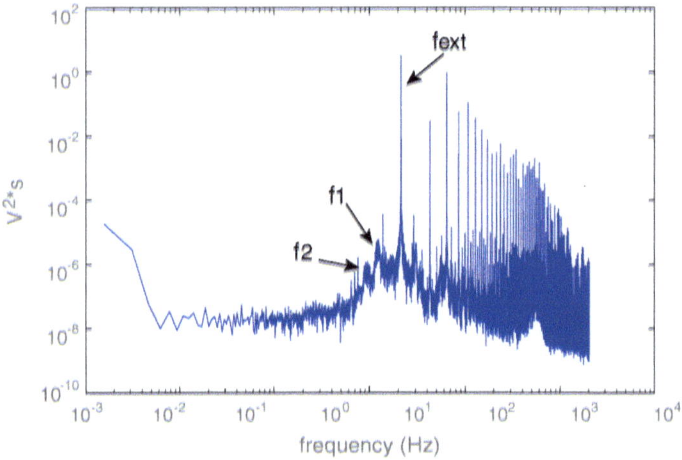

(a) Power spectrum of the output voltage.

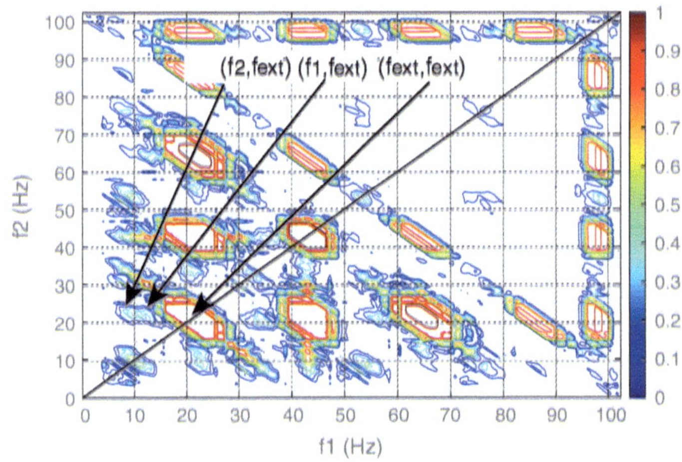

(b) Bicoherence of the output voltage.

Fig. 4 Spectral analysis of the output for an excitation of $a_{rms} = 0.4$ g at 21.5 Hz were recorded across a load resistance of $R_L = 7\,\mathrm{k}\Omega$ for 100 s in order to obtain a sufficient resolution in the low frequency spectrum. Figure 4a and b show the power spectrum and the bicoherence of the signal for the lower acceleration level.

From Fig. 4a, it is evident that there is a main peak at f_{ext} but also at lower frequencies as $f_1 = 12.2$ Hz and $f_2 = 9.4$ Hz are excited. The power spectrum, however, does not have any information on the phases of the components, or on how the modes were excited. With reference to the bicoherence in Fig. 4b, due to its symmetry (Sect. 2), only the lower part of the plot is relevant for the analysis. It is interesting to note

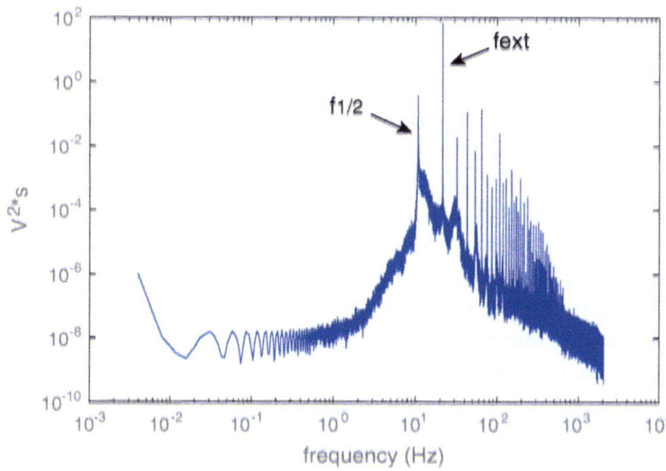

(a) Power Spectrum of the output voltage.

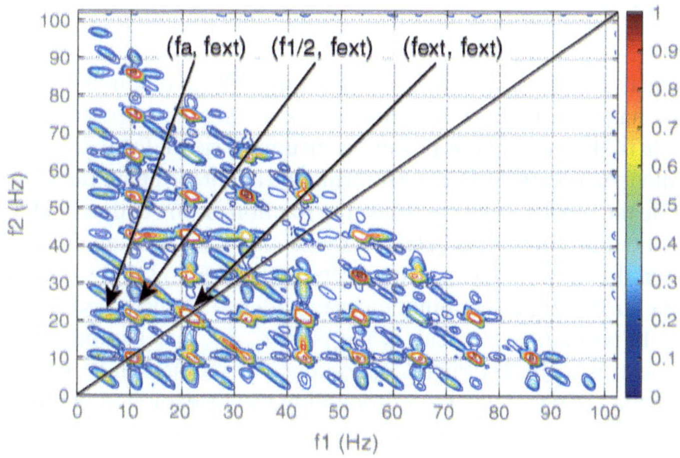

(b) Bicoherence of the output voltage.

Fig. 5 Spectral analysis of the output for an excitation of $a_{rms} = 0.8$ g at 21.5 Hz the presence of peaks in the plot, meaning that the excitation frequency is coupled with the higher harmonics and also with the two lower frequencies (f_1 and f_2). It is important to acknowledge that $f_{ext} = f_1 + f_2$ and so f_1 and f_2 are excited through QPC, even though the coupling is not strong ($b^2 = 0.5$ for the pair (f_{ext}, f_1) and $b^2 = 0.48$ for the pair (f_{ext}, f_2)).

Increasing the acceleration amplitude, the larger mass experienced a stronger repulsive magnetic force and the nonlinearities in the system are expected to increase. Figures 5a and b shows the power spectrum and the bicoherence of the signal for $a_2 = 0.8\,g$.

From the comparison of Figs. 4a and 5a, it is possible to see that the bandwidth of the device is enlarged and that a peak at $f_{1/2} = 10.7$ Hz$= f_{ext}/2$ is present that is evidence of period-doubling. However, it does not contain information on the increase of nonlinearities and, moreover, it is not possible to tell if $f_{1/2}$ is excited independently from f_{ext}. Figure 5b shows the bicoherence of the output voltage and the increase in the number and magnitude of peaks, compared to Fig. 4b. The driving frequency is coupled through QPC to the higher harmonics as in the lower acceleration case, and to $f_{1/2}$; this is evidence of period-doubling due to the hardening effect associated with the magnetic springs. The excitation frequency is also coupled to $f_a = 5$ Hz, a coupling that was not evident in Fig. 4b. Moreover there are peaks at lower frequencies, indicating presence of QPC between $f_{1/2}$, its harmonics and f_a, and between f_a and its harmonics. The greater total number of peaks compared to Fig. 4b indicates the expected increase of nonlinearities, since the larger mass could reach positions closer to the magnets and, consequently, experience a stronger repulsive force.

5 Conclusion

Bispectral analysis was conducted on the output voltage of a nonlinear electromagnetic energy harvester to show the effect of increasing the amplitude of the external excitation. It is shown that the power spectrum carries information about the frequency content of a signal but has no information about the phase interaction between modes, while the bicoherence measures the fraction of power due to QPC. Two tests were carried out at different acceleration levels, but at the same driving frequency. It was shown that more couplings between modes due to QPC were present in the bicoherence for the higher amplitude case, evidence of increased nonlinearities in the system. In particular, the peak at the pair $(f_{1/2}, f_{ext})$ is symptomatic of period-doubling which is a characteristic consequence of the hardening effect introduced by the magnetic springs. Such period-doubling can be exploited in order to enhance the harvested power because, by using a nonlinear potential, it could be possible to excite modes not present in the external acceleration and, hence, widen the bandwith of the response.

Acknowledgements The authors acknowledge the financial support of Science Foundation Ireland under Grant No.10/CE/I1853 and the Irish Research Council (IRC) for funding under their Enterprise Partnership Scheme (EPS). Bell Labs Ireland would also like to thank the Industrial Development Agency (IDA) Ireland for their continued support.

References

1. Barnett, T.P., Johnson, L.C., Naitoh, P., Hicks, N., Nute, C.: Bispectrum analysis of electroencephalogram signals during waking and sleeping. Science **172**(3981), 401–402 (1971)
2. Boco, E., Nico, V., Frizzell, R., Punch, J.: A c-battery scale energy harvester: part b–transducer optimization and modeling. In: ASME 2015 Conference on Smart Materials, Adaptive Structures and Intelligent Systems, pp. V002T07A009–V002T07A009. American Society of Mechanical Engineers (2015)
3. Choi, D., Chang, J., Stearman, R., Powers, E.: Bispectral interaction of nonlinear mode interactions pp. 602–609 (1984)
4. Cottone, F., Goyal, S., Punch, J.: Energy harvester apparatus having improved efficiency (2013). US Patent 8,350,394
5. Hasselmann, K., Munk, W., MacDonald, G.: Bispectra of ocean waves. Time Series Analysis, pp. 125–139 (1963)
6. Kim, Y.C., Powers, E.J.: Digital bispectral analysis and its applications to nonlinear wave interactions. IEEE Trans. Plasma Sci. **7**(2), 120–131 (1979)
7. Kollias, A., Nikolaidis, I.: In-wall thermoelectric harvesting for wireless sensor networks. In: SMARTGREENS, pp. 213–221 (2014)
8. Miller, M.D.: Bispectral analysis of the driven sine-gordon chain. Phys. Rev. B **34**, 6326–6333 (1986)
9. Neri, I., López-Suárez, M.: Efficient nonlinear energy harvesting with wrinkled piezoelectric membranes. Energy Harvest. Syst. 1–5 (2015)
10. Nico, V., Boco, E., Frizzell, R., Punch, J.: A c-battery scale energy harvester: part a—system dynamics. In: ASME 2015 Conference on Smart Materials, Adaptive Structures and Intelligent Systems, pp. V002T07A008–V002T07A008. American Society of Mechanical Engineers (2015)
11. Nico, V., Boco, E., Frizzell, R., Punch, J.: A high figure of merit vibrational energy harvester for low frequency applications. Appl. Phys. Lett. **108**, 013,902 (2016)
12. Pezeshki, C., Elgar, S., Krishna, R.: Bispectral analysis of possessing chaotic motion. J. Sound Vib. **137**(3), 357–368 (1990)
13. Pickavet, M., Vereecken, W., Demeyer, S., Audenaert, P., Vermeulen, B., Develder, C., Colle, D., Dhoedt, B., Demeester, P.: Worldwide energy needs for ict: the rise of power-aware networking pp. 1–3 (2008)
14. Yeh, T., Atta, C.: Spectral transfer of scalar and velocity fields in heated-grid turbulence. J. Fluid Mech. **58**(02), 233–261 (1973)

Experimental Investigations of MR Fluids in Air and Water Used for Brakes and Clutches

Alin Ilie Bosioc, Trandafir Emanuel Beja, Sebastian Muntean, Istvan Borbáth and Ladislau Vékás

1 Introduction

Magnetic fluids have been developed at the beginning of the 1948, when Rabinow [1] developed a clutch device. According with Rabinow, the advantages of using this kind of clutch with magnetic fluids are: the amount of electrical power necessary to control is small and time response is also small. Since then the magnetic fluids have been implemented in many applications. Several important applications can be found in: sealing's (especially in the case of gas transporting installations) [2, 3], sensors [4], dumpers [5, 6].

Magneto-rheological brakes (MRB) are one important research direction concerning these smart materials. Some of the advantages are convenient low-power control (through electrically generated magnetic field) and very good torque to weight ratio. The latter recommends them for applications where small devices able to generate high value of torque are needed, [7, 8]. The electrical control possibility

A.I. Bosioc (✉) · T.E. Beja
Hydraulic Machinery Department, University Politehnica Timişoara, Mihai Viteazu, no. 1, Timişoara, BV, Romania
e-mail: alin.bosioc@upt.ro

T.E. Beja
e-mail: emanuel_beja@yahoo.com

S. Muntean · L. Vékás
Center for Advanced Research in Engineering Sciences, Romanian Academy, Mihai Viteazu, no. 24, Timişoara, BV, Romania
e-mail: seby@acad-tim.tm.edu.ro

L. Vékás
e-mail: vekas.ladislau@gmail.com

I. Borbáth
SC. Roseal S.A., Str. Nicolae Balcescu, no. 5A, Odorheiu Secuiesc, Romania
e-mail: borbathistvan@roseal.eu

L.F.M. da Silva (ed.), *Materials Design and Applications*,
Advanced Structured Materials 65, DOI 10.1007/978-3-319-50784-2_16

(the magnetic field can easily be electrically generated and controlled) is one very important advantage, [9]. According to Park [10], an electronically-controlled brake is easier to adapt in any application by changing the software. Such a device ensures easier integration and what they call "new control features".

The paper presents the design and testing of an underwater-operating MRB. The main challenge in designing the MRB is that needs to work in an improper medium as water. The second section describes the MRB with the mechanical design. The third section presents experimental setup and the control and acquisition program. Fourth section presents the experimental result and the last section draws the conclusions.

2 Experimental Test Rig

A MR brake was built for Laboratory of Hydraulic Machinery from Politehnica University Timisoara in order to slow down the speed and increase the operating regimes from the hydraulic machinery test rig [11–14]. For this purpose a preliminary test rig was designed, developed and built in order to analyze in detail different brake models for other applications and to test different MR fluids available on the market as well as several MR fluids developed and characterized in our laboratory. This method was chosen taking into account the mechanical properties of the MR brake (easy to control, small dimensions, adjustable). The test rig, Fig. 1, consists in a variable speed drive motor, a torque detector, the MR brake and the control and acquisition system.

The main characteristics and technical data about the brake system are presented in Table 1.

The control and acquisition system set the nominal speed of the electric motor and records the following data: (i) the torque value, (ii) electric motor speed, (iii) the voltage and current from the coil of brake or clutch and (iv) the brake temperature, respectively. The temperature is acquired on the outside wall or inside of the brake with ±1.5 °C accuracy. Note that this investigation was performed in air and water. In all measurements the temperature inside MRB close to the MR fluid was below threshold limit. A picture with our preliminary test rig for MR brake applications is presented in Fig. 1.

The existing MRB, presented in Fig. 2, is a bell shaped MRB, with an important advantage: it encapsulates two brakes by doubling the friction surface. The rotating part is the shaft and disc (bell shape). The magnetic circuit is a fixed part and is built using two pieces: the coil and the magnetic casing. To increase the torque of the MRB, the disc radius was increased as much as possible. The magnetic design of the coil ensures a radial magnetic field, in order to use the entire surface of contact where MR fluid is inserted. The entire magnetic circuit is built of magnetized iron.

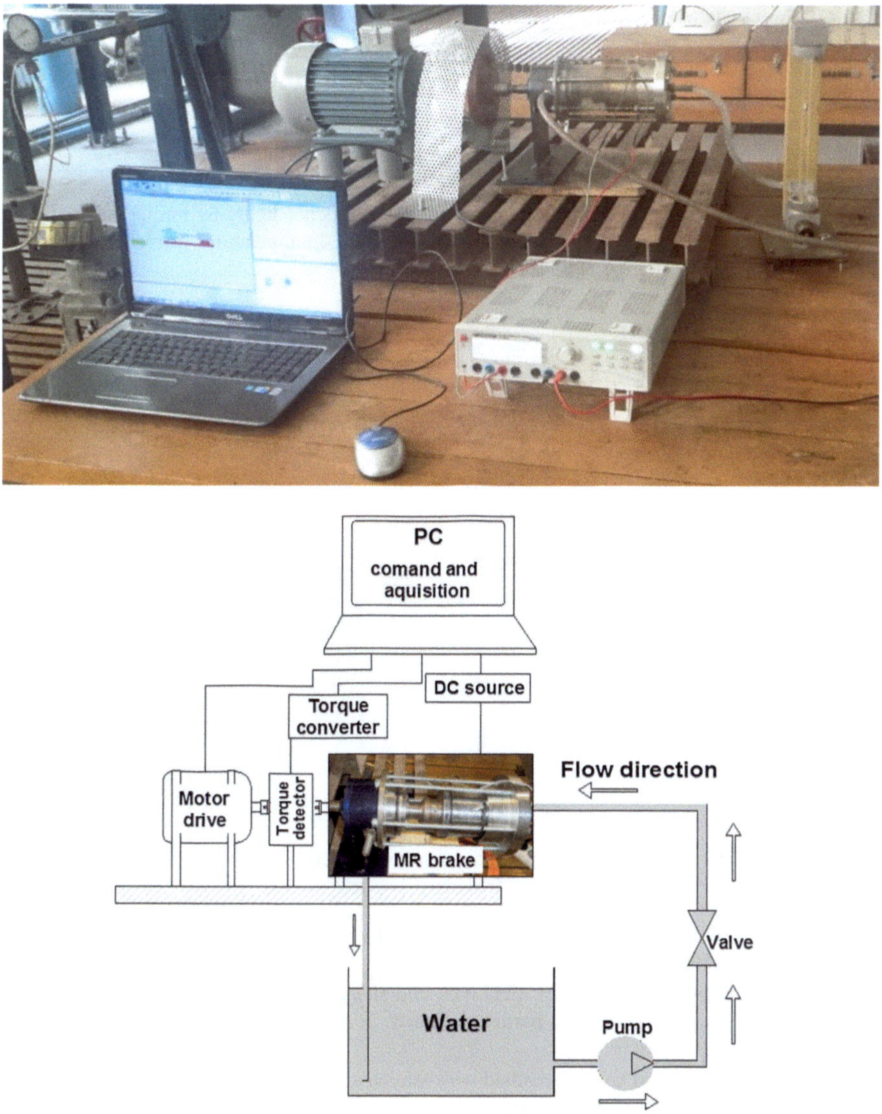

Fig. 1 Experimental test rig for MRB applications, image from laboratory (*up*) and sketch of the test rig with main components and working principle (*down*)

3 MR Fluids and the Experimental Setup

The MR fluids used in the experiments are presented in Table 2.

The SMR 5%, SMR 15% and SMR 25%, have been produced in Laboratory of Magnetic Fluids, Romanian Academy-Timisoara Branch. The SMR fluids based ferrofluid have been specially designed for this type of applications, the team from

Table 1 Main characteristics of the brake system

Component	Main characteristics
Motor drive	0–1000 rpm, 3 kW
Torque detector	0–9000 rpm, 20 Nm
MR brake	0–1000 rpm, design for air/water applications
Control/acquisition system	–10 Hz acquisition frequency –Maintain a constant speed independently by torque variation –Has the possibility to acquire 4 temperatures inside/outside of the brake

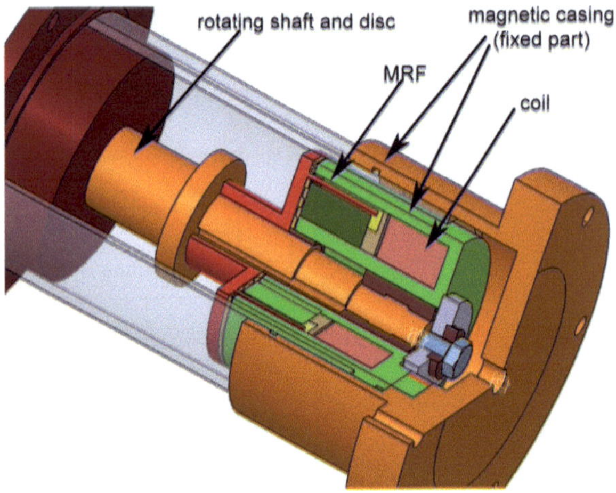

Fig. 2 Sketch of the MRB existing in laboratory

Table 2 Technical data for the MR fluids used in experiments

MR fluid	Technical data	Producer
MRF 336 AG	Silicone oil with micro particles, density 3450 kg/m^3	Lord corporation
SMR 5%	Mineral oil based ferrofluid with micron sized Fe particles with micro particles, density 1230 kg/m^3	Lab. of Magnetic Fluids Romanian Academy–Timisoara Branch
SMR 15%	Mineral oil based ferrofluid with micron sized Fe particles with micro particles, density 3880 kg/m^3	Lab. of Magnetic Fluids Romanian Academy–Timisoara Branch
SMR 25%	Mineral oil based ferrofluid with micron sized Fe particles with micro particles, density 5700 kg/m^3	Lab. of Magnetic Fluids Romanian Academy–Timisoara Branch

Timisoara having a long experience the preparation and characterization of this type of fluids, [15, 16].

The control and acquisition program is presented in Fig. 3. The application allows setting the speed of the motor drive and a smooth start at speeds between 0–1000 rpm. Parallel the program allows monitoring the torque and the speed in real time. At the end of each measurement the acquired data can be saved in different formats. The measurements have been performed in two environments: air and water. For each environment was set a speed between 300 and 1000 rpm with a

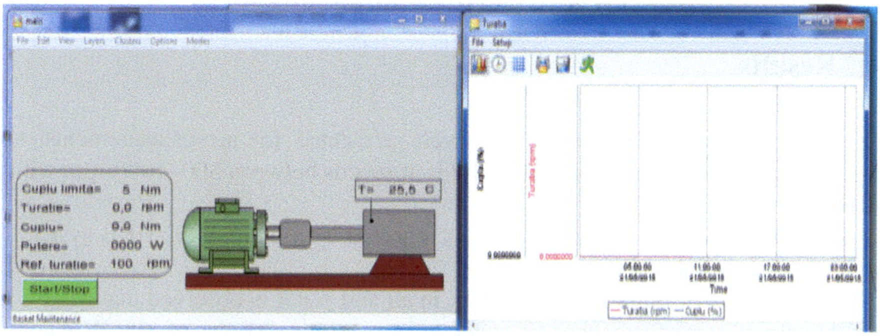

Fig. 3 Control and acquisition program implemented on the MRB test rig

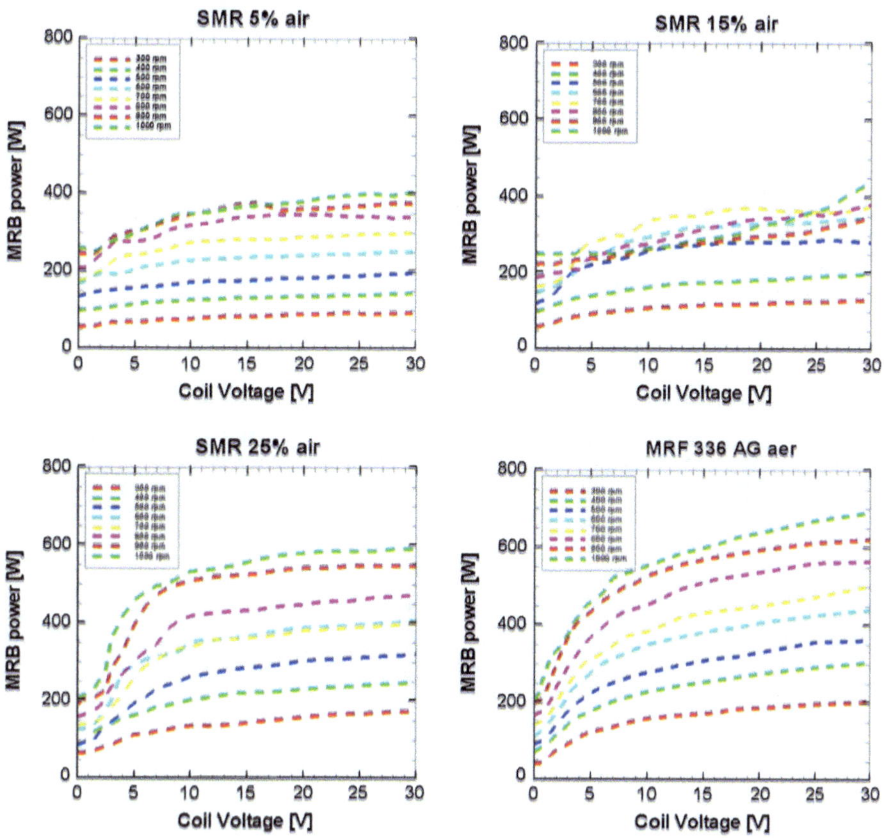

Fig. 4 MRB power for different speeds and coil voltages **in air**

step size of 100 rpm. At each speed the coil voltage was modified between 0–30 V with a step size of 1 V. For each voltage step size was registered: speed, torque, coil voltage, internal and external temperatures.

4 Results

The first experimental investigations were presented for air measurements, see Fig. 4. Have been tested all four MR fluids at speeds between 300–1000 rpm with a step size of 100 rpm.

Same measurement configuration was repeated by mounting all MRB system in water, as presented in Fig. 5.

From the first evaluation of the MRB, in air and water is observed that MRF 336 AG has the best performances. It generates a maximum power of 800 W when is

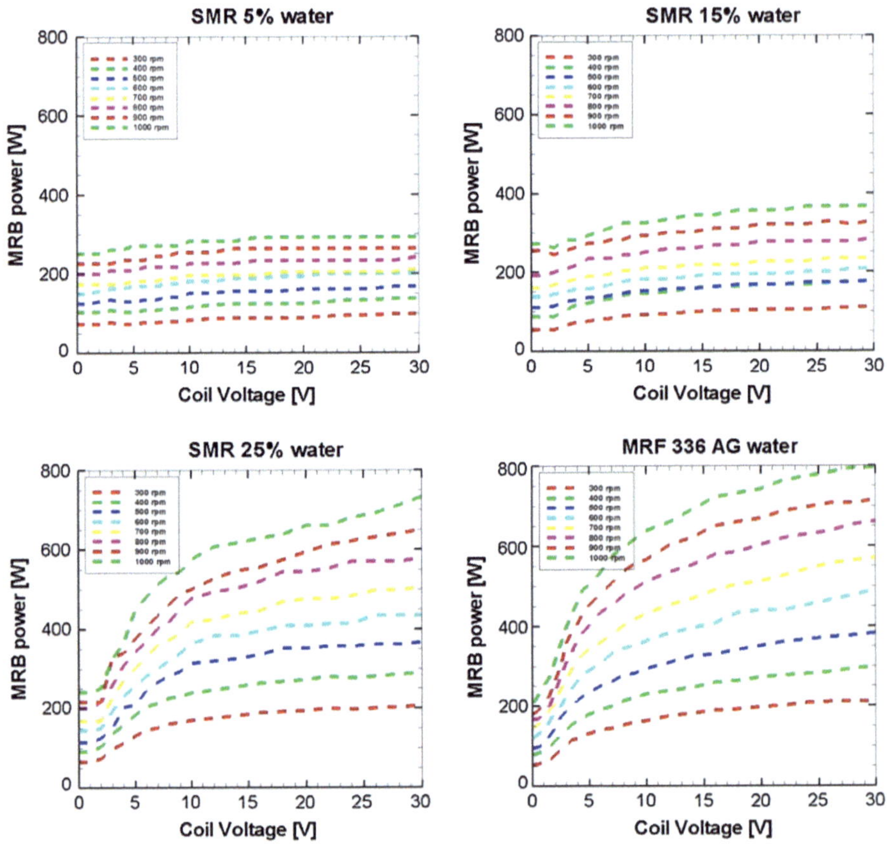

Fig. 5 MRB power for different speeds and coil voltages **in water**

used in water at the maximum coil voltage of 30 V. Even so, SMR 25% has the maximum brake power around 750 W, with 7% less.

A next analysis consist in evaluation of the MRB at two voltages 10 and 30 V for all investigated MR fluids in air and water.

The evaluation presented in Fig. 6 shows clearly the power generated by each type of the MRF. SMR 5 and SMR 15%, even they have different iron composition, generate approximately same power in the brake. Instead, SMR 25% generate approximately same power as the commercial sample.

Keeping into account the fact that the final brake will operate in water, an important consideration is the washing of the MR fluid in time during the operation. Accordingly, Fig. 7 present the recirculated water resulted at the end of the measurements. The visualization of the water shows that MRF 336 AG, even it has good braking characteristics is washed in time. The SMR 25% sample is not washed (in the same time as MRF 336 AG).

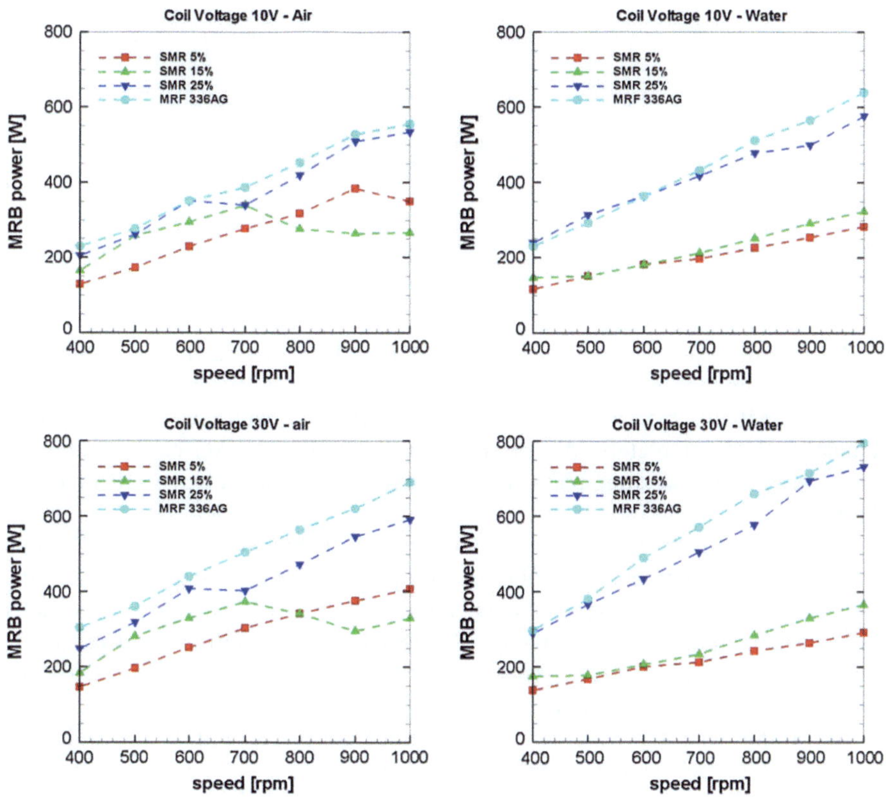

Fig. 6 MRB power for two voltages of the coil (10 V *up*, 30 V *down*) **for air** (*left*) and **water** (*right*)

Fig. 7 Visualization of the recirculated water from hydraulic system at the end of **water measurements**

A last visualization consist in disassemble the MRB after each measurement and observe the amount of MR fluid left on the rotating disc. According with Fig. 8, the amount of MR fluid is same as beginning for SMR 25%, while the MRF 336 AG is almost washed completely. As a result even the MRF 336 AG has good braking characteristics; the washing phenomenon can remove completely the MR fluid from the brake.

A last analysis consists in analysis of external temperature for each MR fluid only for air measurements, see Fig. 9. During the water measurements no cooling system was used. The external temperature at the MRB surface was same as ambient temperature from laboratory. The registered temperature corresponds to a complete measurement at 1000 rpm, and varying the voltage between 0–30 V in a time interval of around 25 min.

According with the variation of the temperature, see Fig. 9, the maximum temperature is registered for MRF 336 AG. Even so, the maximum temperature does not reach the threshold temperature given by the producer.

Fig. 8 Visualization of the MR fluids at the end of the **water measurements** on the rotating disc (bell disc)

Fig. 9 Temperature variation for each MR fluid during the tests

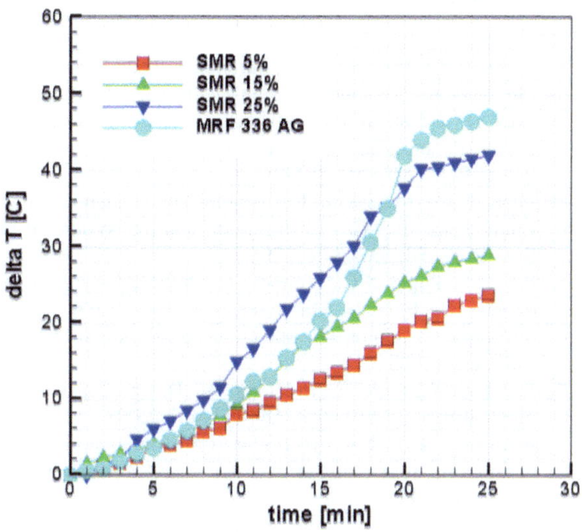

5 Conclusions

The aim of this study is to provide a first experimental evaluation of different MR fluids used for a MRB used in air and water. There have been tested MR fluids from the market and MR fluids developed especially for this type of applications (produced by our laboratory).

The air and water measurements conclude that approximately similar braking characteristics are observed for MRF 336 AG and SMR 25%. The MRF 336 AG has with 15% more braking power in air and with 7% more in water. The major inconvenient of the MRF 336 AG represent the washing phenomenon when is in contact with water. Instead, SMR 25% is more stable. The maximum temperature generated in operation is with 45 °C more than at the beginning of the measurement. The temperature is in the normal working conditions for all MR fluids. For water operation, in order to increase the braking power, it can be produced a new MR fluid with a higher iron concentration.

Acknowledgements Dr. Eng. Bosioc A.I. and Eng. Beja T.E. have been supported by a grant of the Romanian National Authority for Scientific Research and Innovation, CNCS—UEFISCDI, project number PN-II-RU-TE-2014-4-1089. Dr. Eng. Muntean Sebastian and Dr. Phys. Ladislau Vekas have been supported by a grant of the Romanian National Authority for Scientific Research and Innovation, CNCS—UEFISCDI, "*MagNanoMicroSeal*" project number PN-II-PT-PCCA-2011-3.2-0538.

References

1. Rabinov, J.: The magnetic fluid clutch. Trans. AIEE, 1308–1315 (1948)
2. Moskowitz, R.: Dynamic sealing with magnetic fluids. ASLE Trans. **2**(18), 135–143 (1975)
3. Bica, D., Balanean, F., Borbath, I., Boros, T.F., Galffy, D., Vekas, L.: Process for preparing composite magnetic fluids for magneto-fluidic rotary seals, for high pressures. Romania, Patent RO122725-B1 (1999)
4. Popa, N.C., Rousseau, J.J., Siblini, A., Chatelon, J.P., Jamon, D., Royer, F., Robert, S., Choueikani, F.: Gas distribution control system using magnetic fluid sensors. Rom. Rep. Phys. **58**, 337–349 (2006)
5. Liţă, M., Popa, N.C., Velescu, C., Vékás, L.: Investigation of a magnetorheological fluid damper. IEEE Trans. Magn. **40**(2), 469–472 (2004)
6. Sireteanu, T., Ghita, G., Stancioiu, D.: Fluide şi amortizoare magneto-reologice. Ed. Bren, Bucuresti (2005)
7. Karakoc, K., Park, E., Suleman, A.: Design considerations for an automotive magnetorheological brake. Mechatron **18**, 434–447 (2008)
8. Nguyen, Q.H., Choi, S.B.: Selection of magnetorheological brake types via optimal design considering maximum torque and constrained volume. Smart Mater. Struct. **21**, 015012 (2012). IOP Publishing
9. Li, W.H., Du, H.: Design and experimental evaluation of a magnetorheological brake. Int. J. Adv. Manuf. Technol. **21**, 508–515 (2003)
10. Park, E.J., Stoikov, D., Falcao da Luz, L., Suleman, A.: A performance evaluation of an automotive magnetorheological brake design with a sliding mode controller. Mechatron., **16**, 405–416 (2006)

11. Nishi, M., Liu, S.: An outlook on the draft-tube-surge Study. IJFMS **6**(1), 33–48 (2013)
12. Frunzăverde, D., Muntean, S., Mărginean, G., Câmpian, V., Marşavina, L., Terzi, R., Serban, V.: Failure analysis of a francis turbine runner. In: 25th IAHR Symposium on Hydraulic Machinery and Systems, IOP Conference Series: Earth and Environmental Science, vol. 2, no. 012115 (2010)
13. Bosioc, A.I., Muntean, S., Tanasa, C., Susan-Resiga, R.F., Vékás, L.: Unsteady pressure measurements of decelerated swirling flow in a discharge cone at lower runner speeds. In: 27th IAHR Symposium on Hydraulic Machinery and Systems, IOP Conference Series: Earth Environmental Science, vol. 22, 032008 (2014)
14. Bosioc, A.I., Muntean, S., Susan-Resiga, R.F., Vekas, L., Bernad, S. Numerical simulation of the swirl generator discharge cone at lower runner speeds. In: 8th Symposium on Numerical Analysis of Fluid Flow and Heat Transfer, AIP Conference Proceedings, vol. 1558, p. 204 (2013). doi:10.1063/1.4825456
15. Borbáth, T., Borbáth, I., Günther, S., Marinica, O., Vékás, L., Odenbach, S.: Three-dimensional microstructural investigation of high magnetization nano–micro composite fluids using x-ray microcomputed tomography. Smart Mater. Struct. **23**, 055018 (2014). doi:10.1088/0964-1726/23/5/055018
16. Susan-Resiga, D., Vekas, L.: Yield stress and flow behavior of concentrated ferrofluid-based magnetorheological fluids: the influence of composition. Rheol. Acta **53**, 645–653 (2014). doi:10.1007/s00397-014-0785z

Hydrodynamic Investigations in a Swirl Generator Using a Magneto-Rheological Brake

Sebastian Muntean, Alin Ilie Bosioc, Raul Alexandru Szakal, Ladislau Vékás and Romeo Florin Susan-Resiga

1 Introduction

The actual requirements from the energy market enforce that hydraulic turbine to operate far from the best efficiency point. When the hydraulic turbines operates at partial discharge or at overload discharge, downstream the runner (in the draft tube cone), the decelerated swirling flow becomes highly unstable. In these conditions is developed a spiral vortex breakdown, also known in engineering literature as the precessing vortex rope. The flow unsteadiness produced by the vortex rope results in severe pressure fluctuations that hinder the turbine operation or may cause accidents [1]. So during the day, the hydraulic machines (which usually have minimum a couple of tons) are forced to operate in a wide range of operating regimes. Due to these regimes downstream the runner, more precisely in the draft tube cone appear the vortex rope. The vortex rope produces pressure pulsations on the draft tube cone wall and in all hydraulic circuit, additional hydraulic losses and unwanted accidents

S. Muntean · L. Vékás
Center for Advanced Research in Engineering Sciences,
Romanian Academy – Timişoara Branch, Bv. Mihai Viteazu, No. 24,
Timişoara, Romania
e-mail: seby@acad-tim.tm.edu.ro

L. Vékás
e-mail: vekas.ladislau@gmail.com

A.I. Bosioc (✉) · R.A. Szakal · R.F. Susan-Resiga
Hydraulic Machinery Department, University Politehnica Timişoara,
Bv. Mihai Viteazu, No. 1, Timişoara, Romania
e-mail: alin.bosioc@upt.ro

R.A. Szakal
e-mail: raul.szakal@yahoo.com

R.F. Susan-Resiga
e-mail: romeo.resiga@upt.ro

L.F.M. da Silva (ed.), *Materials Design and Applications*,
Advanced Structured Materials 65, DOI 10.1007/978-3-319-50784-2_17

in power plant [1–3]. To investigate the unwanted effects of the hydraulic turbines operated at these regimes in the laboratory was design and realized a swirl generator which mimics a hydraulic turbine operated at only one regime (0.7 from the best efficiency discharge). The swirl generator [2] has stationary and rotating blades for generating swirling flow. The rotating blades (or the runner) rotate at 1020 rpm driven by water. In order to increase the investigated regimes of our swirl generator (from 0.7 to the best efficiency discharge) we design and built a MRB in order to slow down the speed of the runner. So our purpose is to slow down the speed of the runner from 1020 rpm to 690 rpm maintaining the same water discharge to the runner blades. The main challenge in designing the brake is the fact that all system needs to operate underwater conditions. As an advantage is the fact that the MRB system is continuously cooled by water (the temperature of the water is 21 °C) and no cooling system is needed and on another side proper sealing's needs to be used in order to keep water away from MRB system, [2].

Different control methods have been introduced and tested in order to mitigate the vortex rope effects. Kurokawa et al. [4] proposed J-Groove method which consists in mounting radial groves on the draft tube cone. Thike [5] and Falvey [6] have tested different extensions of runner cone to suppress the stagnant region and associated vortex rope. Recently, a new control method was proposed [7] and successfully tested experimentally which involves using an axial water jet to the runner cone [8]. The results clearly showed that using the optimum amount of jet water, the stagnant region and the vortex rope are completely eliminated from the draft tube cone. This technique was tested experimentally on a test rig developed at the Hydraulic Machines Laboratory from the Politehnica University Timisoara, with a test section that mimics the swirling from in a real Francis turbine. Foroutan and Yavuzkurt [9] have reached at a similar conclusion, when tested numerically water injection method on a Francis turbine. The vortex rope effects are present both at partial discharge and at full load discharge.

The experimental test rig developed in laboratory was design to investigate one operating regime. A new solution by using a magneto-rheological brake (MRB) was tested by slowing down the runner speed in order to extend the investigation at several swirling flow configurations. This solution covers a wide range of discharge values from 0.7 up to best efficiency point (BEP) discharge at constant guide vane opening. The paper presents the hydrodynamic investigations performed with the swirl generator equipped with MRB working underwater conditions. The second section describes the mechanical design for the chosen type of MRB. Next section provides the experimental results and the MRB behaviour. The conclusions are drawn in the last section together with the future work.

2 Experimental Test Rig

The experimental test rig was designed to investigate the decelerated swirling flow with vortex rope in a conical diffuser. The test rig, Fig. 1 (left), has a main circuit which is used to generate the swirling flow and the auxiliary circuit which is used to

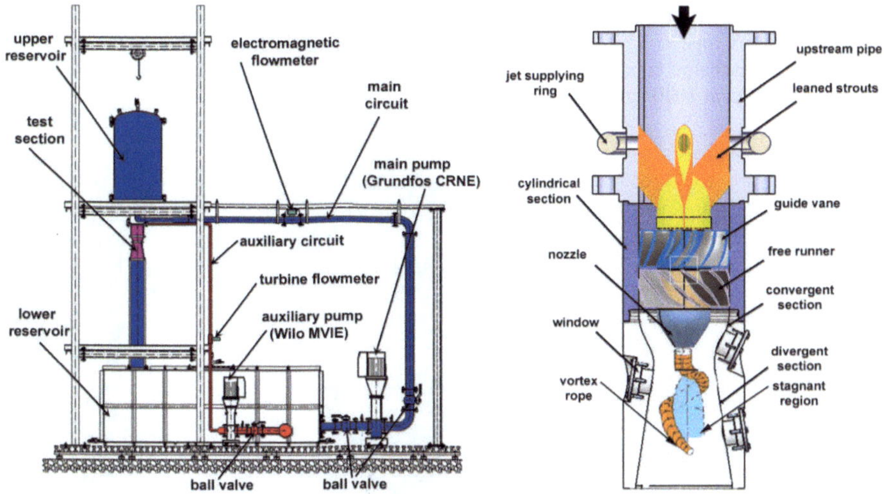

Fig. 1 Experimental test rig for investigation of the decelerated swirling flow (*left*) and sketch of the swirl apparatus with formation of stagnant region and vortex rope (*right*)

investigate new flow control techniques, [8, 10]. The swirl apparatus, Fig. 1 (right), is an original design which reproduces the actual swirling flow encountered in real Francis turbines operated at partial discharge.

Instead of using a hydraulic turbine for the experimental investigations, a swirl apparatus was designed. The swirl apparatus, Fig. 1 (right) has a swirl generator which uses two blade rows.

Upstream is a non-rotating blade (guide vanes with 13 blades) and downstream is a second row of rotating blades (free runner) in order to create at the exit a similar flow as a real Francis runner, [11].

The runner with 10 blades spins at the runaway speed, acting as a turbine near the hub and as a pump near the shroud, with vanishing overall torque. This swirl generator ends with a convergent section leading just downstream the throat to a swirling flow configuration similar to the one downstream a Francis runner operated at 70% the best efficiency discharge [12].

From the visualisation on the test rig, Fig. 2, is observed the cavitating vortex rope. The vortex rope is a vortex sheet which rolls-up as a helical vortex around a central stagnant region. This is the source of the pressure pulsations in the draft tube cone travelling in all hydraulic system when the hydraulic turbine operates far from the best efficiency point regime.

As was mentioned before in a first configuration the runner rotates freely at the runaway speed of 1020 rpm. In order to increase the operating regimes from the experimental test rig, a magneto-rheological brake (MRB) system was designed and installed in order to control the runner speed. As a result, the runner speed is slow down up to 35% using the MRB system providing several swirling flow configurations at the discharge cone inlet covering a wide range (from part load to full load

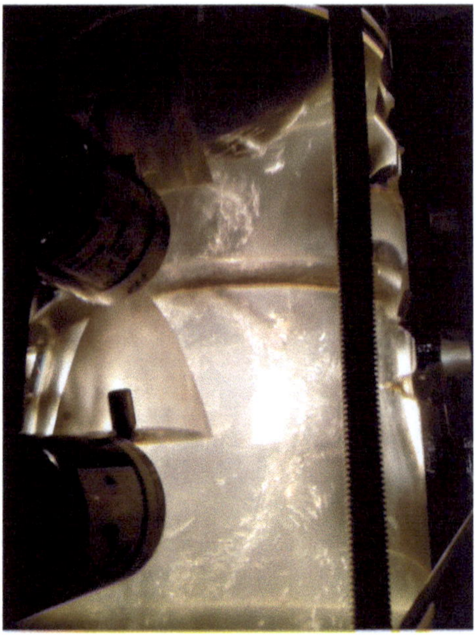

Fig. 2 Visualisation of the cavitating vortex rope developed in discharge cone of the test section at 1020 rpm

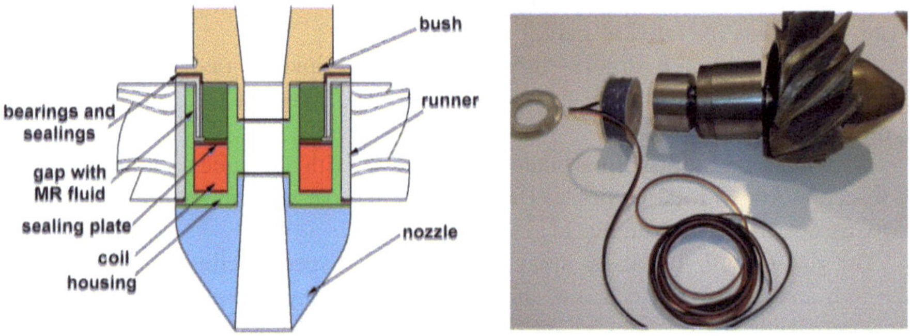

Fig. 3 Sketch of the MRB system (*left*) and picture with mounted pieces (*right*)

conditions). This solution was chosen because of it has several advantages as follows: small and robust, no mechanical parts, fast response time in order to control and low power consuming. In our case all system was especially designed, because was adapted to our geometrical constraints. As a working principle, a magneto-rheological (MR) fluid is mounted between rotating part and fixed one, respectively. The viscosity of the MR fluid is changing in terms of the intensity of the magnetic field generated by the coil leading to slowing down the runner speed. A sketch with the MRB system and assembled components are presented in Fig. 3.

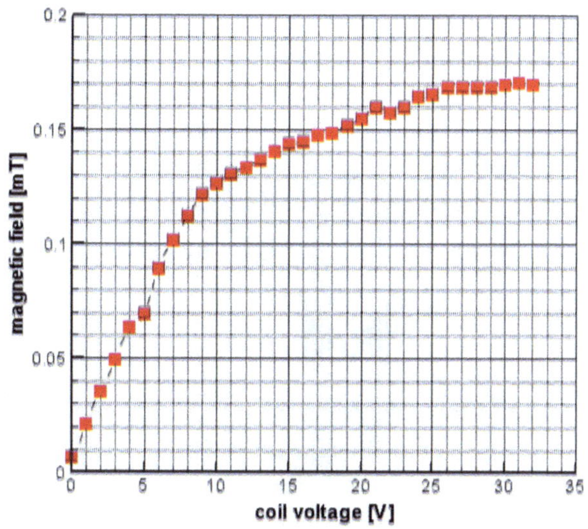

Fig. 4 Magnetic field developed in the gap where MR fluid is inserted

The runner was manufactured so that to be mounted the MRB system. The MRB system consists in iron housing, a coil, seals and bearings and the MR fluid. The housing is made of iron and was design in order to close the magnetic field in the area where MR fluid is mounted. The coil is made by cupper wires and a custom program was used to choose the number of wires and the wire thickness. A series of seals were used to keep the MR fluid separated by water (all system is surrounded by water), also these seals were used as bearings.

The most important component by all system is the MR fluid. We have chosen a concentrated ferrofluid based MR fluid produced in our laboratory with 35% micro sized Fe particles concentration, [13]. For the MR brake controlling the speed of the runner a quantity of 20 ml of MR fluid was necessary. The gap between rotating and fixed parts is 1 mm and time response of the entire system is lower than 0.1 s. Because the MRB system is working in water, no cooling system is necessary. A DC power supply with a maximum electrical voltage of 30 V and a maximum electrical current of 5 A was used in order to control the magnetic field of the MR brake. According with Fig. 4, the magnetic field generated by the coil was measured.

3 Experimental Setup

The investigated regimes are presented in Table 1.

The unsteady pressure was measured in 9 operating points on the wall from the test rig. According with Fig. 5, two pressure transducers are installed on each level being positioned at 180° each other.

Table 1 Speed regimes and the flow rate for experimental investigations

Main characteristic	Speed (rpm)	Flow rate (l/s)
	1020, 990, 960, **924**, 900, 870, **840**, 810, 780, **690**	30

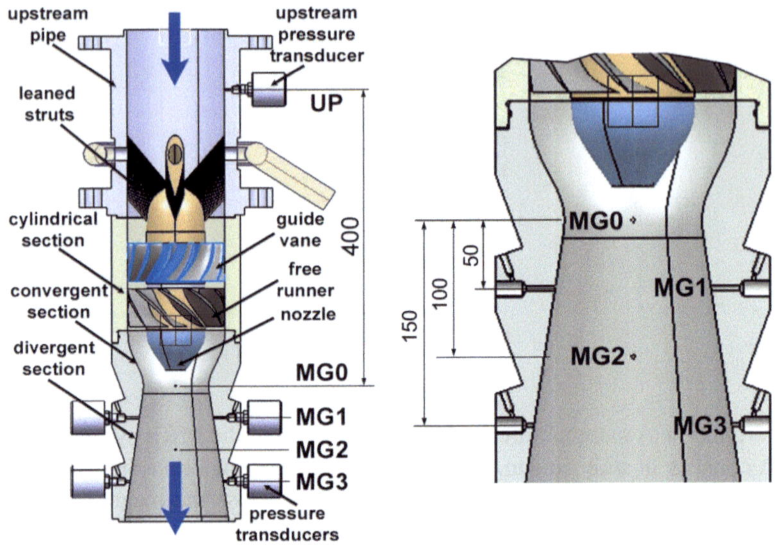

Fig. 5 Meridian cross-section of the test section (*right*), zoom-up of the convergent-divergent test sections with pressure taps (*left*)

The first level corresponds to the throat (MG0) and the next levels are displaced at 50 mm (MG1), 100 mm (MG2) and 150 mm (MG3) with respect to the first one. One more transducer is flush mounted on the wall of the upstream cylindrical part of the test section, see Fig. 5. The position of the upstream pressure transducer is 400 mm with respect to the throat (MG0) level. The transducers measurement range was ±1 bar with a precision of ±0.13%. However, the upper limit of frequency for unsteady pressure transducers is beyond to 50 Hz.

In our investigations a nominal discharge of 30 l/s is selected. In this case, the swirling flow configuration provided by the stay vanes leads to a free runner speed of 1020 rpm. With the MRB, the speed is slow down up to 690 rpm, with a step size of 30 rpm. Accordingly the speed is reduced with 35%.

The dimensionless flux of moment of momentum distribution versus dimensionless discharge was computed for several runner speed values based on experimental data, see Fig. 6 (right). The same distribution is obtained for FLINDT Francis turbine proving the same behaviour for the swirl generator [2], Fig. 6 (left).

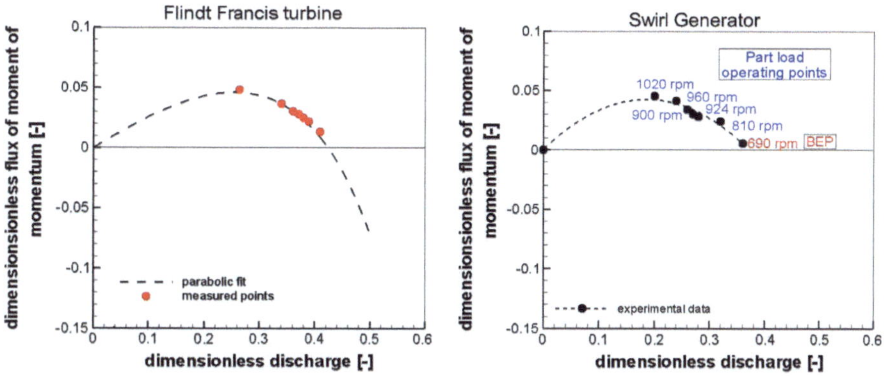

Fig. 6 Dimensionless flux of moment of momentum calculated for FLINDT Francis turbine (*left*) and dimensionless flux of moment of momentum calculated for our swirl generator (*right*)

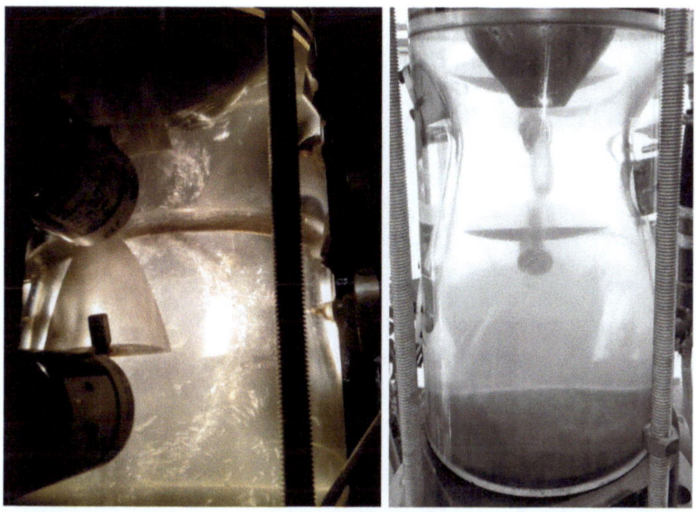

Fig. 7 Visualization of the cavitating flow from the test rig at: 1020 rpm (*right*) and 690 rpm(*left*)

4 Results

A first step in evaluation of the MRB used on the test rig was to visualise the cavitating flow, see Fig. 7.

According with Fig. 7, at the runaway speed (1020 rpm), the vortex rope is a vortex sheet which rolls-up as a helical vortex around a central stagnant region; the vortex rope is the source of pressure pulsations from the draft tube cone. At 690 rpm, the flow in the draft tube cone has an axisymmetric form, while the pressure pulsations have a minimum value. Also the speed of the free runner was measured. Accordingly, at the runaway regime, the variation in runner speed is approximately ±1%, but as the rotational speed decreases, the variation increases up to ±5%. This increase can be related to the shape and size of the magnetic particles clusters within the film of the magneto rheological suspension.

A second analysis consists in evaluation of signal decomposition, keeping into account that on each level two pressure signals are registered on opposite sides. According with Jacob [14], there are two types of draft tube cone pulsations. The plunging type is acting as a water hammer in the length of the cone, while the rotating type is acting in the cross sections of the cone. The rotating type is produced by flow instabilities as vortex rope due to its shape.

The decomposition procedure to obtain rotating and plunging pulsation types is described in Bosioc et al. [8]. Also for pressure pulsations decomposition from the experimental data was using the procedure based on the Parceval's theorem.

According with Fig. 8, the larger amplitudes around 18 Hz for 1020 rpm, 15 Hz for 924 rpm and 10 Hz for 840 rpm correspond to the rotating component.

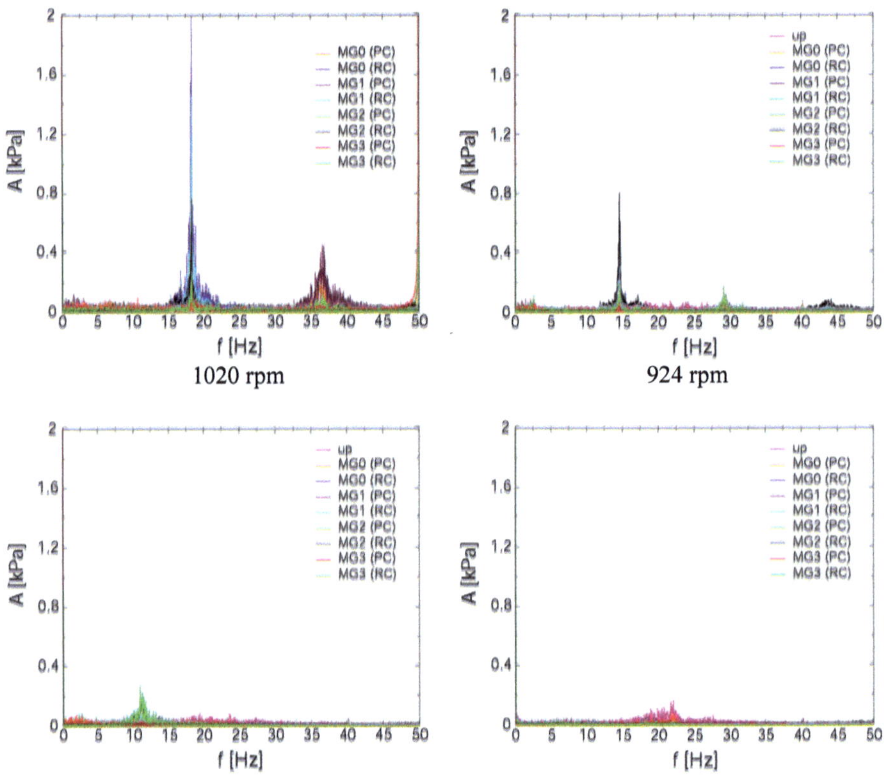

Fig. 8 Fourier spectra of the decomposed unsteady pressure signals at all four levels on the cone wall from MG0 to MG3 at four speeds: 1020, 924, 840 and 690 rpm

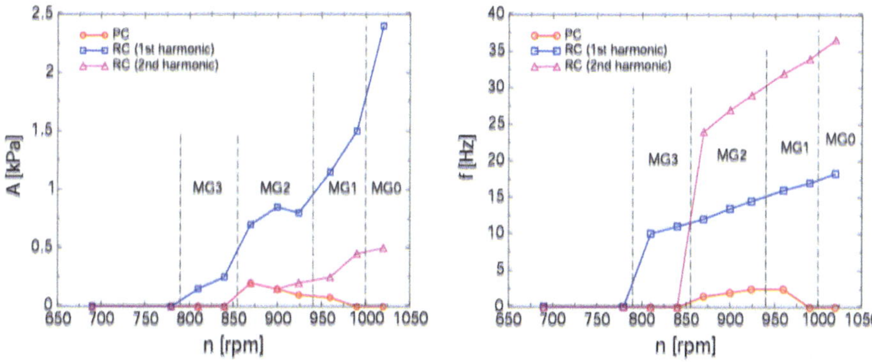

Fig. 9 Amplitude (*left*) and frequency (*right*) of the decomposed unsteady pressure signals at all four levels on the cone wall from MG0 to MG3

This rotating component is due to the existing vortex rope from the draft tube cone. When the runner is slow down with the MRB at 690 rpm, the amplitude is decreased at a minimum value. The main conclusion from these graphs is: when the speed of the runner is reduced, we are going closer to the best efficiency regime. According with this regime, minimum pressure pulsations can be found in the draft tube cone (Fig. 9).

The rotational component of pressure pulsation associated with the vortex rope is significant for part load regime especially in first three levels (MG0, MG1 and MG2).

5 Conclusions

The paper presents experimental measurements of decelerated swirling flow in a conical diffuser at different operating regimes. The runner speed of the swirl generator was controlled by a MR brake system. Accordingly, the dimensionless discharge and dimensionless flux of moment of momentum are calculated, when the runner speed is slow down. It acts as a real runner operating from part load up to BEP regimes at constant guide vane opening. The pressure measurements were made at four levels in the cone, each level having two opposites' pressure transducers. The amplitude, frequency and signal decomposition help to characterize the flow from dynamic point of view. In the case of dynamical analysis, the amplitude associated to the frequency of the rotating component has maximum values in the first part of the cone when is working at partial discharge. This results from the shape of the vortex rope, being with maximum eccentricity in the middle of the cone. At BEP the amplitudes with minimum values are recorded along to the length of the cone.

Acknowledgements This work was supported by a grant series of the Romanian National Authority for Scientific Research, CNCS—UEFISCDI, PNII-ID-PCE-2012-4-0634.

References

1. Nishi, M., Liu, S.: An outlook on the draft-tube-surge study. IJFMS **6**(1), 33–48 (2013)
2. Bosioc, A.I., Muntean, S., Tanasa, C., Susan-Resiga, R.F., Vékás, L.: Unsteady pressure measurements of decelerated swirling flow in a discharge cone at lower runner speeds. In: 27th IAHR Symposium on Hydraulic Machinery and Systems, IOP Conference Series: Earth and Environmental Science, vol. 22, no. 032008 (2014)
3. Frunzăverde, D., Muntean, S., Mărginean, G., Câmpian, V., Marşavina, L., Terzi, R., Serban, V.: Failure analysis of a Francis turbine runner. In: 25th IAHR Symposium on Hydraulic Machinery and Systems, IOP Conference Series: Earth and Environmental Science, vol. 2, no. 012115 (2010)
4. Kurokawa, J., Kajigaya, A., Matusi, J., Imamura, H.: Supression of the swirl in a conical diffuser by use of J-*Groove*. In: Proceedings of the 20th IAHR Symposium on Hydraulic Machinery and Systems, Charlotte, North Carolina, USA (2000)
5. Thike, R.H.: Practical solutions for draft tube instability. Water Power Dam Constr. **33**(2), 31–37 (1981)
6. Falvey, H.T., Cassidy, J.J.: Frequency and amplitude of pressure surges generated by swirling flows. In: Proceedings of 5th Symposium of International Association for Hydraulic Machinery, Equipments and Cavitation, Stockholm, Sweden (1970)
7. Susan-Resiga, R.F., Vu, T.C., Muntean, S., Ciocan, G.D., Nennemann, B.: Jet control of the draft tube vortex rope in Francis turbines at partial discharge. In: Proceedings of the 23rd IAHR Symposium on Hydraulic Machinery and Systems, Yokohama, Japan, F192 (2006)
8. Bosioc, A.I., Susan-Resiga, R.F., Muntean, S., Tanasa, C.: Unsteady pressure analysis of a swirling flow with vortex rope and axial water injection in a discharge cone. ASME J. Fluids Eng. **134**, 081104 (2012)
9. Foroutan, H., Yavuzkurt, S.: Flow in the simplified draft tube of a Francis turbine operating at partial load. Part 2. Control of the vortex rope. J. Appl. Mech. **81**(6), 061011 (2014)
10. Tanasa, C., Susan-Resiga, R.F., Muntean, S., Bosioc, A.I.: Flow-feedback method for mitigating the vortex rope in decelerated swirling flows. ASME J. Fluids Eng. **135**, 061304 (2013)
11. Susan-Resiga, R.F., Ciocan, G.D., Anton, I., Avellan, F.: Analysis of the swirling flow downstream a Francis turbine runner. ASME J. Fluids Eng. **128**, 177–189 (2006)
12. Ciocan, G.D., Iliescu, M.S., Vu, T.C., Nennemann, B., Avellan, F.: Experimental study and numerical simulation of the FLINDT draft tube rotating vortex. ASME J. Fluids Eng. **129**, 146–158 (2007)
13. Susan-Resiga, D., Vekas, L.: Yield stress and flow behavior of concentrated ferrofluid-based magnetorheological fluids: the influence of composition. Rheol. Acta **53**, 645–653 (2014). doi:10.1007/s00397-014-0785z
14. Jacob, T., Evaluation sur modele reduit et prediction de la stabilite de functionnement des turbines Francis, Ph.D. thesis, Laussane, Switzerland (1993)

Part VI
Additive Manufacturing

Direct Digital Manufacturing: A Challenge to the Artistic Glass Production

R. Felismina, M. Silva, A. Mateus and C. Malça

1 Introduction

Currently the high hand-labor costs, long production times, lack of automation and high energy consumption associated with the high temperatures required to produce pieces of glass, are identified as the main inhibitors of growth and economic development of the decorative and utility glass industrial sector. Current production processes impose limitations on the complexity of the geometry and shape of the glass pieces to be produced, which in turn greatly restricts the creativity of designers and consequently, the characteristics of differentiation and innovation presented by new products on the market.

Additive Manufacturing (AM) technology allows the rapid, automated and fully flexible manufacturing of products from templates generated by CAD (Computer Aided Design) in a fast, automated and fully flexible manner. Compared with the classical manufacturing process for artistic glass parts, the AM requires a low volume of material, and offers the potential of higher speeds, greater quality and

R. Felismina · M. Silva · A. Mateus · C. Malça (✉)
Centre for Rapid and Sustainable Product Development, Polytechnic Institute of Leiria, R. de Portugal, 2430-028 Marinha Grande, Portugal
e-mail: candida@isec.pt

R. Felismina
e-mail: raimundo.felismina@ipleiria.pt

M. Silva
e-mail: miguel.r.silva@ipleiria.pt

A. Mateus
e-mail: artur.mateus@ipleiria.pt

C. Malça
Department of Mechanical Engineering, Polytechnic Institute of Coimbra, R. Pedro Nunes, 3030-199 Coimbra, Portugal

L.F.M. da Silva (ed.), *Materials Design and Applications*,
Advanced Structured Materials 65, DOI 10.1007/978-3-319-50784-2_18

lower cost of production. Furthermore, the absence of tools translates into an almost total freedom in the generation of complex geometries, which offers a versatility which is non-existent with current productive systems. AM is also associated with the absence of waste material regardless of any complex shape, material or additive technique. This allows designers to create physical forms of immeasurable complexity from the generation and optimization of three-dimensional digital models (3D) [1–3].

The core of this work is focused on direct digital additive manufacturing of glass components. This requires a comprehensive methodology that integrates several approaches to glass additive manufacturing with design, control, management, information, as well as the networked communication intrinsic to the developments associated with the industrial revolution 4.0. Thus, the global purposes of this work consider the design and development of innovative equipment with a "*hand-made*" stamp as it includes a haptic interface where the artisan and/or the designer can create their pieces directly using their hands, i.e. without using a computational drawing tool. Through the use of appropriate computational tools, this piece will be converted into a digital representation that can be used subsequently by the additive manufacturing equipment to produce quickly and automatically the hand designed piece. Alternatively, the pieces to be produced may have also originated from any computational design tool commercially available.

In short, the equipment includes a trajectory generator composed of a set of four or more axes (x, y, z, ω or more), which are monitored and controlled by software. The system contains three interfaces: (i) a haptic interface, where the craftsman and/or designer may create a piece by hand but which generates a digital

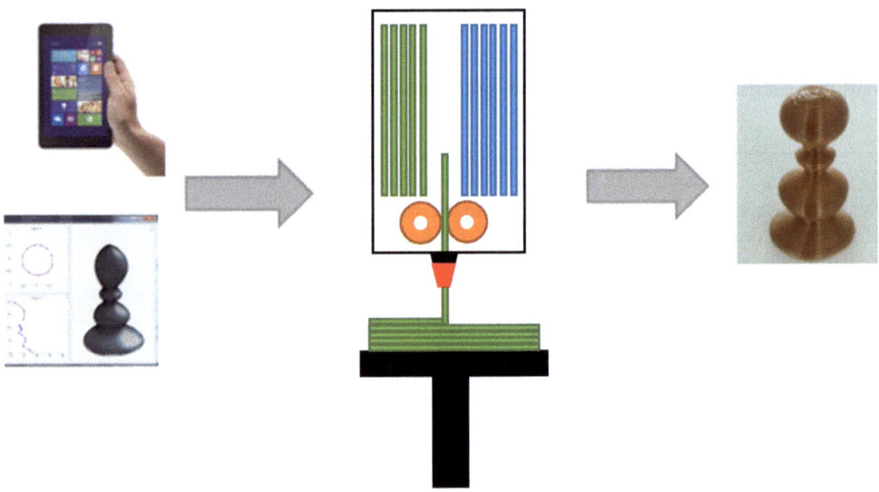

Fig. 1 Illustration of the global direct digital additive manufacturing equipment for 3D glass parts printing

representation which will subsequently be used by the additive manufacturing equipment to produce quickly and automatically the part designed by hand; (ii) a connection interface to the operator, and (iii) other communication with external systems. These are outlined in summary form in Fig. 1.

2 Evolution of the Decorative and Utilitarian Glass Industry in Portugal

Although the glass industry was introduced into Portugal in the second half of the fifteenth century with the founding of a small factory in the town of San Pedro of Vila-Chã, it came to Marinha Grande in 1745, as a result of the transfer of the Royal Glass Factory of Coina by King João. The abundance of the essential raw materials in the region prompted the development of a glass industry in the Marinha Grande region where, currently the glass industry remains active. As a result of the 1755 earthquake in Portugal, there was a great demand for glass due to the need to reconstruct the City of Lisbon. With the support of the Marques of Pombal, in 1769 King José I granted the Stephens brothers, founders of the Royal Navy of the Great Glass Factory, the power to use the wood of the Leiria pine forest to feed the large furnaces, making Marinha Grande the most important center of glass making in Portugal. This led to an significant increase in the number of companies operating in the glass industry in the nineteenth century [4]. Consequent to this expansion, other core businesses developed in Marinha Grande as is exemplified by the moulds industry that is today recognized worldwide for the quality of the products, export-oriented, technological modernization, skilled labor and production flexibility.

In the middle of the twentieth century the container/vessels glass sector, was greatly automated. In contrast, the artistic and utility glass sector witnessed continued stagnation as a lot of handwork, sometimes unskilled, is still needed. In addition, the glass sector has been associated with high energy consumption because of the very high manufacturing temperatures. As a result, the final product has a high commercial value. This strangles the economic growth of companies which have seen a drastic reduction in their profit margins. Reducing costs is seen as the only way to compete in the global market. Moreover, in recent decades we have witnessed an unprecedented invasion of Western markets with products from the countries of Eastern Europe and Asia, where the hand labour and energy costs are lower and this represents a threat to the global economic system.

The production of decorative glass pieces with more complex three dimensional shapes using conventional thermoforming processes, e.g. molding, slumping, fusing, forming, sintering and blowing, also require the production of complex molds, which are difficult to obtain, and this represents a serious obstacle to designers in creating different parts, besides the high costs and times associated with the process of development of new products.

As a consequence of poor economic growth and the absence of innovative business strategies, aspects such as quality, innovation, and design were compromised. This condemns continues to condemn, in the medium term, the survival of traditional companies producing utility and decoration glasses in Portugal. In addition, the disappearance of the artistic glass manufacturing sector compromises the survival of other companies that are involved in the life cycle of such glass products. There is an urgent need to promote within the glass industry the concepts of innovation and technological development so that companies start to compete with their own brands that meet the quality requirements and technical accuracy imposed by the different national markets, and fundamentally the international ones. This work arises in the context to draw together tradition and innovation for the economic strengthening of a potentially productive sector—the glass artistic and utility industry. This industry was the genesis of the major industrial clusters in the Leiria region and hence is of immeasurable historical and cultural importance. It is expected that the implementation of the proposed AM system results in the mass production of pieces in glass, clear or multicolor, without creative boundaries which will lead to the differentiation and affirmation of this glass sector at the regional—Marinha Grande—national and international levels. The new equipment will undoubtedly contribute to the survival of this sector, as well as to its expansion making it more competitive as it meets the demands of today's markets.

3 Glass Additive Manufacturing Technology

To meet the market demands with regard to price, quality and delivery times, it became necessary to resort to the production of models and prototypes that can be used to assess the functional and aesthetic value of the products. AM has, in this respect, an important, powerful and revolutionary role. Since the appearance of the first additive manufacturing systems, research and the search for new applications has not stopped. In fact all industries that develop and produce components and products should use this technology due to the high production speed, creativity, efficiency, affordability, low production cost, environmental sustainability, design optimization and reduced costs that characterize it [1, 2]. In addition, AM technology opens new paths for innovation and offers a range of logistical, economic and technical advantages [1, 3].

AM includes a number of technological processes that enable making three-dimensional (3D) physical models layer-by-layer directly from a digital definition as described in Ref. [5]. These processes can be used for the production of components on any scale from microscopic objects to large objects. AM technologies can be subdivided according to the type of process and materials involved as well as the functional purposes of each one of the manufactured parts [1, 5]. Notwithstanding different objectives, Selective Laser Sintering (SLS), Selective Laser Melting (SLM), Fused Deposition Modeling (FDM) and 3D printing (3DP) are the most commonly processes with glass AM [5–16]. The SLS and SLM processes were used to work soda-lime glasses, quartz and borosilicate [6–10]. The glass objects printed by these sintering processes are extremely brittle and with an opaque appearance due to light scattering from glass powders caused by incomplete densification [9, 13–16]. The latest SLS experiments with glass, attest to the opacity of the products and their poor mechanical properties. The internal porosity of the parts produced leads to dispersion of light, thereby limiting transparency in the application of this technique [10]. The FDM technique allows for 3D objects with good mechanical, durable and fully functional properties. The 3D object is built layer by layer by the addition of polymeric filaments which are melted and extruded through a nozzle that moves in the x-y plane [11]. The major drawback of this process is the need of structural supports during manufacturing. These supports are made simultaneously with the part to be built using a less resistant material, e.g. a wax or water-soluble polymer [17]. The FDM technique has been used to obtain translucent ABS parts and to extrude bone scaffolds from bioactive glass [11, 12].

Based on extrusion techniques, an AM glass-printing process called G3DP (Glass 3D Printing) has been developed by researchers in the Mediated Matter Group at the Massachusetts Institute of Technology (MIT) Media Lab in collaboration with the Glass Lab at MIT [14–16]. In this 3D printing process, molten glass is loaded into a hopper on the top of the device after being gathered from a conventional glassblowing kiln. In operation, both the hopper and the nozzle, through which the glass is extruded, have to be maintained at very high temperatures. The energy involved to maintain the glass in the molten state before and during the extruded process remains high when compared with conventional procedures to mold glass. Furthermore, this equipment is not able to build objects with smooth surfaces [15, 16]. Another recent development of new AM equipment to print glass came from Israel with the Micron3DP. This printer is also based on the FDM technology, but despite reaching temperatures around 1600 °C, only the manufacturing of small scales parts are allowed [18]. As with the GD3P, the Micron3DP equipment is not able to manufacture smooth surfaces. In both types of equipment the glass filament temperature and the geometry are crucial parameters to ensure sufficient adhesion between the new layer and the part already printed. This is a real challenge together with other processing parameters such as the glass viscosity, flow and feed adjustment rates to produce quality parts.

Recently, The Netherlands Organisation for applied scientific research TNO developed a printing process using powders to set color images in glass. This printing process involves a projection of a color, in the form of glass powder, which is applied to a flat glass sheet. Thereafter, the glass sheet is heated in an oven to melt the glass powder with the underlying glass. Using only 3 basic colors, it is possible by their blending to provide any desired color. Furthermore through this fusion process it is also possible to create embossed structures on the glass sheet [19].

Regarding the use and marketing of additive technologies applied to the production of decorative and utility pieces glass with high detail, high precision and excellent finishing little has yet been done. The development and implementation, in a production line, of efficient equipment for glass additive manufacturing to obtain a mass production of glass parts differentiated with high added value is absolutely crucial for the survival of this Portuguese economic sector. In addition, a drastic reduction in the level of waste raw material during the manufacturing process is expected as well as a significant saving of energy and production time since each product line can continuously manufacture various geometries. With these objectives the equipment, which is described in the next section, was designed and developed.

4 Equipment Conceptualization and Design

The AM equipment developed in this work allows the production by extrusion, layer by layer, of glass parts with a countless combination of colors and shapes through an automatic procedure using glass in the form of rods, as shown in Fig. 2. The equipment designed includes a trajectories generator composed of a set of four or more axes (xyzw or more), which is monitored and controlled by a processing and control software that, in turn, includes a connection interface to the operator and another interface that establishes the communication of the device with other computer exterior systems. The equipment operator can set and control the most appropriate operating conditions to each specific production.

As shown in Fig. 2, the glass in the form of rods of different colors and diameters (2) are stored in the reservoir (1). The rods are driven by an automated material feeder (3) to the extruder nozzle, where a heat source (4) provides the glass molten state, which will be extruded onto the construction platform (5). The equipment also comprises an automatic change system of building platform

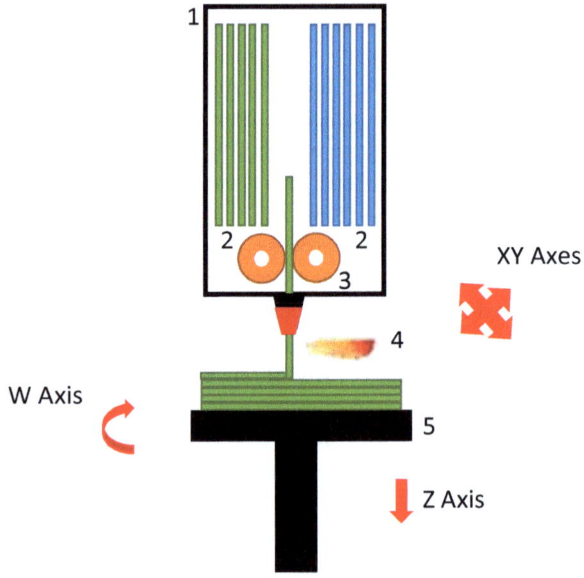

Fig. 2 Illustration of color 3D glass printer with 4 axles

(palletizing) with motion in the axis direction or another (Z or other), i.e., in a vertical or other direction.

The construction platform (5), after complete part production, is replaced by another platform leading to the start of a new production cycle. The platform that contains the finished glass part is placed in a transport system, a conveyor belt, to be brought to an oven where it will be subject to a programmed heat treatment. Note that the construction platform (5) is heated to a temperature suitable for production, which contributes to the total energy costs associated with the production process. If the glass rods (2) are colorless it is possible through a glass powder projection system (6), aided by an additional heat source (7), to incorporate a multitude of colors to the processed glass as shown in Fig. 3. This equipment has been submitted to a provisional patent whose reference is 20161000017437.

A rudimentary prototype of the equipment described above was constructed. The idea of building this prototype was to test the feasibility of the operation principle and mainly the effectiveness of the heat source used to melt the glass. Figure 4 illustrates images of the prototype system printing a 3D glass part from a rod glass that is molten using a combustion torch powered through a mixture of oxygen and propane, A significant energy loss was observed with this kind of heat source. To overcome this drawback a new system to melt the glass was designed. Figure 5 displays the scheme of the new system that will be built. The efficiency of induction resistances or high temperature resistances for the heat source will be evaluated.

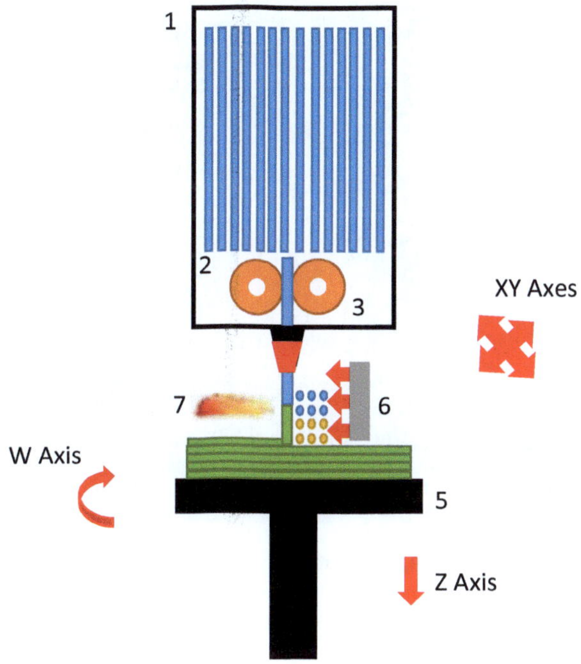

Fig. 3 Illustration of 3D glass printer with the glass powder projection system

Fig. 4 Images of the rudimentary prototype developed printing a 3D glass part

Fig. 5 Design of the new system to melt the glass

From Fig. 4 it was concluded that as reported for the Micron3DP and GD3P devices, for the equipment developed there is sufficient adhesion between layers and the part already printed. Uneven surfaces are also observed. Nevertheless, the thickness of each layer can be reduced to a dimension comparable to the glass rod diameter. An auxiliary system, using hot air pressured has been proposed to increase the surface smoothness. The energy consumption involved to melt the glass is clearly reduced.

Since the viscosity of the glass changes very quickly with temperature, a very accurate procedure to control the temperature of the glass is required. In addition, the success of the manufacturing process of the desired customized glass part depends on the optimisation of other processing parameters such as the flow and feed rates, which are de as a function of the complexity and dimensions of the part.

Much work is still required to develop and understanding of the influence of the extrusion parameters on the parts produced as well as the physical, chemical and mechanical characterization of the glass parts printed with this equipment. Despite this, an important contribution to the development of 3D printing glass parts has been made. With the additive manufacturing equipment described here it is possible to produce custom parts and with greater freedom of geometries in a short time and at perfectly acceptable costs due to the drastic reduction of hand labor and energy consumption that are associated with the glass industry. This is a highly competitive advantage over current conventional and traditional processes used not only in utilitarian glass pieces but also, and above all, in the production of decorative pieces, thereby opening up new creative horizons for designers.

5 Conclusions

This work presents the development of novel AM equipment that combines the current technology of additive manufacturing and the art of working glass by hand. From the equipment implementation viewpoint, a lower production complexity, greater constructive reliability and high adaptability to the different design solutions can be expected. Additionally, the manufacturing process of stylized glass parts is automated, which leads to productivity increase and a considerable reduction on the intervention of manpower and on the energy consumption. This equipment will be able to produce translucent and multicolored three-dimensional art pieces. The equipment will, in an automated fashion and from glass in the form of rods, provide the production of glass components that can have numerous combinations of colors and shapes, since the machine operator can set and control the most appropriate operating conditions to each specific production. For the manufacture of multicolor glass parts, the equipment can, additionally, include one or more heat sources that allow the fusion of the glass powders projection of the different colors on successive layers deposited.

In short, the equipment described here will open a range of possibilities of studies on the application of additive manufacturing to the glass area to produce decorative and utilitarian objects in materials with high melting points without the need to resort to traditional ovens, and thus, contribute to a drastic reduction in energy bills. With the implementation of glass additive manufacturing equipment and due to the absence of drastic reduction of waste and of raw materials, a significant saving of energy and production time is expected. With the implementation of automation in the production process, a certain number of production lines that lead to the mass production of different glass parts with high added value becomes possible, and this undoubtedly contributes to the growth and development of this economic sector.

References

1. Gibson, I., Rosen, D.W., Stucker, B.: Development of additive manufacturing technologies. In: Additive Manufacturing Technologies: Rapid Prototyping to Direct Digital Manufacturing, pp. 17–40. Springer Science+Business Media, LLC (2010)
2. Antoniw, S., McCarthy, N., Pacey, E., Parkin, B., Shelton, P.: Additive Manufacturing: Opportunities and Constraints, 1st edn. Royal Academy of Engineering, London (2013)
3. Wohlers, T., Caffrey, T.: Wohlers Report 2014. Wohlers Associates, Fort Collins (2014)
4. Mendes, J.M.A.: A concentração da indústria vidreira na Marinha Grande. Repercussões socioeconómicas. Revista de História (1988)
5. ASTM. Standard Terminology for Additive Manufacturing Technologies (2012)
6. Khmyrov, R., Grigoriev, S., Okunkova, A., Gusarov, A.: On the possibility of selective laser melting of quartz glass. Phys. Proc. **56**, 345–356 (2014)

7. Klocke, F., McClung, A., Ader, C.: Direct laser sintering of borosilicate glass. In: Solid Freeform Fabrication Symposium Proceedings, Austin, TX, 3–5 Aug 2004, pp. 214–219 (2004)
8. Lee, I., Manthiram, A., Marcus, H.L.: Selective laser sintering of alumina-zinc borosilicate glass composites using monoclinic HB02 as a binder. In: Solid Freeform Fabrication Symposium Proceedings, Austin, TX, 7–9 Aug, pp. 46–54 (1995)
9. Fateri, M., Gebhardt, A.: Selective laser melting of soda-lime glass powder. Int. J. Appl. Ceram. Technol. **12**(1), 53–61 (2015)
10. Luo, J., Pan, H., Kinzel, E.C.: Additive manufacturing of glass. J. Manuf. Sci. Eng. **136**(6), 061024 (2014)
11. Ahn, S.H., Lee, C.S., Jeong, W.: Development of translucent FDM parts by post-processing. Rapid Prototyp. J. **10**, 218–224 (2004)
12. Fu, Q., Saiz, E., Tomsia, A.P.: Bioinspired strong and highly porous glass scaffolds. Adv. Funct. Mater. **21**(6), 1058–1063 (2011)
13. Marchelli, G., Prabhakar, R., Storti, D., Ganter, M.: The guide to glass 3D printing: developments, methods, diagnostics and results. Rapid Prototyp. J. **17**(3), 187–194 (2011)
14. Klein, S., Simske, S., Adams, G., Parraman, C., Walters, P., Huson, D., Hoskins, S.: 3D printing of transparent glass. In: Proceedings of NIP & Digital Fabrication Conference, Society for Imaging Science and Technology, Quebec City, Canada, 9–13 Sep, pp. 336–337 (2012)
15. Klein, J.: Additive manufacturing of optically transparent glass, M.Sc. thesis, Master of Science at the Massachusetts Institute of Technology (2015)
16. Klein, S., Simske, S., Adams, G., Parraman, C., Walters, P., Huson, D., Hoskins, S.: 3D printing of transparent glass. HP Technical report (2012)
17. Yan, Y., Li, S., Zhang, R., Lin, F., Wu, R., Lu, Q., Xiong, Z., Wang, X.: Rapid prototyping and manufacturing technology: principle, representative technics, applications and development trends. Tsinghua Sci. Technol. **14**, 1–12 (2009)
18. Information on http://micron3dp.com/blogs/news/34473924-breakthrough-in-3d-printing-glass
19. Information on https://www.tno.nl/en/focus-area/industry/flexible-free-formproducts/additive-manufacturing/glass-powder-printer

Post-process Influence of Infiltration on Binder Jetting Technology

Edwin Ocaña Garzón, Jorge Lino Alves and Rui J. Neto

1 Introduction

When Binder Jetting technology (BJ) was originally developed it was called 3DP, although the term was never trademarked [1].

Printed parts by 3DP, especially in ceramic powders, are not strong enough to be used as functional parts such as those made by other rapid prototyping (RP) technologies, so it requires post-processing (PP) to improve the desired strength or final characteristics. The mechanical strength is a factor in the design of the vast majority of printed pieces and even more in brittle materials such as plaster. For functional components, PP is a critically important aspect of 3DP, but it is often overlooked in product literature and by the media.

Complete 3DP process involves 3DP plus PP, to which is a considered a challenge by some authors [1, 2]. PP encompasses all of the activities that occur from the time a build is complete (green body) to the time a 3DP part is ready for its intended use (final part). Green strength refers to the initial strength after printing and before any PP to increase the mechanical properties. After printing, a typical green body can be 30–75% vol. powder, 10% vol. binder, and the rest void space [3].

PP is not only the typical removing of support material, there is a considerable post-processing for various AM technologies. In BJ, the result, immediately after

E.O. Garzón (✉)
Departamento de Ciencias de la Energía y Mecánica (DECEM),
Universidad de las Fuerzas Armadas—ESPE, Sangolquí, Ecuador
e-mail: emocania@espe.edu.ec

E.O. Garzón · J.L. Alves · R.J. Neto
Faculty of Engineering, INEGI, University of Porto, 4200-465 Porto, Portugal
e-mail: falves@fe.up.pt

R.J. Neto
e-mail: rneto@inegi.up.pt

L.F.M. da Silva (ed.), *Materials Design and Applications*,
Advanced Structured Materials 65, DOI 10.1007/978-3-319-50784-2_19

printing, is a bound green part embedded in loose powder [4], therefore, the first PP is a minimum post-printing bed manipulation or depowered to handle the print part. Next, some optional PP steps may include coating, sintering, infiltration [5], polishing, or machining; depending on the desired improvement of final characteristics, which can be; surface finish, reduction of stress concentrations, increase of strength, or highlight bright colours. The two most common PP steps are sintering and infiltration [4].

Infiltration is a way to achieve high density parts without the large shrinkage associated with sintering to full density [6]. Infiltration occurs when the selected liquid is drawn into the open pores of the printed part through capillary action and solidifies. Both low and high temperature infiltration are possible, depending on the part material and binding mechanism [5]. The first occurs at or slightly above room temperature, and the second at temperatures between −6 and 10 °C of the infiltrates melting point [3].

Examples of infiltrates include molten wax, varnish, lacquer, cyanoacrylate, polyurethane, and epoxy [7]. This variety of materials and PP methods suggests that a deeper analysis of to this stage is worthwhile. Moreover, there are no significant contributions in the literature concerning this issue.

Successful realization of a specific 3DP process involves not only the printing process itself, but also powder and binder material system [4] understanding along with printing process details (printing parameters) and PP, both of which play a major role in determining the final mechanical properties of the parts produced.

The lack of understanding in the interplay between the powder properties and the final mechanical properties raised the need for a systematic investigation of printing powder for 3DP [8] and PP.

The aim of this study is to analyse the main influential parameters in the post-process of infiltration on the flexural and tensile mechanical properties of printed parts by binder jetting technology, infiltrated with materials and methods suggested by 3D Systems, performing a previous characterization of the powder and a green body specimen.

1.1 Main Parameters on Binder Jetting and Post-process of Infiltration

Good quality of a printed piece with a specific 3DP process involves not only the printing process itself, but also a combination of a powder and binder material system along with process details for printing and PP, both of which play a major role in determining the mechanical properties of the printed parts [9]. Figure 1, summarizes the main parameters of both BJ and the process of infiltration that will affect the final characteristics of a printed functional piece.

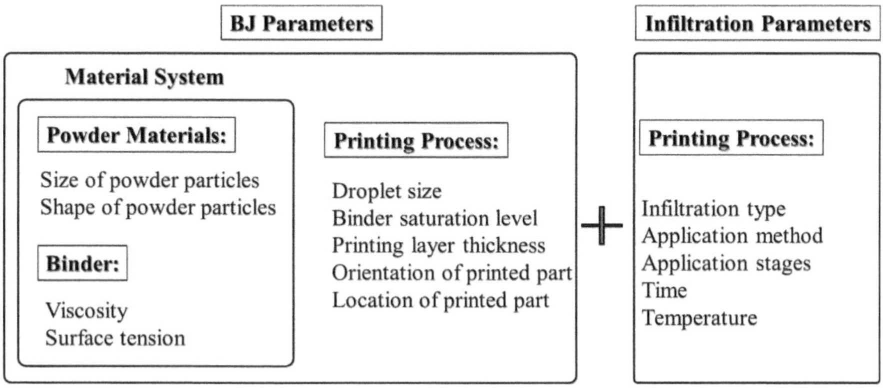

Fig. 1 Key parameters on Binder Jetting and process of infiltration

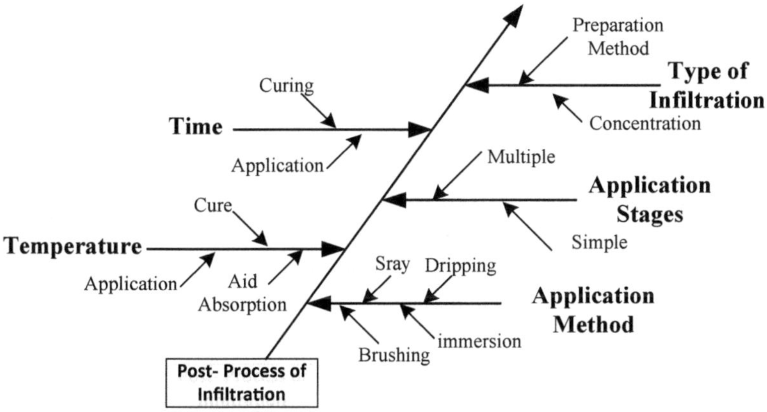

Fig. 2 Fishbone diagram with parameters of post-process of infiltration (adapted from [10])

Figure 2 shows the parameters on post-process infiltration as a fishbone scheme, such as, time, temperature, infiltrate type, application stages, and application method.

The main property of the powder is the depositability, which depends on the size and shape of the particles. For dry state, preferred particles are less than 20 microns, while particles smaller than 5 microns can be deposited both dry and wet [11]. Fine powders (~1 μm) tend to agglomerate due to the Van der Waal's forces and moisture effect [12].

To exploit the advantages of powders with small and large particles, multimodal powders are used. Larger particles allow the powder mixture to be easily spread in dry state, while the smaller particles fill the interstices among the large particles, to increase the density of the printed part [13].

Table 1 Powder characteristics and the influence on printing (adapted from Utela [4])

	Advantages	Disadvantages
Size		
Large particles > 20 μm	Can be spread dry, large pores facilitate fluid migration in bed	Largest particle establishes minimum layer thickness
Small particles < 5 μm	Increased sinterability, lower surface roughness, thinner layers, smaller minimum features	Difficult to spread, agglomerate, may require slurry deposition
Shape		
Spherical	Tend to flow well, low internal friction	
Faceted/anisotropic [14]	Potentially increased packing ratio	Inhibits spreading

In both dry and wet deposition methods, the particle shape is less important than the size, but spherical powders are preferred for dry deposition because they tend to have a better flow [14] and have low internal friction [9]. Table 1 shows the powder characteristics and the influence on printing.

There are different ways for binding the powder. Some of the common selection criteria include (i) binder location (in-liquid vs. in-bed), (ii) binder residue in the final part, and (iii) binder material limitations. It is also important to review the binding methods, some of the most common are: organic liquids, in-bed adhesives, hydration systems, acid/base systems, inorganics, metal salts, solvents, phase-changing materials, and sintering aids/inhibitors [4, 5].

The most relevant parameters of the liquid binder material are: the viscosity and surface tension [4]. The spread characteristics are important because a liquid that wicks out considerably from the impact area results in a rougher surface texture. Increased liquid binder viscosity delays the spreadability [14], while a viscosity reduction is achieved by lowering the solids loading or adding dispersants [15].

The parameters inherent to the printing process are: droplet size [16], binder saturation level [17], printing layer thickness, orientation and location of the printed part (see Fig. 3), which influence the strength and surface quality of the 3D printing process.

The saturation level is the ratio of ink to bed pore volume and depends on the droplet size, droplet spacing, layer thickness, and bed packing density [5]. The saturation level needs to be high enough that the ink penetrates to the previous layer to bind the part in the vertical direction [9], but not so high that excess ink wicks away from the impact zones and roughens the surface finish by binding extra powder [18].

Part orientation relative to the print-head travel path affects printing. The test parts should either be designed to examine this behaviour, or the same test part should be printed at various orientations [4].

Several studies have been conducted to optimize printing parameters with plaster-based materials for commercial 3DP powders, to get best strength and

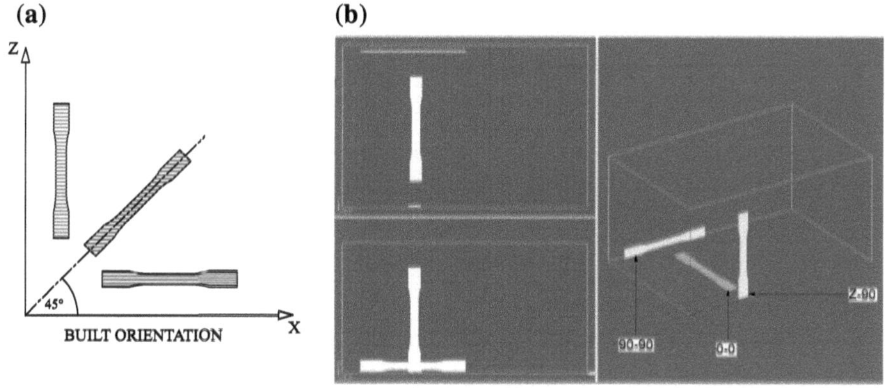

Fig. 3 **a** Three built orientation relative to Z (adapted from [19]), and **b** view of three built orientation in 3D Edit Pro 2.0 software (adapted from [20])

accuracy of green specimens, in some cases performing a characterization of the base material (see Table 2).

Vaezi et al. [21], studied the effects of layer thickness and binder saturation level on flexural and tensile strengths, integrity, surface quality, and dimensional accuracy in 3DP. They proved that the layer thickness and binder saturation level combined together have a significant effect on strength, integrity, and dimensional accuracy of green body part.

Zañartu et al. [22], studied the effect of layer thickness, jetted binder volume per layer, type of binder and temperature on mechanical properties of green body parts, made with an experimental 3DP equipment based on the spiral growth manufacturing (SGM) device.

Butscher [8], conducted a study in powder based three-dimensional printing of calcium phosphate structures for scaffold engineering where he analysed relevant powder properties such as powder particle size, flowability, wettability and compaction rate, and determined the optimal values of the mentioned parameters for printing accuracy.

The major printing parameters examined by Asadi et al. [23], were layer thickness, delay time of spreading the next layer, and build orientation of the specimens, to optimize them in the manufacture of a prototype scaffold with materials based on calcium sulphate.

Feng et al. [24], conducted tests with printed samples with cementitious powders, and verified that the materials are orthotropic. They observed that the mechanical properties and failure depends on inter-layer and inter-strip, and further suggest that the behaviour of 3D printed structures depends strongly on the printing direction.

On the other hand, there are studies that have been reviewed how the print parameters interact to mechanical properties of PP infiltrated specimens, printed in commercial 3DP plaster materials, which are summarized on Table 3. In general, these studies do not perform characterization of base materials.

Table 2 Previous studies of the influence of different parameters in BJ green parts in plaster based materials

		Materials				Printing parameters				Green body Study	Mechanical effects				
		Powder	Binder	Characterization		Binder saturation	Layer thickness	Built orientation	Built position		Tensile	Flexion	Compression	Roughness	Comments
				Chemical comp.	Particle size										
Vaezi, Chua	2011	ZP102	zb56			x	x			x	x	x			Thinner layers produce stronger components
Zanartu, Ramos	2010	zp 131	zb60	x		x	x			x					
Butscher A	2013	zp 130	zb58	x	x					x			x		Determined optimal values for printing accuracy
Asadi, Solati, et al.	2015	zp150	zb63	x	x					x			x		Evaluated mechanical properties, accuracy and porosity
Feng	2015	zp150	zb60						x	x			x		Printing direction has influence on the load bearing capacity

Table 3 Previous studies of influential parameters in printed and infiltrated parts by BJ with plaster materials

		Materials		Printing Parameters				Green body Study	Post- process				Mechanical effects				
		Powder	Binder	Binder Saturation	Layer thickness	Built Orientation	Built Position		Infiltration				Tensile	Flexion	Compression	Roughness	Comments
									Type	Time	Meth	Absor					
Ipens, Urbanic	2015	zp150	zb59			x			x	x		x	x		x		Infiltrantes influence on compression and tensile by print orientation
Galeta, Kladarie, Karakasic	2013	zp 130	zb56		x	x			x				x				Thinner layers produce stronger components
Gharaie, Morsi & Massood	2013	zp 150	zb63		x	x			x				x				Thinner layers and orientation position in "Y" > Strength
Frascati	2007	ZP 130	zb58			x	x		x			x	x	x			
Suwanprateeb, Jintamai	2006	biomat.mix								x	x	x		x			Full immersion infiltration
Pilipovic, Ana	2009	ZP102	zb56						x				x	x		x	Compared 3DP and polyjet parts

Ipens et al. [25, 26], studied printing parts infiltrated with polyurethane resin, cyanoacrylate, epsom salt and epoxy resin, previously analysing time and depth of infiltration by immersion, they reviewed the effect of type of infiltrate, print part orientation on the compressive, tensile and flexural strength. Concluding that for most brittle materials, the ultimate compression strength is much larger than the ultimate tensile strength. The compressive strength trends are more consistent in behaviour compared to the tensile and flexural results, they also show that infiltrates can significantly improve the mechanical characteristics, but performance degradation can also occur, which occurred with the Epsom salts infiltrates.

Galeta and Karakasic Mirko [27], evaluate the application of traditional infiltrants and other affordable alternative infiltrants used instead of genuine infiltrants. The results of the tensile test revealed that the strength of 3D-printed samples comes mainly from the infiltrants, but it may be additionally increased by selecting the best combination of the other two processing factors. The strength of the samples infiltrated with alternative infiltrants was equivalent to that obtained with genuine infiltrants, thus confirming the use of alternative infiltrants. They also observed that thinner build layers produced stronger parts.

Gharaie et al. [19], investigated how tensile properties of parts fabricated by 3D Printing is affected by 3D printing build orientation, and by post-processing methods of infiltration process and drying of parts. Results obtained for various parameters are compared to investigate the optimum procedure to achieve the highest tensile strength.

Previous studies of the influence of different parameters in BJ green parts in plaster based materials.

Additional contribution of this research is conducting a study using different materials of the ones used in the reviewed studies, even with similar characteristics, but that are new to the 3DP market. Therefore, the general structure of this study consists of characterization of powder base material and printed parts, application of infiltration and then, the influence of the type of infiltration on mechanical properties of the same parts that were analysed and discussed.

2 Materials and Methods

2.1 General Methodology

The methodology established for this experimental work consists on the following steps: (i) study of the particle size distribution and chemical composition of powder base material, (ii) analysis of the chemical composition of printed specimens, (iii) PP: infiltration of specimens, (iv) flexural and tensile tests, to verify the influence of infiltration type on the mechanical strength.

2.2 *Fabrication of 3D Printing Specimens*

2.2.1 Materials for Printing

The printing material is VisiJet® PXL Core (3D Systems, USA) with an appropriate water based binder VisiJet® PXL clear (3D Systems, USA). These materials are recommended to concept models and prototypes. This study did not use the option of colour printing.

The 3D printer, Projet 660 Pro printer (3D Systems, USA) was used, and this is equipped with 3D Edit Pro 2.0 software, automated power loading recycling and removal. It has built-in the same camera a compressed air gun to clean the parts, and drying them at 70 °C. The printer has a resolution of 600 × 540 dpi and layer thickness of 0.1 mm, and its 1520 inkjets allows to have a vertical build speed of 28 mm/hour, inside the build chamber of 254 × 381 × 203 mm.

2.2.2 Printing Parameters

All specimens were designed in Solidworks software, and saved in stl format. The printing parameters, binder saturation, layer thickness and printer orientation of specimen were set on the basis of two criteria; first, the flexibility of varying parameters within the printer range, and the second, discard those parameters that produce the best and the worst mechanical strength, opting for the ones that reflect the average mechanical resistance.

The set parameters were: binder saturation level: 37–42%, printing layer thickness: 0.01 mm, and printing orientation: 0°, relatively to the machine X axis, as it is shown in Fig. 4. Five samples were printed for each condition.

Currently there is no specific standard for tensile and flexural tests for materials based on plaster materials manufactured by AM processes, therefore it was decided to apply ASTM standards; for flexural test, ASTM D790-10 [28] (not technically

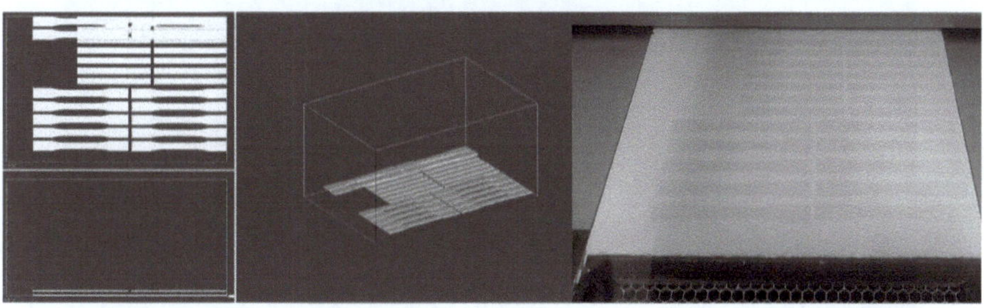

Fig. 4 Print orientation and position of the printed test specimens for tensile and flexural tests

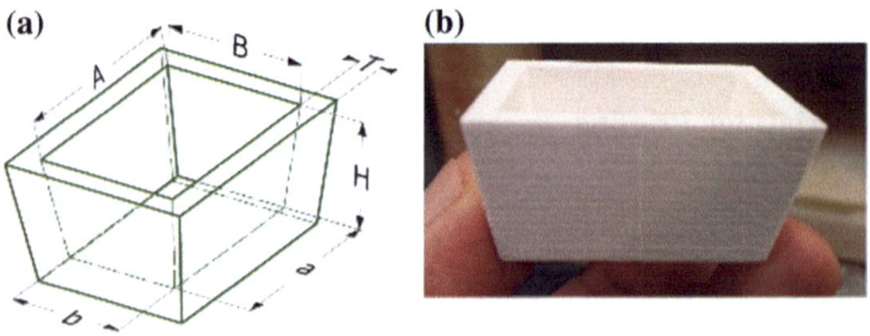

Fig. 5 Green body specimen for microscopic analysis. **a** CAD model, **b** *green* body printed part

equivalent to ISO 178), and for tensile tests ASTM D638-10 [29] (technically equivalent to ISO 527-1).

The chemical composition analysis, used a printed sample of a previous study (see Fig. 5), with internal dimensions of 35 × 25 × 20 mm (A × B × H) and 30 × 20 (a × b) and 4 mm wall thickness.

2.2.3 Drying and Depowdering

After printing, and before the infiltration, the models were dried in stages; first in the printer's chamber bed for 2 h at 55 °C, then in the contiguous chamber where they were further cleaned with compressed air, and dried at 37 °C for 1 h (see Fig. 6).

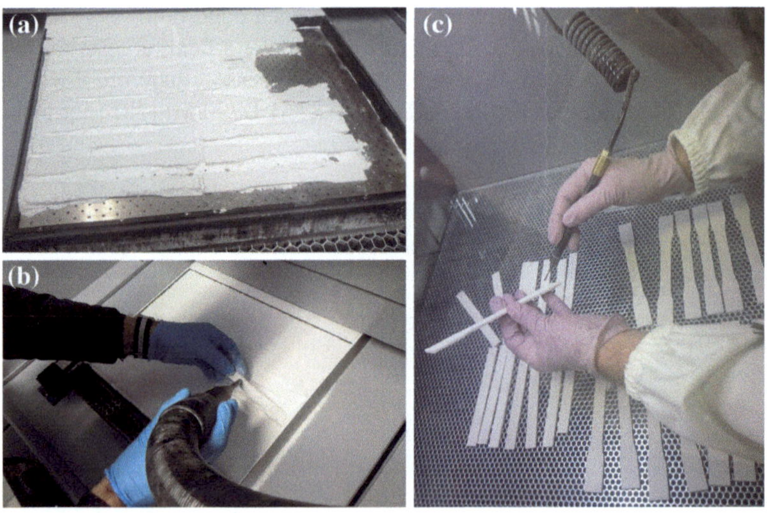

Fig. 6 Depowdering with **a** automatic vacuum system, and compressed air in **b** printing chamber, and **c** clean part chamber

The samples were weighed in a micro-scale JS-100XV (Jennings, Phoenix USA) with 100 g capacity. and accuracy of 0.01 g and measured at room temperature (~23 °C) with a digital calliper with accuracy of 0.01 mm (Mitutoyo, Japan), after drying and before infiltration (green state). This method of weighing and measuring was repeated after infiltration curing time. The percentage of absorbed infiltrate was only calculated for flexural samples, due to the simple geometry. Equation (1) was used to determine the percentage of the absorbed infiltrate, and is based on weight change [30]:

$$\text{Infiltrant absorption } (\%) = \frac{\text{Weight after curing of infiltrate (g)} - \text{Green body weight (g)}}{\text{Green body weight (g)}} \text{x100} \tag{1}$$

For each specimen, the dimensional control of the width and thickness was done in three different points as suggested in references [28, 29], measured in the middle and ends of the samples, to obtain the mean value.

2.3 *Infiltration*

The mechanical tests used samples with four types of infiltrates, three of them recommended by the supplier of the equipment, 3DSystems; Epoxy Resin (E), Cyanoacrylate (CY) and magnesium sulphate (MS), and the fourth was water (W). The last group of specimens were not infiltrated, which will be the green bodies referential.

The E infiltrate (StrengthMax, USA) is a two parts organic mixture, with a low viscosity. CY infiltrate (ColorBond™, USA) is a fast-curing organic mixture. The MS hepta-hydrate or Epsom Salt (Blue Cross Lab, USA) has as active ingredient MS pentahydrate 100%, according to their respective MSDS.

Each type of infiltration has its own processing procedures, which included mixture ratios, application methods and time. Before each infiltration, the respective mixtures were prepared. Infiltrate E mixture was prepared in the ratio of 5 parts in resin and 2 of hardener, and manual mixture for 5 min. MS mixture was prepared with 200 ml of magnesium sulphate, to which was added 250 ml of hot water at 60 °C to enable its dissolution, and manually shaken for 5 min. In the case of cyanoacrylate and water, no premix is required.

For E, MS and W infiltrates, the same methodology was followed by manual application, it proceeded to cover the specimens with a thin brush with light layers allowing the specimen to absorb the infiltrates and be saturated. For CY, a dropper bottle was used instead of a brush, because of its almost instantaneous drying. Several thin layers where applied to permit absorb the liquid until its saturation. These steps are shown in Fig. 7.

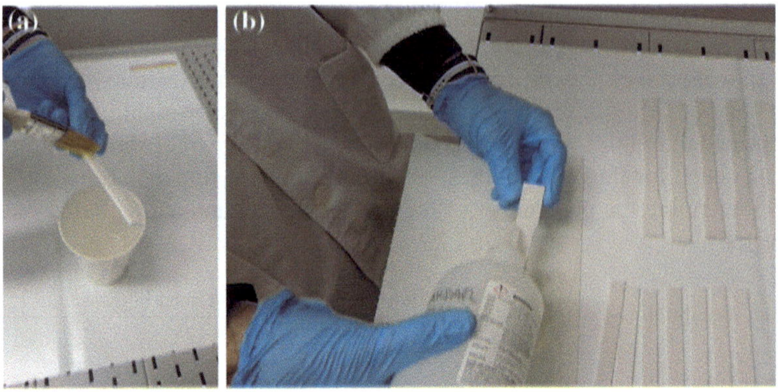

Fig. 7 **a** Flexural specimen infiltrated with Epoxy, and **b** Tensile specimen infiltrated with cyanoacrylate

Subsequent to the infiltration, and prior to the mechanical strength test, all the samples were allowed to dry at room temperature for a minimum of 24 h to obtain the final strength, as recommended in E[1] and CY[2] user guide.

2.4 Mechanical Testing

Three-point bending test was conducted according to ASTM D790-10 standard in a Universal testing machine (TIRAtes 2705, Germany) with a load cell of 5 kN. The samples were supported by two roods of 5 mm radius with a span of 64 mm, and upper load rood press on the middle. The crosshead load speed was 2 mm/s.

Tensile Tests were conducted according to the ASTM D638-10 standard in a Universal test machine (INSTRON 4507, England), using a 200 kN load cell and a 5 mm/min speed, with a distance between extensometers of 50 mm. Figure 8 shows the samples under the flexural and tensile tests.

Both tests were carried out at room temperature (~22 °C), and at least five specimens were tested. The machines software provides the Force (*F*) applied at different time's intervals (*t*) with the respective strains (*D*).

To get the stress, in bending test (support span = 64 mm > 51.2 mm), the following Eq. (2) was applied:

---

[1]3D Systems, I. *StrengthMax User Guide*. 2013 [cited 2016 apr. 29]; Available from: https://3dscentral.3dsystems.com/attachments/2239_22-95042%20StrengthMax%20User%20Guide.pdf.

[2]3D Systems, I. *ColorBond User Guide*. 2013 [cited 2016 apr. 29]; Available from: https://3dscentral.3dsystems.com/attachments/2239_22-95041%20ColorBond%20User%20Guide.pdf.

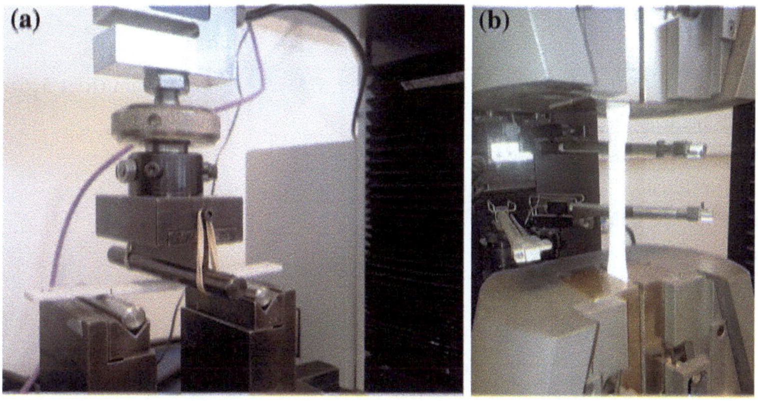

Fig. 8 Flexural and tensile tests

$$\sigma_f = \left(\frac{3PL}{2bd^2}\right)\left[1 + 6\left(\frac{D}{L}\right)^2 - 4\left(\frac{d}{L}\right)\left(\frac{D}{L}\right)\right] \quad (2)$$

where, σ_f is the stress in the outer fibres at midpoint [MPa]; P, the load at given point on the load-deflection curve [N]; L, the support span [mm]; b, the width of beam tested [mm]; d, is the depth of the tested beam [mm]; and D, maximum deflection at the centre of the beam [mm]. The Flexural Strength ($\sigma_{f\,M}$) or Maximum flexural stress sustained by the test specimen during a bending test, was calculated according to Eq. (2), when P is the load at break point.

For tensile analysis, the maximum tensile strength [MPa] of samples infiltrated with E and CY was calculated by dividing the maximum load [N], applied by the machine at break point, by the average original cross-sectional area in the gage length segment of the specimen in [mm^2].

2.5 Characterization of Powder Material and Green Body Printed Part

2.5.1 Powder Characterization

the powder material was characterized, to determine, (i) chemical composition, and (ii) grain size distribution. Two analyses were conducted using a scanning electron microscope (Phenom ProXL, USA) with fully integrated Energy Dispersive X ray spectrometry (EDS).

Detailed chemical powder composition, obtained from a micro volume via a spot analysis, was performed using an EDS analytical technique, for which first it proceeded to charge the dry particles, scattering them into the microscope chamber

Table 4 Parameters for wet and dry analysis at 2 bar by LD

	Dry analysis	Wet analysis
Particle name	Gypsum (RI 1.525)	Gypsum (RI 1.525)
Dispersant name	-	Ethanol
Particle absorption index	0.100	0.100
Weighted residual	0.35%	0.35%
Particle refractive index	1.525	1.525
Dispersant refractive index	-	1.360
Laser obscuration	2.02%	6.09%

through a proprietary venting/loading mechanism, with a beam acceleration voltage of 15 kV.

Particle size distribution of powder was performed in wet and dry condition by laser diffraction (LD), the most popular particle process due to speed, ease of use, and flexibility. For dry analysis pressures of 0, 1, 2, 3 and 4 bar were used, while for the wet analysis, ethanol was used as a dispersant. According to the theory of Lorenz-Mie used by ProSuite software, it is required to enter the refractive index of the particle and its suspension, as well as the absorption of the particle. Table 4 summarizes the parameters used for wet and dry at 2 bar.

2.5.2 Green Body Printed Part Characterization

To analyse the elemental chemical composition of the printed sample shown in Fig. 9, whose chemical composition is a mixture of powder and binder, the elemental mapping was obtained with scanning electron microscope technology (SEM) supported by EDS.

Elemental mapping was obtained in the microscope (Phenom ProXL, USA), which displays the distribution of elements throughout the sample. The elements

Fig. 9 SEM images of a 3D printed green body (*centre*), mapping 1 (*left*), mapping 2 (*right*)

were mapped with a beam acceleration voltage of 15 kV, and an acquisition real time of 265 s. The real time mapping algorithm shows live build-up of these selected elements. A typical microstructure is depicted in Fig. 9.

3 Results and Discussion

3.1 Powder Characterization

With the help of the dedicated software package Element Identification (EID) used to control the fully integrated EDS detector, it was not necessary to change the use of external software packages or references to obtain a result of the elemental chemical composition.

Two types of representative particles were identified, differing in size and appearance, large ones and smaller particles (defined as spot 1 and spot 2). The same particles were considered to analyse the chemical powder composition, and the respective spectrum was obtained, as shown in Fig. 10. One can see that the particle 1 has a similar composition to Calcium Sulphate Semi-hydrate particles ($CaSO_4 \cdot 0.5\ H_2O$), (Fig. 10b) [31], known as Plaster of Paris or Gypsum Plaster [32].

Particle 2, in turn, contains calcium sulphate, with own additives used for plasters such as Polyacrylonitrile (PAN), also known as Creslan 61, which is a synthetic, semi-crystalline organic polymer resin, with a linear formula $(C_3H_3N)_{n.}$ [33].

This powder material can also be compared to 3DP powder ZP131 ($CaSO_4 \cdot 0.67H_2O$), whose average grain size is approximately 40 μm, and has a composition of carbon (14.35%wt; 23.14%at), oxygen (46.02%wt; 55.73%at), aluminium (0.10%wt; 0.07% at), sulphur (16.20%wt; 9.79%at) and calcium (23.33%wt; 11.28% at) [22].

The most common approach for expressing (LD) results is to report the D10, D50 and D90 values, based on a volume distribution [34], in order to Lorenz-Mie theory be used.

Figure 11, shows the overlay curve of particle size distribution of 3D printing VisiJet PXL core powder, in % volume density. It is shown for each test performed to dry particles at pressures of 0, 1, 2, 3 and 4 bar, and testing to wet particles and their mean and standard deviation.

This distribution of particles in volume is not symmetrical. It has a larger deviation to the left, i.e. the greater relative standard deviation up to RSD-D × 10 = 17.2 (RSD, relationship between average standard deviation average), being the smaller RSD-D × 90 = 3.33; this means that there are a high number of fine particles of up to 7.12 μm in diameter, but their volume is only 10% (D × 10), while 50% reach diameters up to 36.1 μm (D × 50), and 90% reach a diameter of up to 77.3 μm.

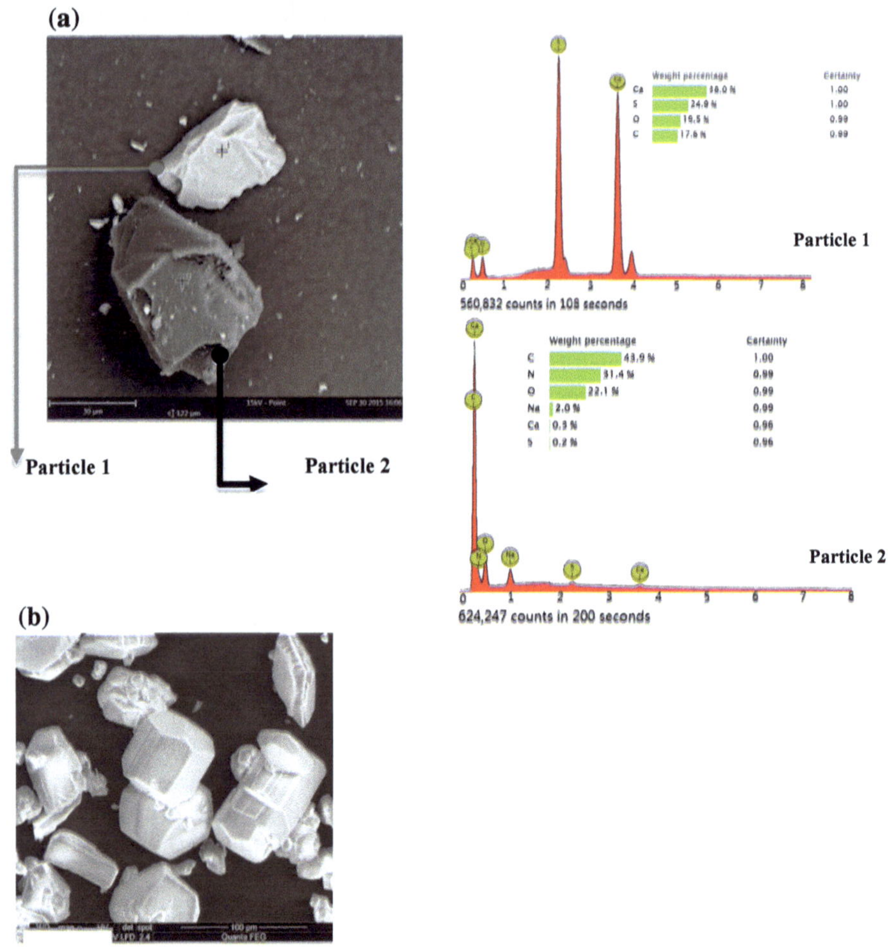

Fig. 10 **a** Chemical analyses (weight %) of powder particles 1 and 2, and comparison with **b** crystals of α-$CaSO_4 \cdot 0.5H_2O$ SEM microphotographs (adapted from [31])

On the other hand, the size of representative particles, 1 and 2, as shown in Fig. 12, was also estimated (graphically) in SEM images, where particle 1 is around 45 μm, while 2 is around 75 μm.

3.2 *Green Body Printing Part Characterization*

In mapping 1 and respective spectrum, Fig. 13a, a structure with a small concentration of sulphur was observed, typical of calcium sulphate, as already mentioned. There is also a high carbon content with 26.6%, which could correspond to plaster

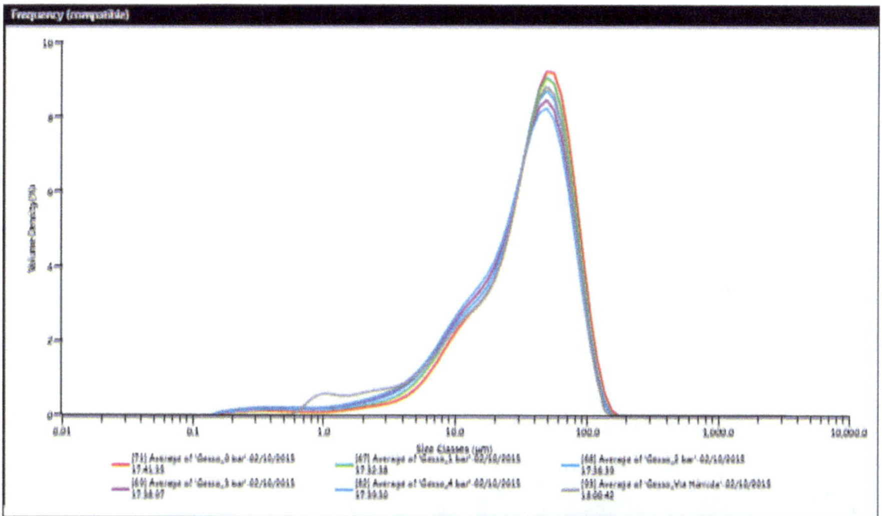

Record Number	Sample Name	Dv (10) (μm)	Dv (50) (μm)	Dv (90) (μm)
71	Average of 'Gesso_0 bar'	9.29	39.6	81.0
67	Average of 'Gesso_1 bar'	7.85	37.6	78.8
68	Average of 'Gesso_2 bar'	6.94	35.8	77.1
69	Average of 'Gesso_3 bar'	6.51	34.5	75.8
82	Average of 'Gesso_4 bar'	6.16	33.3	74.0
93	Average of 'Gesso_Via Húmida'	5.98	35.7	75.8
Mean		7.12	36.1	77.3
1xStd Dev		1.26	2.25	2.57
1xRSD (%)		17.6	6.23	3.33

Fig. 11 Overlay particle size distribution of 3D printing VisiJet PXL core powder, in volume density (%)

Fig. 12 Large representative particles of dry VisiJet PXL core powder

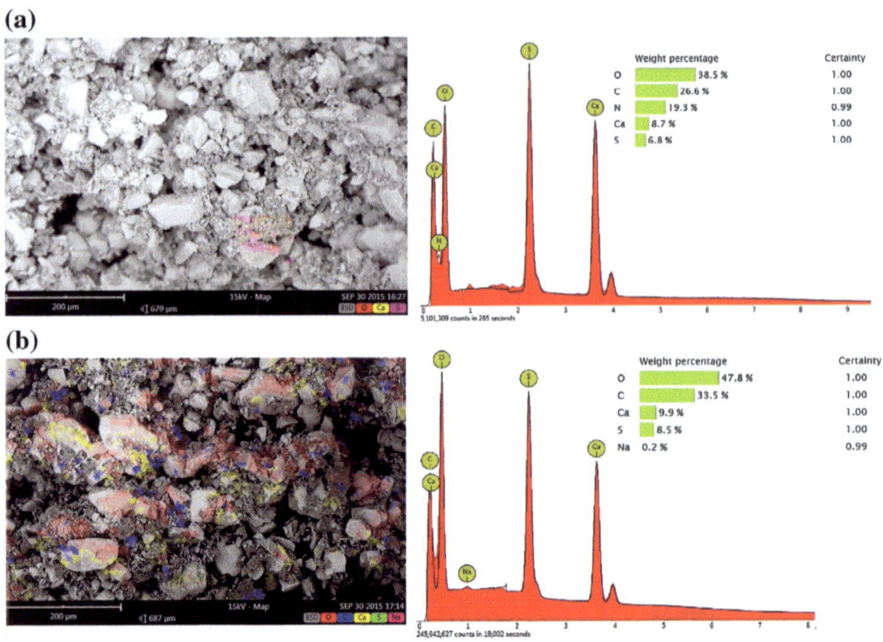

Fig. 13 Elemental chemical analysis of a green body specimen by SEM image and respective spectrum according to **a** mapping 1 **b** mapping 2

additives for 3DP [4, 5, 35]. In mapping 2 and respective spectrum, Fig. 13b, a structure with high concentrations of oxygen and carbon with 47.8 and 33.5%, respectively, is observed.

3.3 Mechanical Properties

Due to low mechanical strength for tensile test of samples with some types of infiltrations, only types E and CY were tested. In the other cases, samples were broken outside the area, invalidating the test.

The results obtained for both tests are presented in Figs. 14 and 15. It can be seen that they are lower than the ones indicated by the equipment supplier. Similar results were also obtained by Chaves [36] for E and CY infiltrates.

For flexural strength, it can be concluded that even for samples infiltrated with E (16.45 ± 1.15 MPa) and CY (14.68 ± 1.15 MPa), their standard deviation does not intersect each other Fig. 14a and its zoom in (b), i.e., there is a statistical significance between them and the other infiltrates. In the case of infiltration with MS, the values obtained were too low, 0.66 MPa, suggesting a possible failure in the application method (see Fig. 14).

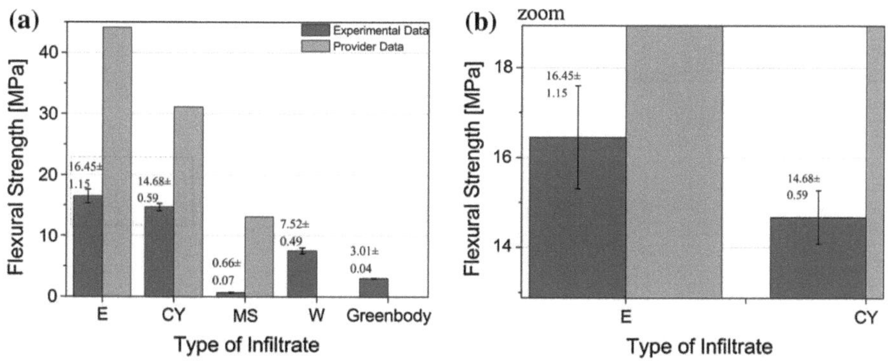

Fig. 14 Maximum flexural strength of infiltrated specimens compared with manufacturing data, and **b** zoom of a part

Fig. 15 Maximum tensile strength of infiltrated specimens compared with manufacturer data

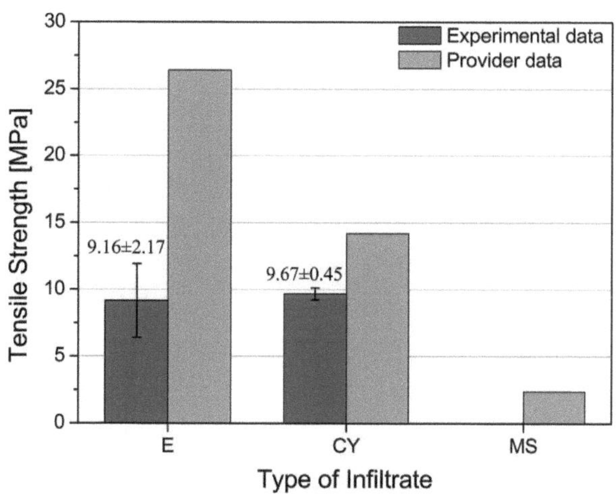

Compared to the green body part (referential), the flexural strength of the samples infiltrated with E increased by 447%, CY by 388%, W by 150%, and only with MS decreased by 78%.

Water as infiltrate allowed a considerable increase in the flexural strength of the samples, but during infiltration, they lost consistency and while drying, they were deformed.

Analysing the tensile strength, as shows Fig. 15, samples with infiltrate E showed a slightly lower average strength of 9.16 ± 2.17 MPa, regarding 9.67 ± 0.45 MPa for CY. Such broad standard deviation in E, statistically, means no relevant influence of these two types of infiltrates on the effect of tensile strength.

Tensile tests demonstrated that the epoxy resin is less resistant than the Cyanoacrylate; which is contrary to the results presented by Ipens [25], Galeta [27]

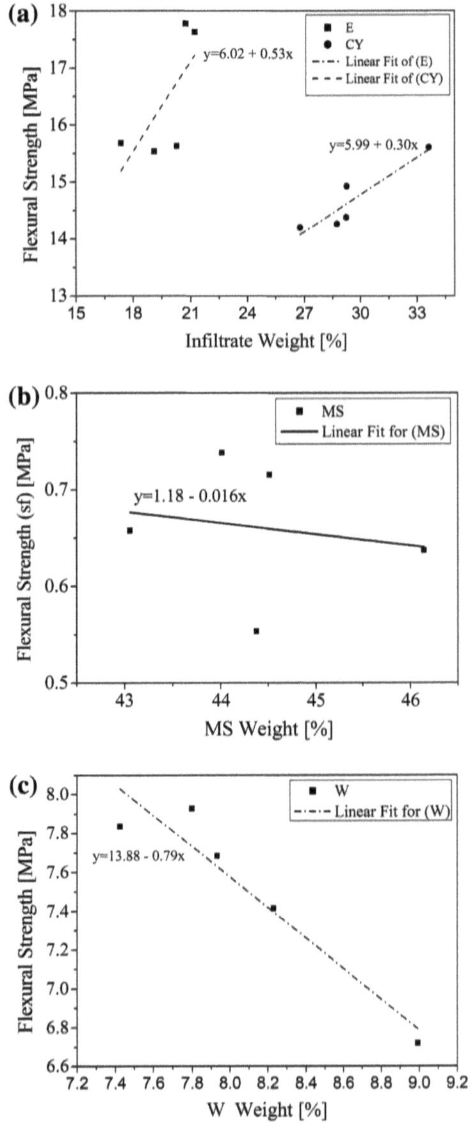

Fig. 16 Weight of dry infiltrant absorbed by specimens (%) versus Flexural strength: **a** (E) and (CY), **b** (MS) and **c** for (W)

and Frascati [20] who claimed that with epoxy better strength was achieved compared to some Cyanoacrylates. On the other hand, in bending analysis, this study agrees with Frascati's results [20], i.e., epoxy resin, enhances the flexural strength.

Equation (1) was applied in order to calculate the infiltration absorption of several samples. Figure 16, shows this absorption tendency, through linear regressions with equations of type $y = a + bx$, where "a" is the Y axis interception, and "b" is the slope of the line.

Figure 16a, indicates a directly proportional relationship, i.e., the more infiltrate absorbed with (E) and (CY) in samples, the stronger their flexural strength becomes, in particular with (E) where the equation of the linear fit, shows a greater slope.

Figure 16b, c, shows an inverse proportional relationship, i.e., the more infiltrate absorbed with (MS) and (W) in samples, the weaker their flexural strength becomes, especially with (W) where the equation of the linear fit, shows a greater slope. In other words, an excessive amount of infiltrate MS and W may reduce flexural strength in samples, suggesting that the use of a better dosage can avoid saturation.

4 Conclusions

To achieve a proper selection of the 3DP post-process, in addition to its application methodology and dosing, at the beginning of the product design, the morphology and composition of the base material and their printing parameters should be known.

The VisiJetPXL core powder used, can be considered multimodal powders, due to its large particles content Dx(90) = 77.3 μm, and its small particles content Dx (10) = 7.12 μm, and even particles with diameters smaller than 0.5 μm. So it exploits the advantages of powders with small and large particles. It was also determined that the powder has a high calcium sulphate content of type α (α-$CaSO_4 \cdot 0.5\ H_2O$), and any synthetic semi-crystalline organic polymer resin content, so it could be considered an additive plaster.

Large increases in flexural strength were obtained, 447, 338 and 150% for epoxy resin, cyanoacrylate and water, respectively, which suggest a strong influence of PP in this technology. The results highlight the importance of this type of study, which deserves further research.

Infiltrating with water, in spite of improving considerably the flexural strength, deforms the specimens, denoting that the infiltrates should not react or change the composition of the base material, but the infiltrates should, basically, fill the interstices between the powder grains bonded with binder during printing (green body porosity), and not modify the accuracy of the test specimens in the PP stage.

On both experimental bending and tensile tests, the strength obtained compared to those displayed by the material supplier were lower.

Large standard deviations in the mechanical strength of the infiltrated specimens, compared with the standard deviations of the green body, characteristic of manual processes (brush and drip), suggest, as a future work, the establishment of methodologies that allow homogeneous infiltration (dip with time control) and controlled dosage, as well as the use of alternative infiltration, which will expand options.

Acknowledgements Authors gratefully acknowledge funding of Project to Universidad de las Fuerzas Armadas ESPE, Ecuador, and SAESCTN-PII&DT/1/2011 co-financed by Programa Operacional Regional do Norte (ON.2—O Novo Norte), under Quadro de Referência Estratégico Nacional (QREN), through Fundo Europeu de Desenvolvimento Regional (FEDER).

References

1. Wohlers, A.: Wohlers Report, vol. 20. pp. 1–1315. Wohlers Associates Inc. (2015)
2. Gibson, I., Rosen, D., Stucker, B.: Additive Manufacturing Technologies: 3D Printing, Rapid Prototyping, and Direct Digital Manufacturing. Springer New York (2014)
3. Liu, J., Rynerson, M.: Method for article fabrication using carbohydrate binder
4. Utela, B., Storti, D., Anderson, R., Ganter, M.: A review of process development steps for new material systems in three dimensional printing (3DP). J. Manuf. Process. **10**(2), 96–104 (2008). doi:10.1016/j.jmapro.2009.03.002
5. Utela, B.R., Storti, D., Anderson, R.L., Ganter, M.: Development process for custom three-dimensional printing (3DP) material systems. J. Manuf. Sci. Eng. **132**(1), 011008 (2010). doi:10.1115/1.4000713
6. Feenstra, F.K.: Method for making a dental element
7. Bredt, J.F., Anderson, T.C., Russell, D.B.: Three dimensional printing material system and method. In. Google Patents (2003)
8. Butscher, A.: Powder based three-dimensional printing of calcium phosphate structures for scaffold engineering. ETH Zurich (2013)
9. Sachs, E.M., Haggerty, J.S., Cima, M.J., Wiliams, A.P.: Three dimensional printing techniques, United States Patent No
10. Upcraft, S., Fletcher, R.: The rapid prototyping technologies. Assembly Autom. **23**(4), 318–330 (2003)
11. Sachs, E.M.: Powder dispensing apparatus using vibration. In. Google Patents (2000)
12. Sachs, E.M., Cima, M.J., Caradonna, M.A., Grau, J., Serdy, J.G., Saxton, P.C., Uhland, S.A., Moon, J.: Jetting layers of powder and the formation of fine powder beds thereby. In. Google Patents (2003)
13. Liu, J., Ryneson, M.L.: Blended powder solid-supersolidus liquid phase sintering. In. Google Patents (2006)
14. Cima, L.G., Cima, M.J.: Preparation of medical devices by solid free-form fabrication methods. ISBN 0736-5845
15. Greil, P.: Polymer derived engineering ceramics. Adv. Eng. Mater. **2**(6), 339–348 (2000)
16. Shirazi, S.F.S., Gharehkhani, S., Mehrali, M., Yarmand, H., Metselaar, H.S.C., Adib Kadri, N., Osman, N.A.A.: A review on powder-based additive manufacturing for tissue engineering: selective laser sintering and inkjet 3D printing. Sci. Technol. Adv. Mater. **16** (3), 033502 (2015). doi:10.1088/1468-6996/16/3/033502
17. Lowmunkong, R., Sohmura, T., Suzuki, Y., Matsuya, S., Ishikawa, K.: Fabrication of freeform bone-filling calcium phosphate ceramics by gypsum 3D printing method. J. Biomed. Mater. Res. B Appl. Biomater. **90**(2), 531–539 (2009). doi:10.1002/jbm.b.31314
18. Monkhouse, D.C., Kumar, S., Rowe, C.W., Yoo, J.: Rapid prototyping and manufacturing process. In. Google Patents (2003)
19. Gharaie, S.H., Morsi, Y., Masood, S.H.: Tensile Properties of Processed 3D Printer ZP150 Powder Material, pp. 813–816. Trans Technical Publications 2013
20. Frascati, J.: Effects of position, orientation, and infiltrating material on three dimensional printing models. University of Central Florida Orlando, Florida (2007)
21. Vaezi, M., Chua, C.K.: Effects of layer thickness and binder saturation level parameters on 3D printing process. Int. J. Adv. Manuf. Technol. **53**(1–4), 275–284 (2011). doi:10.1007/s00170-010-2821-1

22. Zañartu Apara, G., Ramos Grez, J.: Characterization of the mechanical properties of samples fabricated by an experimental SGM device. Rapid Prototyping J. **16**(5), 356–364 (2010). doi:10.1108/13552541011065759
23. Asadi-Eydivand, M., Solati-Hashjin, M., Farzad, A.: Abu Osman, N.A.: Effect of technical parameters on porous structure and strength of 3D printed calcium sulfate prototypes. Robot. Comput.-Integr. Manuf. **37**, 57–67 (2016). doi:10.1016/j.rcim.2015.06.005
24. Feng, P., Meng, X., Chen, J.-F., Ye, L.: Mechanical properties of structures 3D printed with cementitious powders. Constr. Build. Mater. **93**, 486–497 (2015). doi:http://dx.doi.org/10.1016/j.conbuildmat.2015.05.132
25. Impens, D., Urbanic, R.J.: Assessing the Impact of post-processing variables on tensile and compression characteristics for 3D printed components. IFAC-PapersOnLine **48**(3), 652–657 (2015). doi:http://dx.doi.org/10.1016/j.ifacol.2015.06.156
26. Impens, D.: An Experimental Approach to Assess the Impact of Post Processing Variables on the Mechanical Characteristics of 3D Printed (Powder Binding Process) Parts. University of Windsor (2015)
27. Galeta Tomislav, K.I., Karakasic Mirko: Influence of processing factors on the tensile strength of 3d-printed models. Mater. Tehnol. **47**(6), 781–788 (2013)
28. International, A.: ASTM D790-2010 Standard test methods for flexural properties of unreinforced and reinforced plastics and electrical insulating materials. In: ASTM International, West Conshohocken. (2010)
29. ASTM: ASTM D638-2010 Standard test method for tensile properties of plastics. In: West Conshohocken (PA): ASTM International. (2010)
30. Suwanprateeb, J.: Comparative study of 3DP material systems for moisture resistance applications. Rapid Prototyping J. **13**(1), 48–52 (2007). doi:10.1108/13552540710719217
31. LeŠKeviČIenĖ, V.: α-hemihydrate gypsum from flue gas desulphurization gypsum. Mater. Sci./Medziagotyra **19**(2) (2013)
32. Bobby, S.S.: A preliminary investigation of gypsum bonded moulds by three dimensional printing. Int. J. Res. Eng. Technol. **3**(06), 501–507 (2014)
33. Gupta, A.K., Paliwal, D.K., Bajaj, P.: Melting behavior of acrylonitrile polymers. J. Appl. Polym. Sci. **70**(13), 2703–2709 (1998)
34. Spierings, A.B., Herres, N., Levy, G.: Influence of the particle size distribution on surface quality and mechanical properties in AM steel parts. Rapid Prototyping J. **17**(3), 195–202 (2011). doi:10.1108/13552541111124770
35. Viñas, J.L.G.: Nuevas aplicaciones de recursos yesíferos: desarrollo, caracterizaci6n y reciclado. Universidad Politécnica de Madrid (2005)
36. Chaves, A.A.S.: Impressão 3D baixo custo versus impressão em equipamentos de elevado custo. Master University of Porto (2015)

Development of Plaster Mixtures Formulations for Additive Manufacturing

D.E. Caetano, J.L. Alves, R.L. Neto and T.P. Duarte

1 Introduction

For the production of short or even unity series in manufacturing models for processing sanitary ceramics, companies typically use machined polyurethane and epoxy resins. This solution, although efficient, is expensive due the price of these resins that costs approximately 15–20 €/kg, and the waste produced during the machining. These values become more evident with large size/volume pieces, since they require the machining of large quantities of material, high machining time and tooling costs [1].

Thus, it is necessary to find a flexible and economically competitive solution and the production of these moulds or models with gypsum plaster will bring a great reduction in terms of material costs, since the cost of gypsum rounds between 17 and 50 cent/kg. Allied to this lower cost, a big reduction of waste material can be possible too if an additive manufacturing (AM) process is employed, which allows a creation of freeform components with less material to be machined [1].

AM has some disadvantages concerning the surface quality of the parts produced and high initial investment in equipment. Besides that, there is also the problem associated with a production time when compared with conventional processes for machining a piece with the same geometry [2, 3].

D.E. Caetano (✉) · J.L. Alves · R.L. Neto · T.P. Duarte
INEGI, Faculdade de Engenharia, Universidade do Porto, Rua Dr. Roberto Frias 400, 4200-465 Porto, Portugal
e-mail: dc.danielcaetano@gmail.com

J.L. Alves
e-mail: falves@fe.up.pt

R.L. Neto
e-mail: rjn@fe.up.pt

T.P. Duarte
e-mail: tdp@fe.up.pt

L.F.M. da Silva (ed.), *Materials Design and Applications*,
Advanced Structured Materials 65, DOI 10.1007/978-3-319-50784-2_20

The incorporation of an AM system associated with a machining process, such CNC in the same equipment is currently possible for metallic materials, for example, “INTEGREX i-400 AM” developed by Mazak, associates the process of Laser Metal Deposition (LMD) with a CNC equipment [2].

In addition to equipment for metals, there are several hybrid devices directed to polymers, for example, the “FABTotum” [4], the “ZMorph 2.0S” [5] or the “WASP EVO” [6], where the first one is included in the range of 3D low cost printers.

Although this kind of hybrid equipment for metallic materials or polymers, the use of gypsum in a multi-tasking machine according to the Done-in-one concept [2] is still an unexplored field, and there are some problems associated with the control of extrusion and setting time as well as machining possibilities. The gypsum plaster itself does not fulfil the necessary requirements to be machined in a CNC machine, since it generates significant amounts of dust, and the parts do not have enough mechanical strength to allow the machining of sharp edges. However, it is possible to add certain additives to the gypsum formulations, like binders or plasticizers, in order to improve the resistance to fragmentation of the sharp edges during machining [1].

Summing up, this paper is inserted within an ongoing research project in INEGI which main objective is to create one machine that has the possibility to manufacture high size parts, by AM and the subsequent CNC machining, using non-conventional materials such as gypsum, silica sand or thermosetting resins. Thus, this research work focuses on the study of gypsum plaster formulations able to be used in the new hybrid equipment combining AM and machining.

2 Gypsum Properties

Used since ancient times on the form of binder between stone blocks, for example, the pyramids of Egypt, the plaster, extracted from mineral gypsum, is currently a fairly produced worldwide material, having its application expanded to many areas such as construction, medicine, agriculture, artistic sculpture or in the casting process of moulds/models for the sanitary ceramics. The plaster used in this last application is referred as “Plaster of Paris”, and is a white, grey or pink powder, essentially consisting of hemihydrate calcium sulphate that is obtained by controlled dehydration of calcium sulphate ($CaSO_4$) hydrate by removal of about 75% of crystallization water. This reaction can be expressed in Eq. (1), where “Q” is the energy provided in the process [7, 8]:

$$CaSO_4 \cdot 2H_2O + Q \rightarrow CaSO_4 \cdot \frac{1}{2}H_2O + \frac{3}{2}H_2O \tag{1}$$

By dehydration it is possible to obtain two types of hemihydrate, the α-hemihydrate and β-hemihydrate plaster, whose difference is mainly due to the method used in their production. The α-hemihydrate is obtained by a wet process such as autoclaving and β-hemihydrate is obtained under a scarce water vapour process [9].

Concerning hydration, it is possible to reverse de above process passing from one type of hemihydrate, α-hemihydrate or β-hemihydrate, to dihydrate by water addition, expressed in the exothermic reaction (2) [8], that is the focus of this work:

$$CaSO_4 \cdot \frac{1}{2}H_2O + \frac{3}{2}H_2O \rightarrow CaSO_4 \cdot 2H_2O + Q \quad (2)$$

The water reaction with the calcium sulphate hemihydrate forms a slurry which becomes successively more viscous until it hardens completely. This curing time, called "setting time", is usually on the order of 30 min, and varies depending on the type of hemihydrate and consistency of mixture. This setting time is generally lower when using the β-hemihydrate, about 25 min, while the value for α-hemihydrate is approximately 40 min [8, 10].

This hardening is done by meshing of the needles of gypsum crystals during the transition phase, and since the two types of hemihydrates have different setting times, its microstructure will also be slightly different. For the α-hemihydrate, and as a result of a longer setting time, the crystals formed have higher degree of entanglement and interconnection, creating a more consistent and homogeneous structure. On the other hand, as the short β-hemihydrate plaster setting time, the crystals are smaller and stacked, instead of interlocked, as they grow more rapidly at the beginning of hydration, forming shorter and fractured needles, yielding a less consistent and heterogeneous structure [11, 12].

However, the end of setting time of mixtures can be extended to several hours, if retardants are added or reduced to below 3 min if in presence of accelerators [10]. The setting time is the time for complete hardening of the slurry, in other words, when it stops generating heat. Thus, to match the plaster formulations with AM concepts, the use of setting time accelerators is required in order to reduce the extensive time normally associated with de hydration of hemihydrates.

Additives such as potassium sulphate, zinc sulphate or starch are characterized by being a type of chemical accelerator that promotes the formation of sulphate ions [10].

Although these additives reduce the setting time, generally also lead to a reduction of mechanical strength, since in its presence, the crystal morphology of dihydrate will change, giving rise to thinner and shorter needles [9].

Moreover, it is possible to combat this reduction in mechanical strength of plaster formulations using plasticizers. The use of polymers in gypsum formulations leads to an improvement in the mechanical properties. Stav and Bentur [13] have suggested that plaster properties can enhance with incorporation of acrylamide monomers in the inorganic matrix which undergoes polymerization. According to Rubio-Avalos et al. [14], the addition of 10% latex and styrene-butadiene polymer increased the flexural strength of β-hemihydrate from 8.8 MPa to 11.5 MPa.

In short, the addition of superplasticizers contributes to increase flow ability of construction materials such as cement or plaster, for the same ratio water/binder, or allows greater processability for the same water/binder relation. The addition of

0.6% by weight of sulphonated melamine formaldehyde (SMF) led to an increase in compressive and flexural strength by 69% and 42%, respectively, compared to gypsum formulations without any additive, as shown by Pundir et al. [15], and the microstructures obtained are shown on Figs. 1, 2 and 3.

Thus, the above studies, suggest that additives have an important role in the plaster setting time.

Fig. 1 Microstructures of β-hemihydrate without superplasticizer [15]

Fig. 2 Microstructures of β-hemihydrate with 0.4% of superplasticizer (SMF) [15]

Fig. 3 Microstructures of β-hemihydrate with 0.6% of superplasticizer (SMF) [15]

Different types of additives were selected for α and β gypsum and mixtures formulations to seek for the mixture that can reduce the setting time compatible with AM manufacturing and that simultaneously increases the mechanical strength and machinability of the parts.

3 Experimental Work

3.1 Materials and Methods

Different additives were selected and purchased. It is important to note that throughout the experimental work distilled water was the liquid media for the slurry mixtures. Table 1 summarizes all additives acquired and the hemihydrates used in the experimental work.

For better analysis of the results, Table 2 shows the more important properties selected from data sheets of the hemihydrates.

Due to the reduced amounts to mix the hemihydrates with the additives and water of each slurry formulation, a Kenwood KM001 mixer was selected.

Initially, for formulations without additives, after adding the water to the hemihydrate, 90 s soak times were used before start mixing during 180 s.

Subsequently, with the addition of accelerators, these times were reduced to 15 s soak and 15 s mixing, due to the fast increase in viscosity.

The slurry produced for each formulation was casted in a silicone (VTX 950) mould with 170 × 50 × 50 mm.

Table 1 Materials and suppliers

Material type	Designation	Chemical formula	Commercial name	Supplier	City, Country
Accelerator	Potassium sulphate	K_2SO_4	–	VWR International—Material de Laboratório, Lda	Carnaxide, Portugal
Accelerator	Zinc sulfate heptahydrate	$Zn_2SO_4 \cdot 7H_2O$	–	VWR International—Material de Laboratório, Lda	Carnaxide, Portugal
Accelerator	Aluminum sulfate	$Al_2(SO_4)_3$	–	SPD—Sociedade Portuense de Drogas, S.A.	Gondomar, Portugal
Accelerator	Sodium sulphate	Na_2SO_4	–	SPD—Sociedade Portuense de Drogas, S.A.	Gondomar, Portugal
Accelerator	Calcium chloride	$CaCl_2$	–	SPD—Sociedade Portuense de Drogas, S.A.	Gondomar, Portugal
Accelerator	Calcium chloride anhydrous	$CaCl_2$	–	VWR International—Material de Laboratório, Lda	Carnaxide, Portugal
Accelerator	Calcium nitrate tetrahydrate	$Ca(NO_3)_2 \cdot 4H_2O$	–	VWR International—Material de Laboratório, Lda	Carnaxide, Portugal
Accelerator	Calcium carbonate	$CaCO_3$	–	Carban	Leiria, Portugal
Accelerator	Polyethylene glycol	–	Polietilenoglicol 400	VWR International—Material de Laboratório, Lda	Carnaxide, Portugal
Accelerator	Polyvinyl alcohol	–	PVA	INEGI	Porto, Portugal
Retarder	Citric acid	–	–	SPD—Sociedade Portuense de Drogas, S.A.	Gondomar, Portugal
Retarder	Tartaric acid	–	–	SPD—Sociedade Portuense de Drogas, S.A.	Gondomar, Portugal
Binder	Homopolymer of vinyl acetate	–	Polidisp DV	Resiquímica, S.A.	Porto, Portugal
Binder	Styrene-Acrylic Copolymer	–	Polidisp 6660	Resiquímica, S.A.	Porto, Portugal
Binder	Acrylic Copolymers	–	Polidisp 7778	Resiquímica, S.A.	Porto, Portugal
Binder	Copolymers of Acrylic-Veova	–	Polidisp 9957	Resiquímica, S.A.	Porto, Portugal

(continued)

Table 1 (continued)

Material type	Designation	Chemical formula	Commercial name	Supplier	City, Country
Binder	Copolymer of Vinyl Acetate—VeoVa	–	Polidisp 7252	Resiquímica, S.A.	Porto, Portugal
Thickener	Methylcellulose	–	Methocel\® A4C	VWR International—Material de Laboratório, Lda	Carnaxide, Portugal
Thickener	Hydroxyethyl methyl cellulose	–	Walocel MKX 70000 PP 40	DOW wolff cellulosics	Bomlitz, Germany
Wetting Agent	Propylene Glycol	–	–	AMRESCO	Ohio, USA
Wetting Agent	Propylene Glycol	–	–	SPD—Sociedade Portuense de Drogas, S.A.	Gondomar, Portugal
Gypsum	β-hemihydrate	$CaSO_4 \cdot 0{,}5H_2O$	Gesso Cerâmico Extra	SIVAL	Leiria, Portugal
Gypsum	α-hemihydrate	$CaSO_4 \cdot 0{,}5H_2O$	Primopor	SIVAL	Leiria, Portugal

Table 2 Proprieties of acquired plasters (SIVAL's technical sheet)

Product	Gypsum—Primopor (SIVAL)	Gypsum—Cerâmico Extra (SIVAL)
State	α Solid (powder)	β Solid (powder)
Colour	White, light beige	White
Liquid solution pH @ 20 °C	6.0–8.0	6.0–7.0
Density (g/cm^3)	2.75	–
Apparent density of the powder (g/cm^3)	–	0.6–0.7
Initial setting time (min.)	21 ± 2	12 ± 3
Ending setting time (min.)	–	28 ± 5
Flexural strength (MPa)	–	≈ 4.8
Compression strength (MPa)	125–135	–
Linear expansion (%)	<0.19	Max. 0.20
Gypsum/Water ratio (kg/dm^3)	2.70–2.90	1.25–1.40
Residues—300 μm sieve (%)	–	≤ 0.008
Residues—150 μm sieve (%)	–	≤ 0.13
Residues—45 μm sieve (%)	–	≤ 2.3

To determine the setting time temperature (maximum temperature reached during the hardening), a type K thermocouple was placed in the centre of the mould, as shown in Fig. 4.

The samples for mechanical testing, 3 specimens of 160 × 40 × 40 mm, were cast in the silicone mould of Fig. 5.

The flexural strength tests were conducted in accomplishment to BS EN 12859: 2011. Three specimens (Fig. 6) of each formulation were dried until the mass was constant (a constant mass is set when 2 successive weighing's, 24 h apart, differ by less than 0.1%).

The 3 point bend tests were performed in an Instron 4208 machine (Fig. 7), with a 100 kN load cell, applied on the centre of the specimens that were supported in two points with 100 mm span and a feed rate defined by ASTM 647-88: 2004, of 1 mm/min.

The fractured surfaces were analysed to check the porosity, mixture heterogeneities and other defects that can affect the modulus of rupture.

For all formulations, the accelerators, binders or water are expressed in weight percentages relative to the amount of gypsum. The water percentage was selected to allow a homogeneous mixture and fluidity for casting into the moulds.

Fig. 4 Silicon mould for determining the setting time temperature

Fig. 5 Silicon mould for production of samples for 3 point bending test

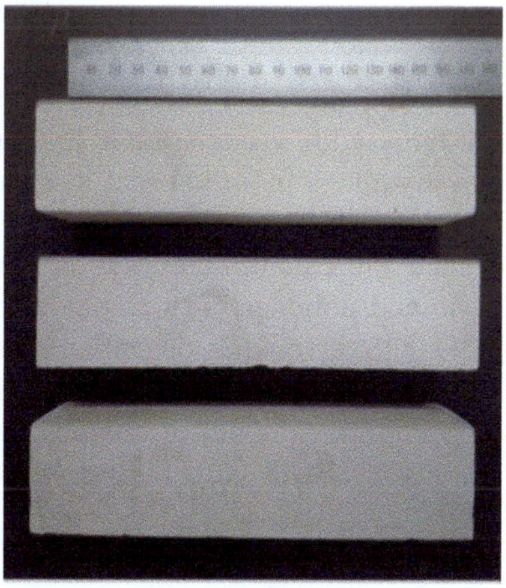

Fig. 6 Test specimens for 3 point bending test

Fig. 7 Instron 4208 testing equipment

Fig. 8 Poor quality samples

During the experimental process, it was possible to analyse the behaviour of mixtures with various additives, and thus, a set of 21 additives and 2 types of hemihydrates, in a total of 130 examined formulations were selected. The selected additives were: potassium sulphate, zinc sulphate, sodium sulphate, calcium carbonate and calcium chloride as accelerators, polidisp 7252, polidisp DV, polidisp 6660 and polidisp 7778 as binders and methocel A4C as thickener.

Due to adversities presented by some formulations, the results with no practical interest, either by the slurry hardened too quickly, heterogeneous mixture or inappropriate amount of bubbles, were excluded. Some of these defects are shown in Figs. 8, 9 and 10.

Fig. 9 Sample with lumps and pores

Fig. 10 Samples with abundant pores

The tested formulations and the results obtained for the bending tests are presented in the following chapter. However, in this paper, only the most interesting mixtures are analysed, which means only two accelerators and two binders.

4 Results

This chapter presents and compares the values from the tests for three "families" of formulations; formulations without additives, formulations with accelerator and formulations with accelerator and binder.

The additives used for comparison are:

- Potassium sulfate (K_2SO_4)—Accelerator
- Calcium carbonate ($CaCO_3$)—Accelerator
- Copolymer of Vinyl Acetate—VeoVa (Polidisp 7252)—Binder
- Acrylic Copolymers (Polidisp 7778)—Binder

4.1 Slurries Temperature Evolution

The Fig. 11 shows the curves for a formulation of each of the "families" of α-hemihydrate formulations and represents the evolution of the temperature inside the slurry during the setting time.

Figure 11 shows that the addition of potassium sulphate leads to a faster temperature increase, reducing the setting time from 45 to 25 min, which represents a reduction of 44%. The addition of Polidisp 7252 delays slightly the speed, reaching

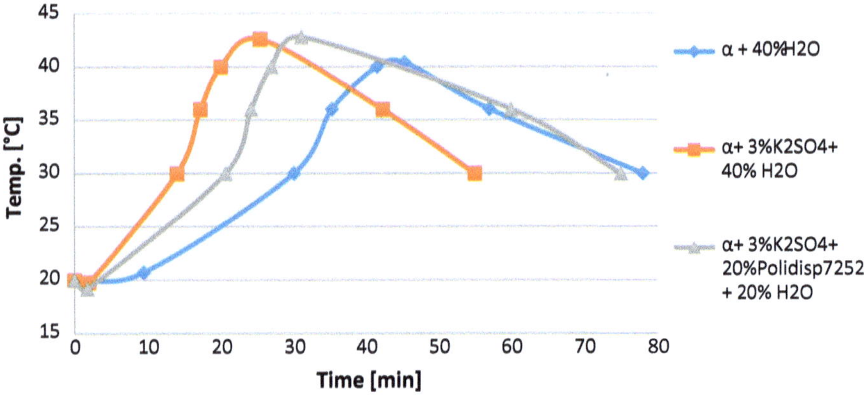

Fig. 11 Temperature evolution as a function of the setting time of gypsum α-hemihydrate

the peak temperature at 31 min, however it still is 14 min faster than the formulation without additives.

Pure β-hemihydrate gypsum setting time is shorter than α-hemihydrate gypsum, and a reduction of 72% (29 to 8 min) can be obtained with the accelerator, as shown in Fig. 12.

Formulation with 50% α-hemihydrate and 50% β-hemihydrate (Fig. 13) have the same tendency has the previous formulations, however with different setting time.

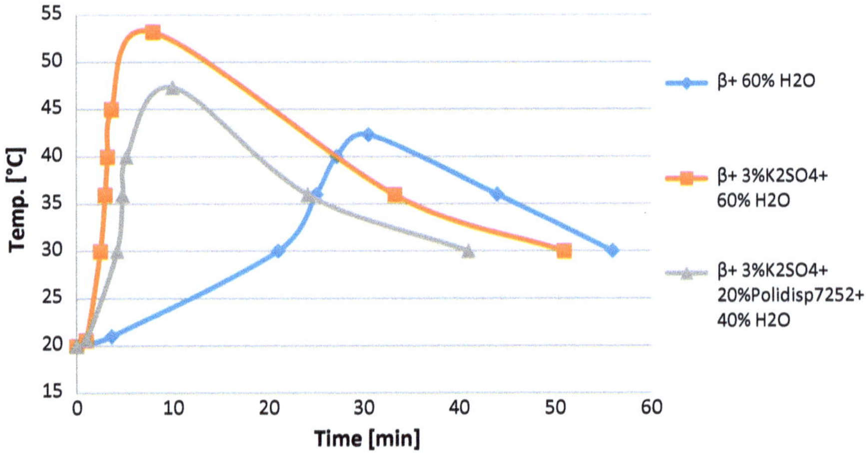

Fig. 12 Temperature evolution according to the setting time for β-hemihydrate gypsum

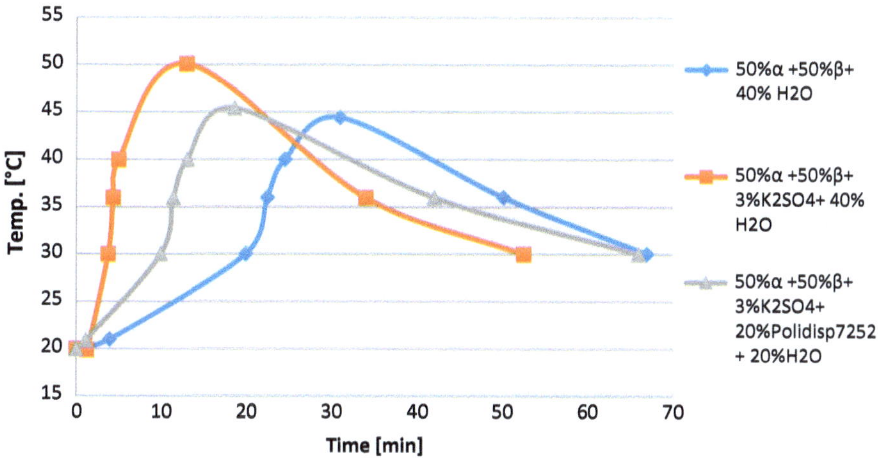

Fig. 13 Temperature evolution as a function of setting time for α-hemihydrate + β-hemihydrate formulation

The results obtained for the three types of formulations show that the setting time is reduced when K_2SO_4 accelerator is introduced in the mixture. Although binder, introduced to increase the mechanical strength of the plaster, has an opposite effect of the accelerator, the setting time is still lower than the one obtained with the pure raw material, which is promising for the combined manufacture process to be implemented.

4.2 Flexural Strength

Figure 14 presents the modulus of rupture and the respective setting time of the various formulations.

In general, the addition of potassium sulphate to gypsum formulations led to a reduction of the setting time of about 45% for the α-hemihydrate gypsum, 74% for β-hemihydrate gypsum and 58% for α + β, however the modulus of rupture is also reduced. This effect in β-hemihydrate is not so notorious.

The combination of an accelerator and a binder, although increasing the setting time, relative to the mixture with only an accelerator, retrieves the lost resistance promoted by the accelerator.

Compared to formulations without additives, the formulations with additives that performed better matching between setting time and flexural strength were the formulations with calcium carbonate ($CaCO_3$) and acrylic copolymer (Polidisp 7778). Regarding formulations with α-hemihydrate, there was a reduction of about

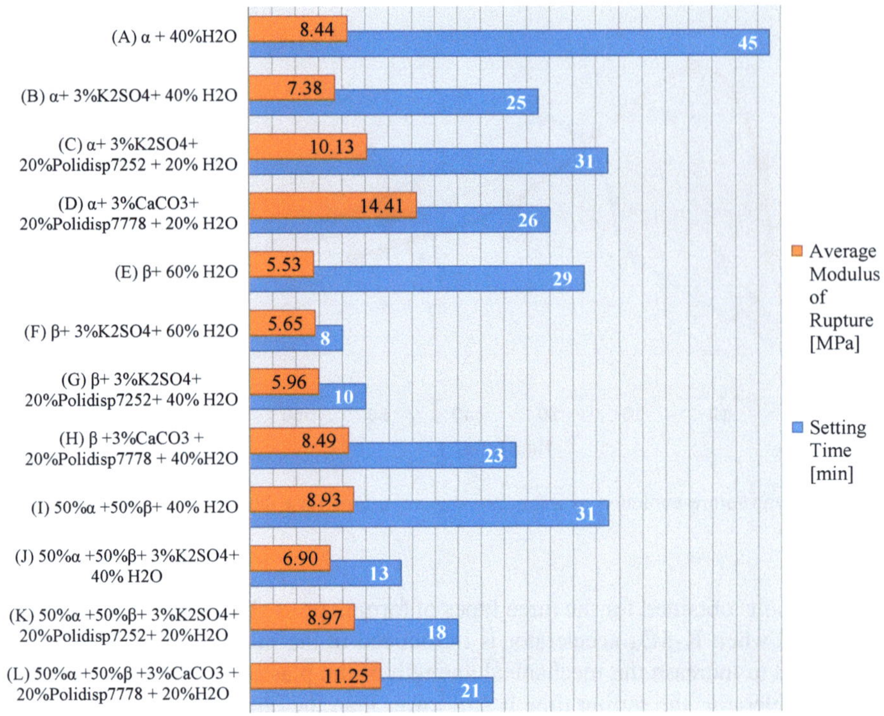

Fig. 14 Comparison of setting times and modulus of rupture from the various formulations

42% of setting time, and a 71% increase in the value of average modulus of rupture (one of the tested samples showed a modulus of rupture value of about 15.48 MPa, which represents a gain of 83%).

For β-hemihydrate with CaCO3 and Polidisp 7778, a 20% reduction in setting time and an increase of 53.5% of modulus of rupture was obtained. For 50% α + 50% β with $CaCO_3$ and Polidisp 7778, promote a 32% reduction in the setting time, and a 26% increase in the average modulus of rupture.

Tables 3, 4 and 5, present the results obtained in 3 point bending with α-hemihydrate, β-hemihydrate and 50% α + 50% β-hemihydrate, respectively.

- It is important to note that in these tables:
- The percentage is always 3% for accelerator and 20% for binder.
- For α-hemihydrate and α + β tables, the percentage of added water is 40% if there is no binder in the formulation and 20% with binder in the formulation.
- For β-hemihydrate table, the percentage of added water is 60% if there is no binder in the formulation and 40% with binder in the formulation.

Table 3 Results for 3 point bending test for α-hemihydrate formulations

#	Formulation		Weight (g)	Max. force (N)	Mr (modulus of rupture) (MPa)	Average Mr (MPa)
1	α	A	384.4	3503	8.18	8.44
		B	384.4	3345	8.03	
		C	385.8	3874	9.11	
2	α + K_2SO_4	A	393.3	3263	7.28	7.38
		B	396.6	3496	7.67	
		C	393.4	3234	7.18	
3	α + Zn_2SO_4	A	395.5	4458	10.04	9.84
		B	397.8	4273	9.70	
		C	395.2	4350	9.77	
4	α + Na_2SO_4	A	397	2367	5.33	4.81
		B	388.5	1756	3.98	
		C	394.5	2282	5.12	
5	α + $CaCl_2$	A	391.2	1992	4.99	4.90
		B	381.2	1979	4.97	
		C	382.9	1895	4.74	
6	α + $CaCO_3$	A	387.6	2617	5.84	6.59
		B	392	2846	6.47	
		C	386.7	3305	7.45	
7	α + K_2SO_4 + Polidisp7252 + Antifoam	A	432.5	4669	10.45	10.13
		B	429.9	4167	9.23	
		C	431.5	4810	10.72	
8	α + Zn_2SO_4 + Polidisp7252 + Antifoam	A	428.6	3659	8.14	7.55
		B	428.8	3291	7.29	
		C	427.2	3259	7.22	
9	α + $CaCO_3$ + Polidisp7252 + Antifoam	A	432.3	4384	9.89	10.20
		B	431	4234	9.56	
		C	432.5	5023	11.16	
10	α + Zn_2SO_4 + Polidisp DV + Antifoam	A	412.9	2217	4.83	5.29
		B	410.8	1997	4.50	
		C	401.6	2937	6.52	
11	α + $CaCO_3$ + Polidisp DV + Antifoam	A	423.9	3232	7.16	7.07
		B	424	3012	6.75	
		C	425.4	3302	7.29	
12	α + $CaCO_3$ + Polidisp 6660 + Antifoam	A	409.1	6027	13.65	13.51
		B	412.5	5860	13.29	
		C	412.7	6003	13.57	
13	α + $CaCO_3$ + Polidisp 7778 + Antifoam	A	423.3	6875	15.48	14.41
		B	424.3	6268	14.24	
		C	420.5	5970	13.51	
14	α + Methocel	A	349	3281	7.41	6.90
		B	349.6	3259	7.07	
		C	347.3	2875	6.22	

Table 4 Results for 3 point bending test for β-hemihydrate formulations

#	Formulation		Weight (g)	Max. force (N)	Mr (modulus of rupture) (MPa)	Average Mr (MPa)
15	β	A	299.7	2465	5.62	5.53
		B	302.6	2492	5.74	
		C	302.6	2284	5.24	
16	β + K_2SO_4	A	306.2	2505	5.67	5.65
		B	305.5	2544	5.77	
		C	305.2	2436	5.50	
17	β + Zn_2SO_4	A	307	1485	3.36	4.57
		B	307	2340	5.29	
		C	306.1	2288	5.05	
18	β + Na_2SO_4	A	297.7	1591	3.63	3.39
		B	302	1571	3.55	
		C	298.9	1349	3.00	
19	β + $CaCl_2$	A	308.4	1075	2.44	2.49
		B	304.6	1200	2.73	
		C	315.5	1024	2.31	
20	β + $CaCO_3$	A	306.5	1706	3.77	3.53
		B	312.2	1258	2.78	
		C	305.5	1787	4.05	
21	β + K_2SO_4 + Polidisp7252 + Antifoam	A	328.1	2595	5.92	5.96
		B	327.4	2568	5.80	
		C	330.2	2715	6.17	
22	β + Zn_2SO_4 + Polidisp7252 + Antifoam	A	328.6	2889	6.46	7.36
		B	327.9	3500	7.74	
		C	328.2	3505	7.87	
23	β + $CaCO_3$ + Polidisp7252 + Antifoam	A	333.7	3998	9.05	8.30
		B	334.1	3554	7.87	
		C	334.4	3579	7.97	
24	β + K_2SO_4 + Polidisp DV + Antifoam	A	302.2	2461	5.63	5.40
		B	303.3	2276	5.20	
		C	300.9	2382	5.38	
25	β + Zn_2SO_4 + Polidisp DV + Antifoam	A	306	3100	6.98	6.50
		B	309.7	2726	6.11	
		C	307.1	2892	6.42	
26	β + $CaCO_3$ + Polidisp DV + Antifoam	A	302.2	2909	6.48	6.61
		B	299.5	3144	7.16	
		C	305.2	2740	6.17	
27	β + K_2SO_4 + Polidisp 6660 + Antifoam	A	326.3	2719	6.26	6.39
		B	324.7	2727	6.18	
		C	326.6	2970	6.73	
28	β + $CaCO_3$ + Polidisp 6660 + Antifoam	A	322.7	2905	6.78	6.66
		B	324.7	2905	6.80	
		C	324.8	2725	6.39	

(continued)

Table 4 (continued)

#	Formulation		Weight (g)	Max. force (N)	Mr (modulus of rupture) (MPa)	Average Mr (MPa)
29	β + K_2SO_4 + Polidisp 7778 + Antifoam	A	327.2	3356	7.58	8.24
		B	325.1	3881	8.82	
		C	326.1	3644	8.30	
30	β + $CaCO_3$ + Polidisp 7778 + Antifoam	A	321.3	3788	8.60	8.49
		B	322.9	3785	8.59	
		C	322.8	3643	8.28	
31	β + Methocel	A	287.3	2358	5.26	4.69
		B	288.5	1626	3.62	
		C	289.8	2368	5.18	

4.3 Fracture Sections of Specimens After 3 Point Bending Test

Comparing the fractured surface sections (40 × 40 mm) of α plaster crystals formulations with β and α + β ones, it is possible to register that pure α-hemihydrate formulations have less occurrence of pores. This is related with the fact that, growing more slowly (higher setting time), allow air bubbles to be released, contributing to more compact formulations that in general have higher flexural strength (see tables above).

The α plaster mixtures containing Polidisp (either 7252 or 7778) acquire a higher viscosity in a shorter period, leading to higher air bubbles entrapment into the slurry, as can be seen in Fig. 15. Figures 16 and 17 show β and α + β sections, where higher amount of porosity can be observed.

The generation of pores (amount and diameter) in the periphery has greater consequence in flexural strength reduction because it is in this region that the specimen is under greater tensile stresses.

Figure 18 shows the fracture of the specimens which formulation had a better flexural strength.

Figure 18a is related to the sample of α plaster formulation and reached a 15.48 MPa modulus of rupture, Fig. 18b is relative to β plaster with a 8.6 MPa and Fig. 18c of α + β with a 14.54 MPa. As $CaCO_3$ is less effective than K_2SO_4 with regard to accelerating the crystallization reaction, i.e., reducing the setting time of the slurry, formulations with this accelerator become more favourable in realising the bubbles introduced during the slurry mixture.

The good flexural strength results for formulations with Polidisp 7778, an acrylic-based binder, is an indicator that this kind of binder is preferred instead of a binder based on vinyl acetate—VeoVa, as Polidisp 7252.

Table 5 Results for 3 point bending test for $\alpha + \beta$ formulations

#	Formulation		Weight (g)	Max. Force (N)	Mr (modulus of rupture) (MPa)	Average Mr (MPa)
32	$\alpha + \beta$	A	386	3351	7.51	8.93
		B	389	4026	8.89	
		C	388.4	4661	10.41	
33	$\alpha + \beta + K_2SO_4$	A	386.6	3737	8.47	6.90
		B	383.6	2464	5.61	
		C	391	2936	6.61	
34	$\alpha + \beta + Zn_2SO_4$	A	379.9	4080	9.20	8.17
		B	381.3	3606	8.04	
		C	379.6	3260	7.27	
35	$\alpha + \beta + Na_2SO_4$	A	386.7	2242	5.10	5.44
		B	384.9	2171	4.92	
		C	385.6	2853	6.29	
36	$\alpha + \beta + CaCO_3$	A	386	3640	8.33	7.99
		B	384.9	3094	7.05	
		C	384.2	3766	8.59	
37	$\alpha + \beta + K_2SO_4$ + Polidisp7252 + Antifoam	A	411.1	3621	8.33	8.97
		B	410.6	3941	9.14	
		C	411.4	4059	9.44	
38	$\alpha + \beta + Zn_2SO_4$ + Polidisp7252 + Antifoam	A	421.7	3782	8.44	7.56
		B	425	3139	6.99	
		C	424.6	3178	7.26	
39	$\alpha + \beta + CaCO_3$ + Polidisp7252 + Antifoam	A	405.5	4499	10.06	10.25
		B	413.4	4969	10.77	
		C	412.6	4513	9.92	
40	$\alpha + \beta + K_2SO_4$ + Polidisp DV + Antifoam	A	388.3	1828	4.24	3.76
		B	388.6	1659	3.80	
		C	389.5	1455	3.22	
41	$\alpha + \beta + Zn_2SO_4$ + Polidisp DV + Antifoam	A	385.2	3820	8.24	7.92
		B	385.1	3893	8.34	
		C	384.8	3329	7.19	
42	$\alpha + \beta + CaCO_3$ + Polidisp DV + Antifoam	A	400.2	2212	4.89	4.64
		B	400.9	1901	4.29	
		C	403.4	2134	4.74	
43	$\alpha + \beta + K_2SO_4$ + Polidisp 6660 + Antifoam	A	410.2	5560	12.50	12.72
		B	412.5	5764	12.66	
		C	411.5	5769	13.00	
44	$\alpha + \beta + CaCO_3$ + Polidisp 6660 + Antifoam	A	404.2	6118	13.71	13.74
		B	406.3	5860	13.09	
		C	401.2	6365	14.43	
45	$\alpha + \beta + K_2SO_4$ + Polidisp 7778 + Antifoam	A	400.6	6035	13.73	13.53
		B	400.1	5820	13.30	
		C	400.2	5911	13.55	
46	$\alpha + \beta + CaCO_3$ + Polidisp 7778 + Antifoam	A	402.6	4852	10.74	11.25
		B	400.4	6526	14.54	
		C	397.1	3750	8.46	

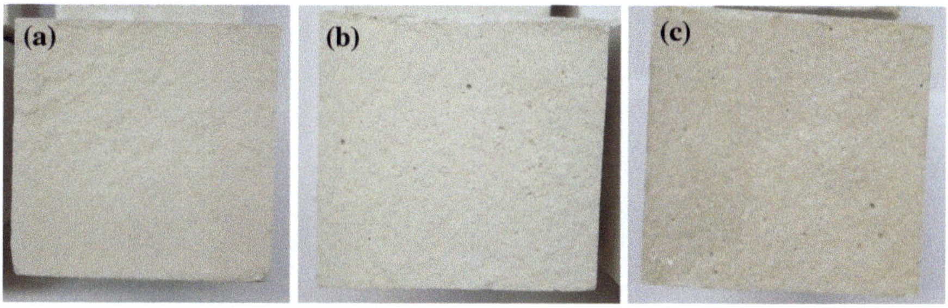

Fig. 15 Fracture surface of the tested specimens of gypsum α-hemihydrate formulations; **a** pure gypsum; **b** $\alpha + K_2SO_4$; **c** $\alpha + K_2SO_4$ + Polidisp 7252

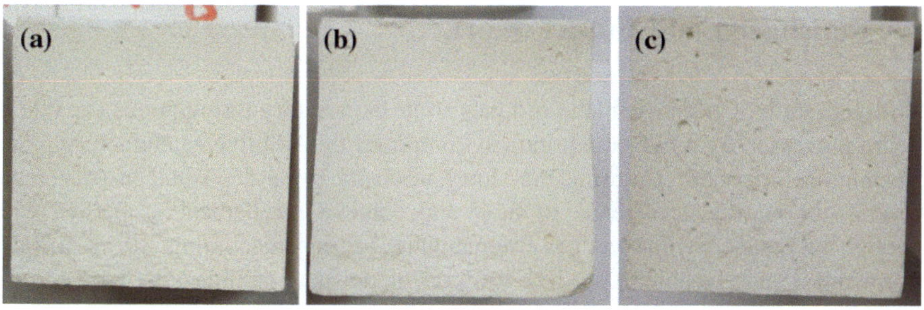

Fig. 16 Fracture surface of the tested specimens of gypsum β-hemihydrate formulations; **a** pure gypsum β; **b** $\beta + K_2SO_4$; **c** $\beta + K_2SO_4$ + Polidisp 7252

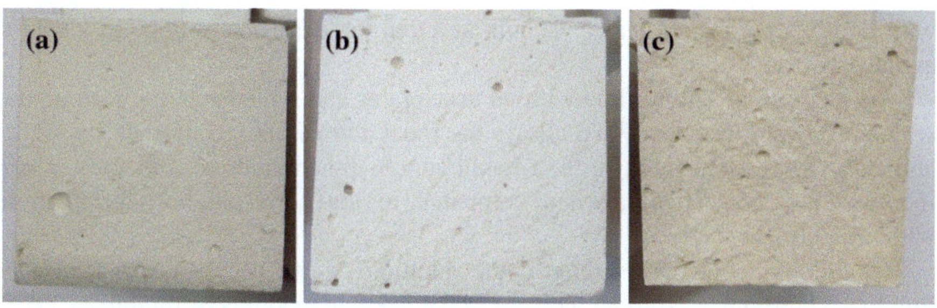

Fig. 17 Fracture surface of the test specimens of gypsum $\alpha + \beta$ formulations; **a** $\alpha + \beta$; **b** $\alpha + \beta + K_2SO_4$; **c** $\alpha + \beta + K_2SO_4$ + Polidisp 7252

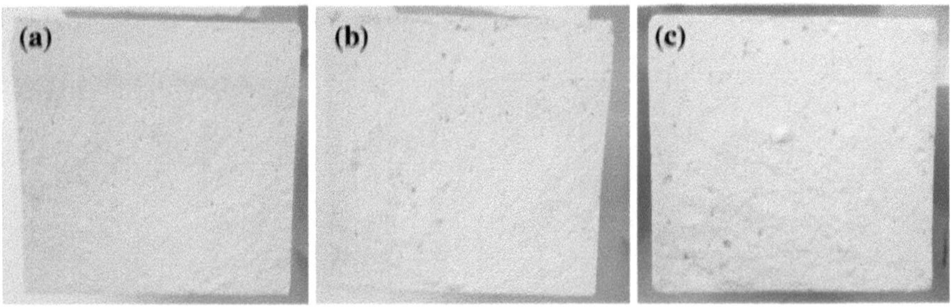

Fig. 18 Fracture surface of the test specimens of gypsum + $CaCO_3$ + Polidisp 7778 formulations; **a** $\alpha + CaCO_3$ + Polidisp 7778; **b** $\beta + CaCO_3$ + Polidisp 7778; **c** $\alpha + \beta + CaCO_3$ + Polidisp 7778

5 Conclusion and Future Work

With this study it was possible to produce some exploratory formulations capable of being processed by a hybrid equipment composed by Additive Manufacturing and Machining processes. Tailoring the slurry viscosity to be deposited in successive layers without the occurrence of flow and achieve mechanical properties with enough resistance to sharp edges fragmentation during machining, seems to be a promising step to be able to produce low cost models and moulds with good surface finishing.

The fast reduction in the setting time and thus the rapid increase in slurry viscosity obstructs the release of air bubbles generated during the mixing of the components, thereby causing pores in the final parts that decrease the mechanical strength of the plaster. Thus, it is advisable that in the future the plaster mixture with liquid is performed under vacuum and that the extrusion is done in a way to avoid porosity.

The increase in setting time with an accelerator and a binder in the formulation with $CaCO_3$ and Polidisp 7778 allows the reduction of the setting time by 42, 20 and 32% for formulations with α-hemihydrate, β-hemihydrate and plaster α + -hemihydrate, respectively, when compared to the corresponding formulations without additives.

Regarding the flexural strength, the addition of the accelerator $CaCO_3$ and Polidisp 7778 binder led to an increase in the average modulus of rupture of 71, 53 and 26% for formulations with α-hemihydrate, β-hemihydrate and plaster α + -hemihydrate, respectively, compared to the respective formulations without additives.

Considering the results of this study, the next objective is making extruding tests of these slurries pastes in order to control the fluidity and thixotropy, and perform adhesion tests between layers of extruded slurries, and do machining tests of gypsum blocks with additives to investigative the surface finishing after machining and analysis of cutting tools wear.

Acknowledgements Authors gratefully acknowledge the funding of Project NORTE-01-0145-FEDER-000022—SciTech—Science and Technology for Competitive and Sustainable Industries, cofinanced by Programa Operacional Regional do Norte (NORTE2020), through Fundo Europeu de Desenvolvimento Regional (FEDER).

References

1. Caetano, D.: Desenvolvimento de Formulações de Pastas de Gesso para Fabrico Aditivo, Master thesis, in MIEM, FEUP, Portugal (2016)
2. Yamazaki, T.: 18th CIRP Conference on Electro Physical and Chemical Machining (ISEM XVIII) Development of A Hybrid Multi-tasking Machine Tool: Integration of Additive Manufacturing Technology with CNC Machining. Procedia CIRP, vol. 42, pp. 81–86 (2016)
3. Ferreira, I.A.: Impressão tridimensional de baixo custo aplicada à pasta de açúcar, Master thesis, in MIEM, FEUP, Portugal (2015)
4. FABtotum web page. (cited 20 June 2016). http://www.fabtotum.com/3d-printers
5. Zmorph blog. (cited 20 June 2016). http://blog.zmorph3d.com/best-all-in-one-3d-printer-in-2016/
6. WASP Web page. (cited 20 June 2016). http://www.personalfab.it/en/products/powerwasp-cnc-mill-lcdsddis/
7. Harper, C.A.: Handbook of Ceramics, Glasses, and Diamonds. McGraw-Hill (2001)
8. Nunes, J.C.O.: Fabrico de peças metálicas utilizando moldações cerâmicas e moldações em gesso, Master thesis, in DEMGI, FEUP (1999)
9. Singh, N.B., Middendorf, B.: Calcium sulphate hemihydrate hydration leading to gypsum crystallization. Prog. Cryst. Growth Charact. Mater. **53**, 57–77 (2007)
10. Kuntze, R.A.: Gypsum: Connecting Science and Technology. ASTM International (2009)
11. Bobby, S.S., Singamneni, S.: Influence of moisture in the gypsum moulds made by 3D printing. Proc. Eng. **97**, 1618–1625 (2014)
12. Song, K., Mitchell, J., Gladden, L.F.: Observing microstructural evolution during plaster hydration. Diffus. Fundam. **10**, 22.1–22.3 (2009)
13. Stav, E., Bentur, A.: Reactions and early-age properties in in-situ polymerized gypsum-acrylamide composites. Mater. Struct. **29**, 366–371 (1996)
14. Rubio-Avalos, J.C., Manzano-Ramírez, A., Luna-Bárcenas, J.G., Pérez-Robles, J.F., Alonso-Guzmán, E.M., Contreras-García, M.E., González-Hernández, J.: Flexural behavior and microstructure analysis of a gypsum-SBR composite material. Mater. Lett. **59**(2–3), 230–233 (2005)
15. Pundir, A., Garg, M., Singh, R.: Evaluation of properties of gypsum plaster-superplasticizer blends of improved performance. J. Build. Eng. **4**, 223–230 (2015)

Part VII
Machining

FE Modal and Harmonic Analysis of Micro Drill with Ultrasonic Horn

Muddu Allaparthi, Mohammed Rajik Khan and Syam Narayana Addepalli

1 Introduction

Micro drilling machining operation is most crucial in electronics and printed circuit board industries for drilling large number of micro holes which are generally required in various computer and electronic products. Hole accuracy, delamination and burr formation are the serious concerns which are destructing the performance of these products. Ultrasonic vibration assistance drilling technologies have shown significant improvements in terms of quality and performance in various industrial applications. This technique is used for high precision machining application of drilling holes in hard, difficult-to-cut and brittle materials. The superimposing of ultrasonic vibration frequency (≤ 20 kHz) through transducer to the drill bit via horn in longitudinal direction has reported many advantages compared to conventional machining [1]. The reduction of torque and thrust forces, low tool wear, burr minimization, good surface finish, etc. are some of its benefits. Researchers like Thomas et al. [2] and Azarhoushang et al. [3] have conducted experiments on drilling of hard materials like Steel and Inconel alloys by applying ultrasonic assistance to conventional drilling. Babitsky et al. [4] have shown the effects of vibration excitation and energy transfer during ultrasonically assisted drilling. Chang et al. [5] have presented a thrust force model for vibration assisted drilling of aluminum 6061-T6. Makhdum et al. [6] have shown the effect of ultrasonically-assisted drilling on carbon-fibre-reinforced plastics. Investigation on forces and torques in conventional

M. Allaparthi · M.R. Khan (✉) · S.N. Addepalli
National Institute of Technology, Rourkela 769008, Odisha, India
e-mail: khanmr@nitrkl.ac.in

M. Allaparthi
e-mail: 513ID1003@nitrkl.ac.in

S.N. Addepalli
e-mail: 214ID1392@nitrkl.ac.in

L.F.M. da Silva (ed.), *Materials Design and Applications*,
Advanced Structured Materials 65, DOI 10.1007/978-3-319-50784-2_21

and ultrasonically assisted drilling of cortical bone have been conducted experimentally by Alam et al. [7]. Ultrasonic assisted machining have some disadvantages in the form of unpredictable resonance and time varying and ill-defined loading conditions at the tip of the drill bit due to vibro-impact [1]. Li et al. [8] have performed experimentations on autoresonant control of ultrasonically assisted drilling to understand the vibro impact regime of drilling. Few researchers have also worked towards the application of ultrasonic assistance to micro drilling processes. Chern et al. [9] have conducted experiments on micro drilling using ultrasonic vibration assistance to work piece material. FE modeling of micro drill and experimentations on high speed ultrasonic vibration assisted micro drilling have been performed by Zhang et al. [10].

Hence, it is known through the past literatures that, the application of ultrasonic vibration assistance to drilling process over conventional drilling has shown good results. But, very few literatures have been found addressing effects of ultrasonic vibration over micro drilling process. This may be due to the complexity of micro drilling machining conditions under the assistance of ultrasonic vibration. The transfer of ultrasonic vibration energy and to determine accurate resonance frequency to match drilling process for efficient micro drilling is one of the main concern.

In the present work, the aim is to perform the Finite Element (FE) analysis of micro drill bit with horn subjected to ultrasonic vibration frequency. FE modal analysis has been presented to show the effect of natural frequency and different modes shapes of stepped and conical horn along with micro drill bit with multiple combination of horn and drill materials. Also, FE harmonic analysis is performed to show the displacement loads on horns with micro drill bit at different amplitudes.

2 Design of Micro Drill Bit and Ultrasonic Horn

For any representation of analysis the geometrical features of the model is most important to find out the results accurately. The design of models with adequate specifications that resembles the quality of standards can meet the task precisely. There are so many commercial CAD software available in the market to represent the models in required format. For the present work, the three dimensional (3D) solid models of micro drill bit and horn are designed in CATIA V6. The micro drill bit considered for our analysis is having a design specification of 0.3 mm diameter with 2 step shank having a point angle 118° and helix angle 35°. The geometrical dimensions of micro drill bit are shown in Fig. 1. The principle function of horn is to amplify the amplitude of vibration energy transmitted from transducer to the cutting tool at the operational range of ultrasonic frequency for effective machining [11]. There are various shapes of horns existing for different ultrasonic applications such as stepped, exponential, cylindrical and conical. In the present analysis, stepped and conical shaped horns are chosen to match the desired ultrasonic frequency with micro drill bit. The governing equation of longitudinally vibrating horn

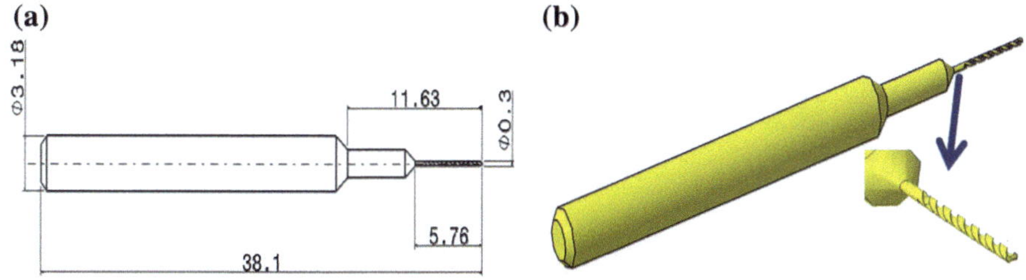

Fig. 1 Micro drill bit. **a** 2D orthographic view (dimensions in mm). **b** 3D view with closed view of flute

with varying circular cross section for one dimensional (1D) continuum is expressed mathematically [11] as,

$$\frac{\partial^2 u(x,t)}{\partial t^2} = c_p^2 \left[\frac{1}{S(x)} \frac{\partial S(x)}{\partial x} \frac{\partial u(x,t)}{\partial x} + \frac{\partial^2 u(x,t)}{\partial x^2} \right] \tag{1}$$

where, x is the position coordinate of any point on the circular cross section of the horn, $u(x,t)$ is the longitudinal displacement of cross section from its position within the undeformed horn. $S(x) = \pi(r(x))^2$ is the cross sectional area of horn, $r(x)$ is the radius of its circular cross section, $c_p = \sqrt{E/\rho}$ is the velocity of longitudinal waves in 1D continuum and E and ρ are the Young's modulus and the density of the horn materials respectively.

The resonant length of the stepped shaped horn can be mathematically [12] expressed as,

$$L = k_1(c_p/4f) + k_2(c_p/4f) \tag{2}$$

To simplify the calculation, the correction factors k_1 and k_2 are assumed to be unity. Hence, $L = c_p/2f$ and the amplitude ratio is given as $(D_1/D_2)^2$, where, D_1 and D_2 are the diameters at the larger and smaller ends of the horn respectively. The resonant length of the conical shaped horn is calculated approximately from the exponential shaped horn length [1] as,

$$\tan \alpha l_2 = \frac{\alpha l_1 (\sqrt{S_1/S_2} - 1)^2 - [\sqrt{S_1/S_2}(\alpha l_1)^2 + (\sqrt{S_1/S_2} - 1)^2] \tan \alpha l_1}{\alpha l_1 [\sqrt{S_1/S_2} \, . \, \alpha l_1 - (\sqrt{S_1/S_2} - 1) \tan \alpha l_1]} \tag{3}$$

where, l_1 and l_2 are the lengths of the conical section and cylindrical cross section of the horn respectively, α is the wave length constant, and S_1 and S_2 are the cross sections of the input and output end of the conical bar.

Considering the above analytical relations, stepped and conical horns are geometrically designed to match the ultrasonic resonance frequency of 20 kHz. Figures 2 and 3 shows the geometrical design of stepped and conical horns. The micro

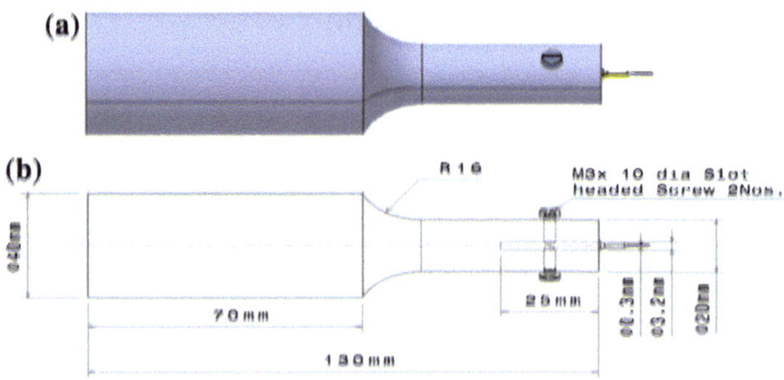

Fig. 2 Assembly of micro drill bit with stepped horn. **a** 3D view and **b** 2D view

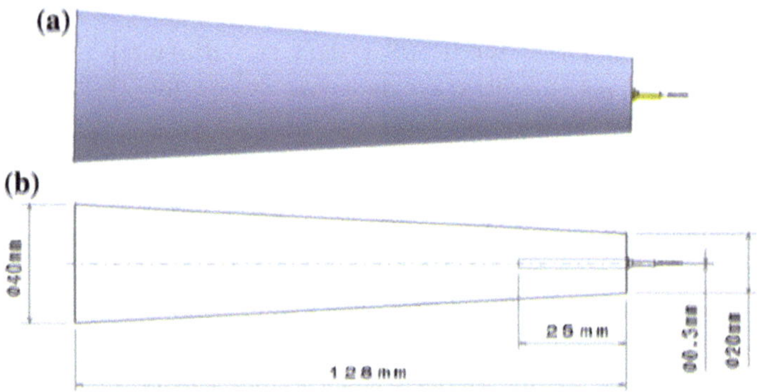

Fig. 3 Assembly of micro drill bit with conical horn. **a** 3D view and **b** 2D view

drill bit is inserted into the horn at lower diameter end. Two slotted cheese head screws (as shown in Fig. 2) have been used to hold the micro drill bit in stepped horn. The assemblies of horn with micro drill bit for stepped and conical shaped horns are shown in Figs. 2 and 3 respectively.

3 FE Modal Analysis

Modal analysis is the basic platform to study the dynamic properties of any materials to determine the natural frequencies and mode shapes of a structure under vibrational excitation. Here, modal analysis is performed to study the characteristics of the micro drill bit under the range of ultrasonic frequency excitation. In the present analysis, the frequency characteristics of stepped and conical horn with micro drill bit are evolved using Block Lanczos solver method. For our analysis, Tungsten carbide (WC) and High speed steel (HSS) are considered as micro drill bit material which has good strength and resistance against different workpiece materials to be drilled. Aluminum 6061-T6, Titanium (Ti-6Al-4 V) and Mild steel are chosen for horn materials which are light in weight and possesses good characteristics under the influence of vibration. The mechanical properties of materials assigned for micro drill bit and horn are shown in Table 1. Bonded contact has been set apart to the assembly of the horn with drill bit. A fine Hex dominant with all quad elements has been assigned to assembly bodies of both the horns. The element size for both the horns and the micro drill are taken as 4 mm and 0.1 mm respectively. Table 2 shows the meshing details of both the horns along with micro drill bit. The mesh generated on stepped and conical horns along with micro drill are shown in Figs. 4 and 5.

A frequency range of 15–30 kHz has been chosen for identifying the natural frequencies and modes shapes of both the horns with drill bit for multiple material

Table 1 Mechanical properties of micro drill bit and horn materials

Mechanical properties	Micro drill bit materials		Horn materials		
	Tungsten carbide	High speed steel	Aluminium 6061-T6	Titanium	Mild steel
Density (kg/m^3)	15600	8160	2700	4510	7870
Young's Modulus (GPa)	690	210	69	120	205
Poisson's ratio (ν)	0.31	0.30	0.33	0.34	0.3
Speed of sound (c_p) (m/s)	6220	5141	5091	5158	5103

Table 2 Elements and nodes generated on horn and drill bit

Mesh type	Meshing body	No of elements	No of nodes
Hex dominant (All Quad)	Stepped horn + micro drill with screws	19329	59051
	Conical horn with micro drill	11851	39373

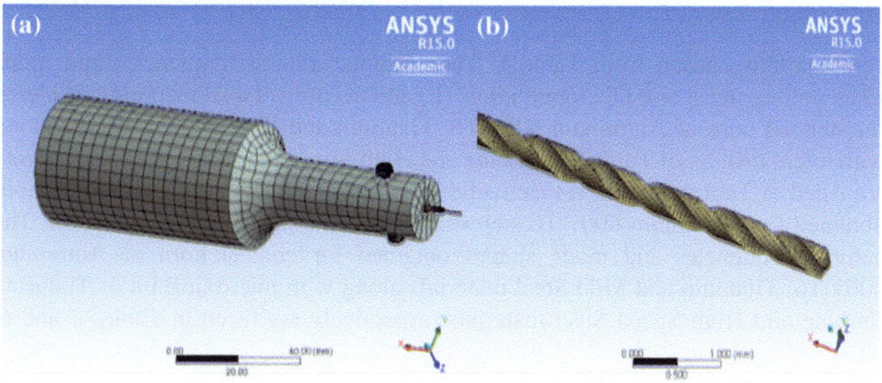

Fig. 4 Meshing of **a** stepped horn with micro drill. **b** Flute part (enlarged view)

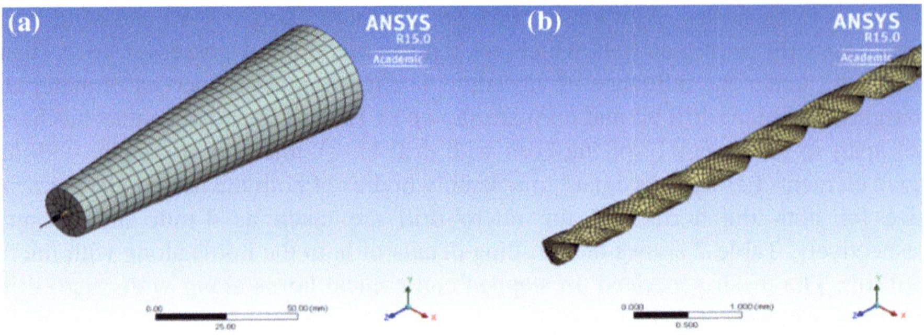

Fig. 5 Meshing of **a** conical horn with micro drill. **b** Flute part (enlarged view)

Table 3 Natural frequencies obtained for stepped horn with tungsten carbide micro drill bit

Mode no and mode shape	Stepped horn materials		
	Aluminium 6061-T6 (Hz)	Titanium (Hz)	Mild steel (Hz)
1 (T)	17308	17386	17639
2 (T)	17452	17470	17689
3 (L)	20008	20431	20845
4 (Tw)	22087	21729	22199
5 (T)	25619	25785	26182
6 (T)	25813	25905	26261

Table 4 Natural frequencies obtained for stepped horn with high speed steel micro drill bit

Mode no and mode shape	Stepped horn materials		
	Aluminium 6061-T6 (Hz)	Titanium (Hz)	Mild steel (Hz)
1 (T)	17437	17462	17676
2 (T)	17586	17549	17721
3 (L)	20326	20625	20956
4 (Tw)	22086	21728	22196
5 (T)	25736	25858	26220
6 (T)	25937	25982	26298

combinations chosen from Table 1. The obtained natural frequencies and mode shapes along transverse (T), longitudinal (L) and twisting (Tw) directions considered for stepped horn of Aluminum 6061-T6, Titanium and Mild Steel materials along with micro drill bit of Tungsten Carbide and High Speed Steel materials respectively are listed in Tables 3 and 4 respectively. Figure 6 shows the different mode shapes obtained for Aluminum 6061-T6 stepped horn with tungsten carbide drill bit. The natural frequencies and mode shapes obtained for conical horn of Aluminum 6061-T6, Titanium and Mild Steel materials along with micro drill bit of Tungsten Carbide and High Speed Steel materials respectively are listed in Tables 5 and 6.

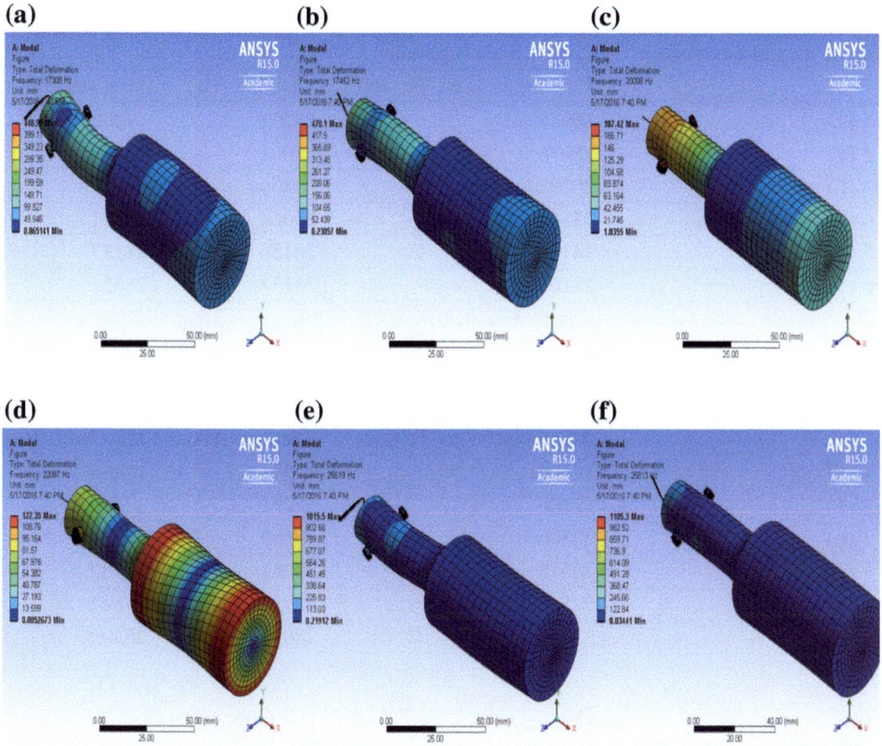

Fig. 6 Mode shapes obtained for Aluminum 6061-T6 stepped horn with tungsten carbide micro drill bit at different natural frequencies **a** 17308 Hz, **b** 17452 Hz, **c** 20008 Hz, **d** 22087 Hz, **e** 25619 Hz, **f** 28813 Hz

Figure 7 shows the different mode shapes obtained for Aluminum 6061-T6 conical shaped horn with tungsten carbide drill bit.

From the above results, it has been observed that both the horns with micro drill bit are well set for the ultrasonic frequency range. For both the chosen horn and micro drill bit material combinations, initially transverse bending modes have been observed at frequencies below ultrasonic range (<20 kHz). The longitudinal mode is seen with in the desirable ultrasonic range above 20 kHz (<100 kHz). Later the twisting and transverse modes shapes are observed beyond the desirable frequency range (>20 kHz) as shown in Figs. 6 and 7 respectively.

The assembly of stepped and conical shaped horns of Aluminium 6061-T6 with Tungsten carbide micro drill bit shows closeness to ultrasonic frequency response longitudinally in comparison to others. Further the results obtained from modal analysis with longitudinal modes are used to perform the harmonic analysis to verify the dynamic frequency responses with respect to material combination for stepped and conical shaped horn with drill bit assembly.

Table 5 Natural frequencies obtained for conical horn with Tungsten carbide micro drill bit

Mode no and mode shape	Conical horn materials		
	Aluminium 6061-T6 (Hz)	Titanium (Hz)	Mild steel (Hz)
1 (T)	16975	17001	17208
2 (T)	16976	17001	17208
3 (L)	20298	20404	20646
4 (Tw)	25517	25064	25581
5 (T)	28101	28136	28537
6 (T)	28102	28137	28537

Table 6 Natural frequencies obtained for conical horn with High speed steel micro drill bit

Mode no and mode shape	Conical horn materials		
	Aluminium 6061-T6 (Hz)	Titanium (Hz)	Mild steel (Hz)
1 (T)	17106	17291	17465
2 (T)	17106	17291	17465
3 (L)	20506	20685	20715
4 (Tw)	25518	25261	25782
5 (T)	27556	27720	27892
6 (T)	27761	28034	28306

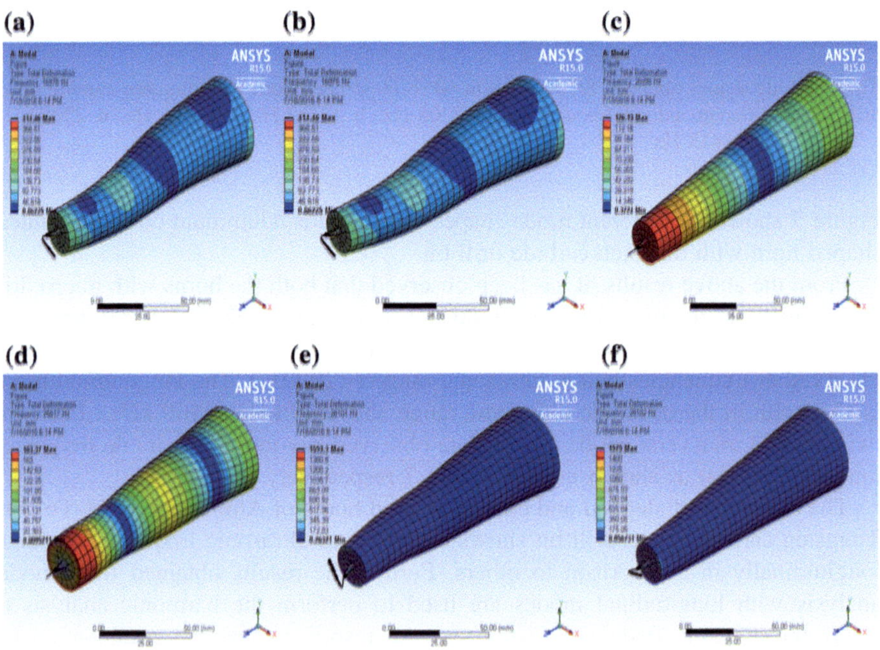

Fig. 7 Mode shapes obtained for Aluminum 6061-T6 conical horn with tungsten carbide micro drill bit at different natural frequencies **a** 16975 Hz, **b** 16976 Hz, **c** 20298 Hz, **d** 25517 Hz, **e** 28101 Hz, **f** 28102 Hz

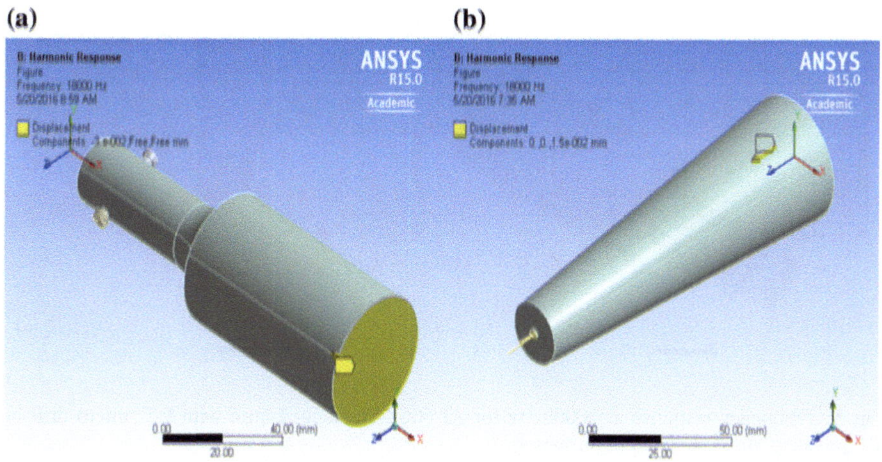

Fig. 8 Displacement boundary applied on back end **a** Stepped horn, **b** Conical horn

4 FE Harmonic Analysis

Harmonic analyses are used to determine the steady-state response of a linear structure to loads that vary sinusoidally with time. For the ultrasonic vibration assistance systems the transmission of vibration energy may vary for different cross-sectional structures. For the present analysis, the response of vibrating bodies i.e. horn with micro drill bit considering stepped and conical shaped horns under desirable ultrasonic frequency range are simulated using Full method analysis. Here, frequency responses are considered for longitudinal modes only. For the present work, the displacement loading boundary conditions are assumed to verify the frequency response and amplification factors of the Aluminum 6061-T6 horn material with tungsten carbide micro drill bit for both the horn shapes. The displacement loads of 15 and 30 μm are applied on horn larger end diameter in axial direction as shown in Fig. 8. Frequency response is considered in terms of stress and total deformation from the face of the micro drill bit chisel edge. Figure 9 shows the stress distribution and total deformation formed on aluminum 6061-T6 stepped horn with tungsten carbide micro drill bit at a frequency of 20008 Hz. Figure 10 shows the stress distribution and total deformation in aluminum 6061-T6 conical horn with tungsten carbide micro drill bit at a frequency of 20298 Hz. The frequency responses in terms of stresses and total deformation at longitudinal modes of 30 μm amplitude for Aluminum 6061T6 stepped and conical horn along with tungsten carbide micro drill bit are shown graphically using Figs. 11 and 12.

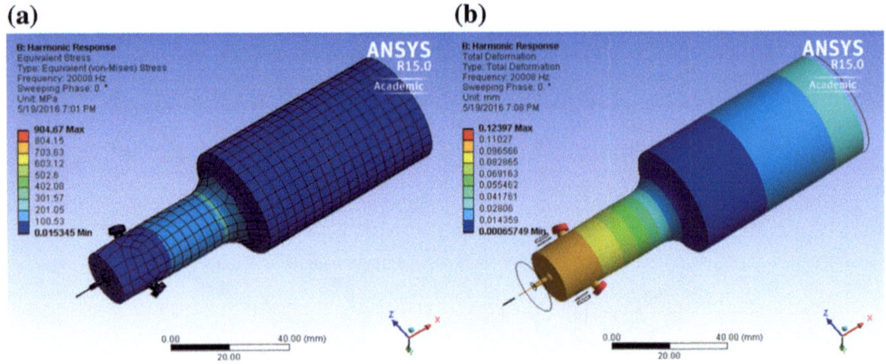

Fig. 9 Frequency response at 20008 Hz for Al 6061-T6 stepped horn with WC micro drill bit. **a** Equivalent stress. **b** Total deformation

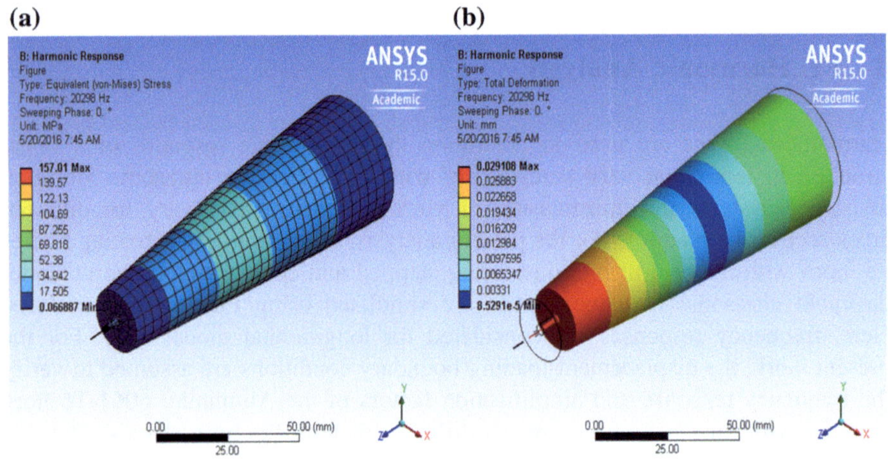

Fig. 10 Frequency response at 20298 Hz for Al 6061-T6 conical horn with WC micro drill bit. **a** Equivalent stress. **b** Total deformation

5 Discussion of Results

For the modal analysis, the natural frequencies obtained for stepped and conical shaped horn with varying material combinations have matched well with the specified frequency range 15–30 kHz. The initial frequencies for all the material combinations have shown transverse mode shapes below the ultrasonic frequency. The most essential longitudinal mode shape has been formed above the ultrasonic vibration frequency range and latter followed by twisting and transverse modes. Aluminum 6061-T6 material with combination of tungsten carbide drill bit for stepped shaped horn have attained very close to ultrasonic tuning frequency

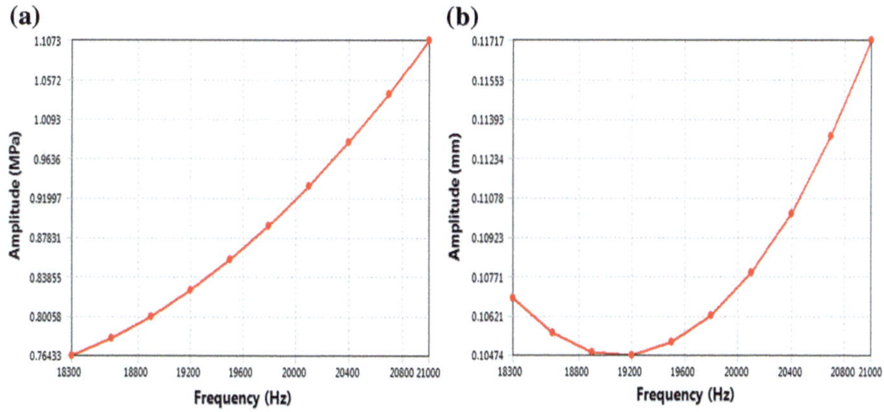

Fig. 11 Frequency versus amplitude of Al 6061-T6 stepped horn with WC micro drill bit at 30 μm. **a** Equivalent stress. **b** Total deformation

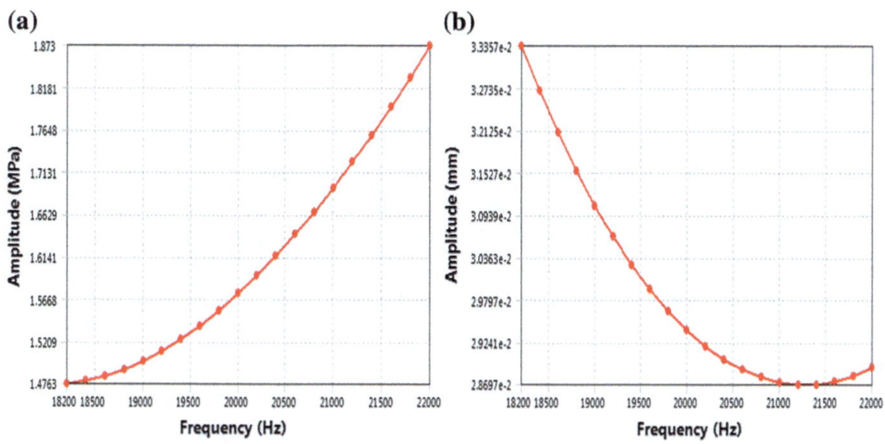

Fig. 12 Frequency versus amplitude of Al 6061-T6 conical horn with WC micro drill bit at 30 μm. **a** Equivalent stress. **b** Total deformation

(20008 Hz). Titanium and Mild steel materials with combination of WC and HSS drill bit materials for stepped horn have also shown the similarities with varying frequencies closer to Al 6061-T6. In case of Conical shaped horn, the Al 6061 T6 material with combination of WC drill bit showed 20298 Hz frequency which is less than the other material titanium and mild steel. In usual practice there will be vibration energy losses in between the assembly of transducer with horn and drill bit due to material internal structures and boundary conditions. Harmonic analysis have been performed for the stepped and conical horn of Al 6061-T6 material with WC micro drill bit to verify the frequency responses. The equivalent stress distribution and total deformation for both the horns at desirable longitudinal modes 20008 Hz for stepped horn and 20298 Hz for conical horn have been shown in Figs. 9 and 10 respectively. The equivalent stress acting on Al 6061-T6 materials for stepped horn and conical horn are well set with in the elastic limit range without exceeding the material property limits. The total deformation occurred in stepped horn is 0.096 mm and for conical horn is 0.022 mm. As compared to stepped shaped horn, the conical shaped horn is showing lower deformation of material. This is due to variations in material properties, speed of sound and horn length etc. The amplification factors for stepped and conical horn form the input end to output end are satisfied to the amplification ratio i.e. greater than 1. From the results, the Al 6061-T6 stepped horn with WC micro drill bit can be recommended for ease of manufacturing and can be suitable for matching the ultrasonic frequency of 20 ± 500 kHz tuning generator.

The present horns are designed after several repeated modal analysis conducted to match the resonance frequency with different material combinations while varying the horn length 2 to 3 mm above the theoretical calculations.

6 Conclusion

Modal and harmonic analysis have been performed for stepped and conical shaped horns along with micro drill bit considering different material combinations essentially which are used in present industrial applications. The designed dimensions of both the horns satisfied the longitudinal modes with the ultrasonic vibrational frequency of 20 kHz. The equivalent stresses and total deformation has been tested for both the horns at longitudinal modes. Both showed good results with in the material elastic limits for horns. The present simulation results will be helpful to conduct proper experimentation to achieve efficient machining and surface finish. In future, more no of material combinations for horns which are light in weight can be analyzed to determine the ultrasonic vibration frequency range.

References

1. Astashev, V.K., Babitsky, V.I. In: Babitsky, V.I., Wittenburg, J. (eds.) Ultrasonic Processes and Machines-Dynamics, Control and Applications. Springer, Berlin, New York (2007)
2. Thomas, P.N.H., Babitsky, V.I.: Experiments and simulation on ultrasonically assisted drilling. J. Sound Vib. **308**, 815 (2007)
3. Azarhoushang, B., Akbari, J.: Ultrasonic-assisted drilling of Inconel 738-LC. Int. J. Mach. Tools Manuf. **47**, 1027 (2007)
4. Babitsky, V.I., Astashev, V.K., Meadows, A.: Vibration excitation and energy transfer during ultrasonically assisted drilling. J. Sound Vib. **308**, 805 (2007)
5. Chang, S.F., Bone, G.M.: Thrust force model for vibration-assisted drilling of aluminum 6061-T6. Int. J. Mach. Tools Manuf. **49**, 1070 (2009)

6. Makhdum, F., Phadnis, V.A., Roy, A., Silberschmidt, V.V.: Effect of ultrasonically-assisted drilling on carbon-fibre-reinforced plastics. J. Sound Vib. **333**, 5939 (2014)
7. Alam, K., Mitrofanov, A.V., Silberschmidt, V.V.: Experimental investigations of forces and torque in conventional and ultrasonically-assisted drilling of cortical bone. Med. Eng. Phys. **33**, 234 (2011)
8. Li, X., et al.: Experimental analysis on autoresonant control of ultrasonically assisted drilling. Mechatronics (2015). http://dx.doi.org/10.1016/j.mechatronics.2015.05.006
9. Chern, G.L., Lee, H.J.: Using workpiece vibration cutting for micro-drilling. Int. J. Adv. Manuf. Technol. **27**, 688 (2006)
10. Zhang, Z., Babitsky, V.I.: Finite element modeling of a micro-drill and experiments on high speed ultrasonically assisted micro-drilling. J. Sound Vib. **330**, 2124 (2011)
11. Nad, M.: Ultrasonic horn design for ultrasonic machining technologies. Appl. Comput. Mech. **4**, 79 (2010)
12. Seah, K.H.W., Wong, Y.S., Lee, L.C.: Design of tool holders for ultrasonic machining using FEM. J. Mater. Process. Technol. **37**, 801 (1993)

Optimization of Machining Parameters to Minimize Surface Roughness in the Turning of Carbon-Filled and Glass Fiber-Filled Polytetrafluoroethylene

Muhammet Emre Sanci, Serdar Halis and Yavuz Kaplan

1 Introduction

Surface roughness is an important characteristic in industrial applications and involves friction, lubrication, heat transmission, corrosion resistance and wear. It is acknowledged that surface roughness determines the longevity and reliability of machine parts. Surface roughness is also an indicator of the amount of energy and other resources consumed during machining [1–5]. Thus, the determination of surface roughness is an essential factor in industrial applications.

One of the best thermoplastic polymers is Polytetrafluoroethylene (PTFE), which exhibits high thermal stability as well as good chemical resistance and dielectric properties. Because of its excellent mechanical properties and low friction coefficient, PTFE is the preferred engineering plastic for many applications and processing techniques. It is used in the production of seals, bearings, O-rings, electrical insulators, valve bodies and laboratory instruments requiring chemical resistance. Additionally, it is used in non-stick surfaces, engine parts, and for applications in the biotechnology and medical fields. At the same time, PTFE is used to coat automotive parts such as clutches, valves, etc. [6–9].

Glass fiber and carbon fillers can increase the mechanical properties of PTFE. However, fillers can lead to machining difficulties. Polymer composites differ from metals during machining processes owing to the time–temperature dependence of

M.E. Sanci (✉)
Department of Mechatronics Engineering, Pamukkale University, Denizli, Turkey
e-mail: msanci@pau.edu.tr

S. Halis
Department of Automotive Engineering, Gazi University, Ankara, Turkey
e-mail: serdarhalis@gazi.edu.tr

Y. Kaplan
Department of Manufacturing Engineering, Pamukkale University, Denizli, Turkey
e-mail: ykaplan@pau.edu.tr

L.F.M. da Silva (ed.), *Materials Design and Applications*,
Advanced Structured Materials 65, DOI 10.1007/978-3-319-50784-2_22

the polymers and additional structural fillers. The quality of a machined surface can be determined by taking into account the parameters of surface roughness, cutting speed, feed rate and depth of cut. Good machinability and superior product quality with minimum surface roughness values can then be obtained [1, 10].

Artificial neural networks (ANNs) are artificial intelligence nonlinear mapping systems that can solve problems of modeling and predicting experimental data. An ANN is commonly designed in a multi-layer form that includes an input layer, a hidden layer and an output layer. Jeyakumar et al. developed the response surface method (RSM) and ANN-based prediction models to determine the surface roughness of Al6061/SiC_p. The ANN model was found to perform better than the RSM model in determining the optimum machining parameters for minimum surface roughness [11].

In this study, an ANN-based prediction model was developed to determine the optimum cutting parameters (cutting speed, feed rate, and depth of cut) in terms of surface roughness in the turning of 25% carbon-filled and 25% glass fiber-filled PTFE. The performance of the ANN model was compared with the experimental results in order to determine its efficiency.

2 Materials and Methods

2.1 Measuring of Surface Roughness

The experiments were conducted using the MAHR MARSURF PS 1 mobile roughness-measuring instrument. The Korloy WNMG-NC5330 TiN-coated carbide insert was selected as the cutting tool. The machining parameters included two cutting speeds (150 and 200 m/min), three feed rates (0.1, 0.2, and 0.3 mm/rev) and three depths of cut (1, 2 and 3 mm) and the response considered was the average surface roughness (Ra). The experiments were carried out using commercially available pure (unfilled) PTFE, 25% carbon-filled PTFE and 25% glass fiber-filled PTFE in the form of cylindrical specimens with a diameter of 50 mm. The values reported in the study were taken from readings at different points on the circumference of the workpiece samples.

2.2 Production of PTFE

The PTFE was supplied by the APAMEYA Company. The PTFE powder and filler materials were mixed mechanically in the extruder to produce cylindrical samples 50 mm in diameter, as shown in Fig. 1.

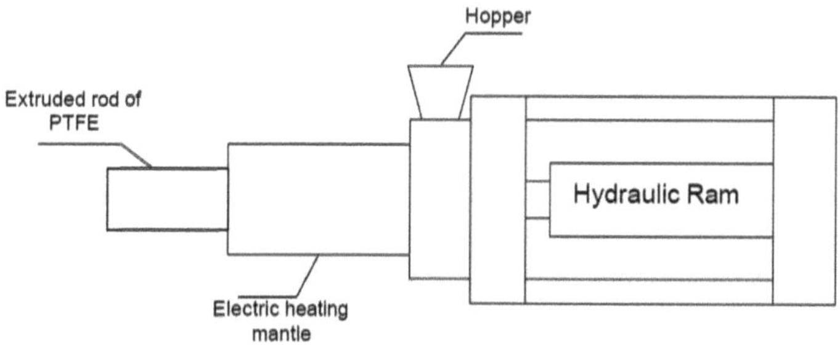

Fig. 1 Schematic illustration of PTFE sample production

2.3 *Artifical Neural Network*

The architectural approach was proposed to deal with the optimization of machining parameters in order to minimize surface roughness in the turning of carbon-filled and glass fiber-filled polytetrafluoroethylene. This approach is based upon nonlinear autoregressive models with exogenous input called NARX recurrent neural networks. The NARX model is one of the powerful class of models greatly suited for modeling non-linear systems, especially in time series. Control systems are one of the principal application fields of the NARX dynamic neural network, which contains recurrent feedback from several layers of the network to the input layer [12–14]. The architecture of the NARX neural network is given in Fig. 2.

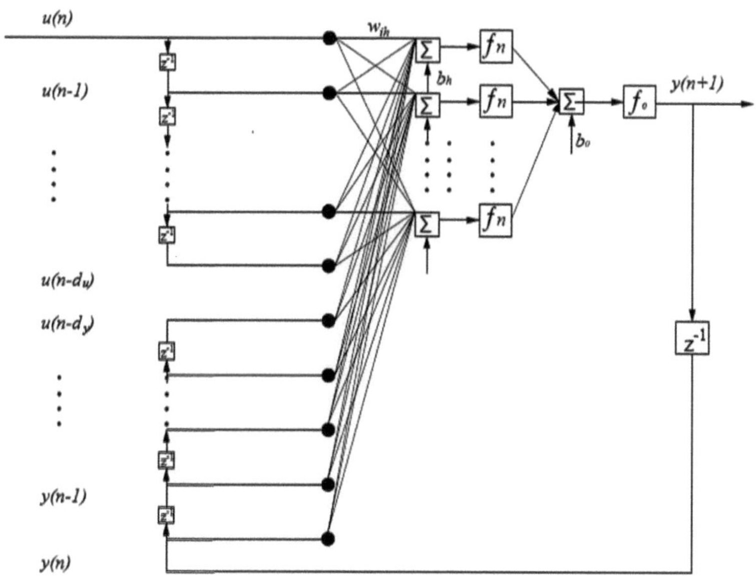

Fig. 2 Architecture of the NARX neural network

3 Results and Discussion

3.1 Experimental Results of Surface Roughness

The surface roughness tests were conducted on a CNC turning machine under dry conditions. The effect of feed rate and depth of cut on surface roughness in the turning of PTFE at a cutting speed of 150 m/min is shown in Fig. 3.

It can be seen that the surface roughness of pure (unfilled), 25% carbon-filled and 25% glass fiber-filled PTFE increases with the increase of feed rate in all depth of cut parameters. Therefore, an increasing feed rate resulted in higher surface roughness values, while a low feed rate was assumed to produce a better surface finish. In addition, the highest Ra value (4.4 μm) was observed on pure (unfilled) PTFE in turning at a feed rate of 0.3 mm/rev and a depth of cut of 3 mm. However, under these conditions, the effect of depth of cut on the surface roughness is very complex and it does not exhibit a regular behavior. Figure 4 presents a comparison of the surface roughness values of pure (unfilled), 25% carbon-filled and 25% glass fiber-filled PTFE.

Figure 4 reveals that the highest Ra values were obtained on pure PTFE in turning at all machining parameters. The lower Ra values were observed on carbon-filled and glass fiber-filled PTFE. This indicated that the carbon and glass fiber fillers led to a better surface finish with the preferred experimental parameters. In particular, the carbon fillers provided good machinability and superior product quality due to minimum surface roughness values. The effect of cutting speed on surface roughness in the turning of PTFE at a depth of cut of 1 mm is shown in Fig. 5.

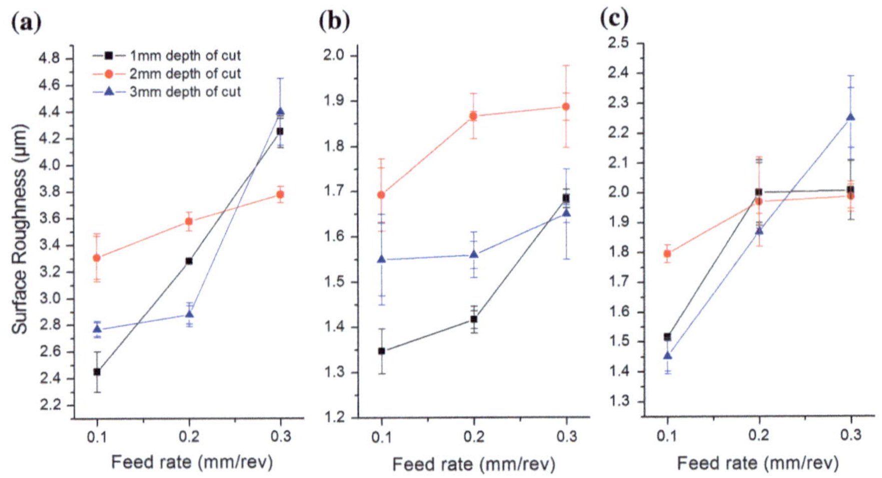

Fig. 3 Effect of feed rate and depth of cut on surface roughness in the turning of PTFE: **a** Pure (unfilled), **b** 25% carbon-filled, **c** 25% glass fiber-filled

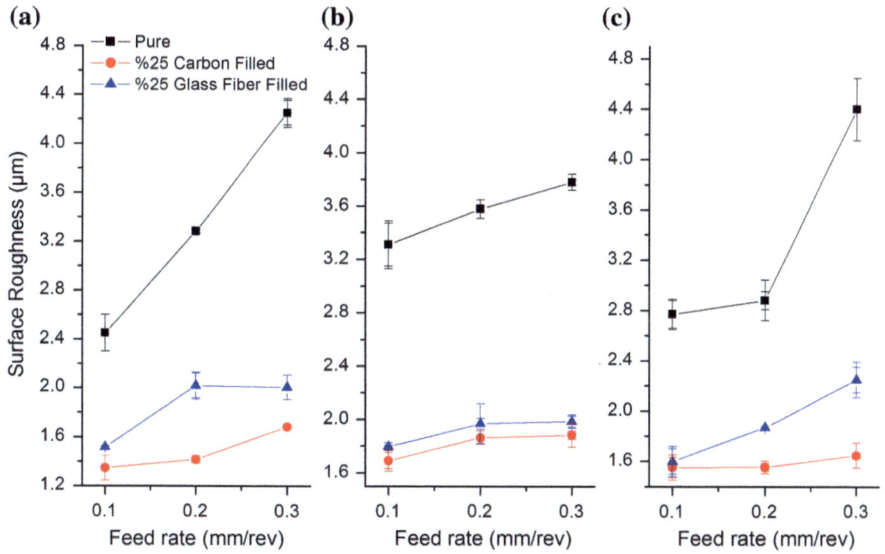

Fig. 4 Effect of feed rate and filled materials on surface roughness in the turning of PTFE: **a** 1 mm depth of cut, **b** 2 mm depth of cut, **c** 3 mm depth of cut

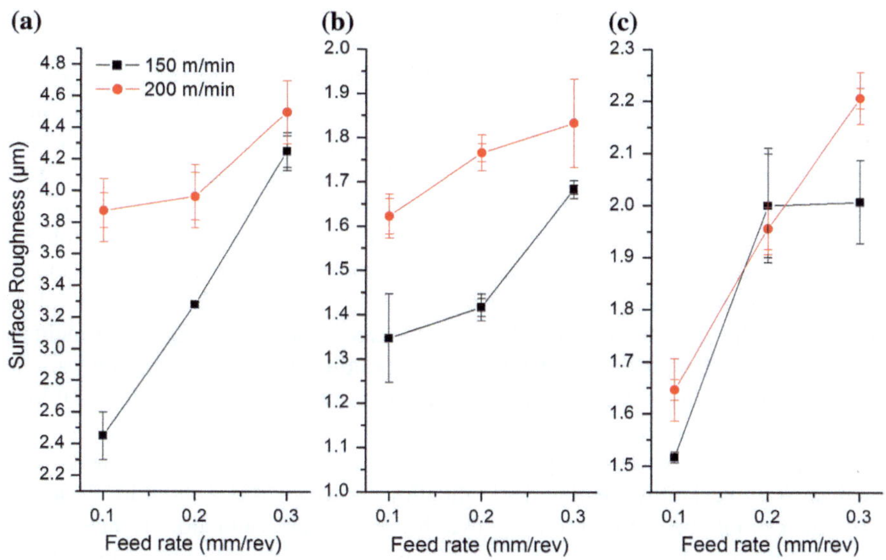

Fig. 5 Effect of feed rate and cutting speed on surface roughness in the turning of PTFE: **a** Pure (unfilled), **b** 25% carbon-filled, **c** 25% glass fiber-filled

It can be seen that the surface roughness values of pure, 25% carbon-filled and 25% glass fiber-filled PTFE generally decrease in all samples with the increase of cutting speed. Thus, a higher cutting speed resulted in lower surface roughness values, and a low feed rate was assumed to produce a better surface finish.

3.2 *Comparison of ANN Predictions with Experimental Results*

The dataset was taken from the experimental work in order to analyze the surface roughness during the turning of pure (unfilled), 25% carbon-filled and 25% glass fiber-filled PTFE. The MATLAB 2015 Neural Network toolbox with the NARX model was used for the experiments. The NARX network was capable of using multi-time series input and multi-time series output applications. In the current study, The Levenberg Marquardt (LM) algorithm, which is fast and consumes less memory [15], was used for training the algorithm.

The hyperbolic tangent sigmoid transfer function was used for activation of the function in the hidden layer as well as for the output layer. The learning algorithm used was the back propagation algorithm, which minimizes the total mean square error of the output computed by the network via a gradient-descent method. Data obtained from the experiments (machining parameters and surface roughness values) were used at the network learning stage. During network learning, the output of the network was compared with the desired output. The learning process is iterative and was stopped early to improve generalization by an increase in the mean square error of the validation samples. In sum, the data from 18 experimental trials were measured in order to build the neural network for each of the three experimental samples. In total, data from 54 experimental trials were used for the neural network modeling study, as shown in Table 1. It was necessary to decide on the number of neurons based on trial and error. This was accomplished by gradually increasing the number of neurons and observing the results of the change on the predicted values. As a result, the structure of the network was selected as 3-9-1 (Fig. 6). It included three input neurons in the input layer (corresponding to three machining parameters), one hidden layer with nine neurons and one output neuron in the output layer (corresponding to surface roughness).

No specific rule was employed for determining the number of data items to be used for training and for testing and validation; however, the general idea was that more data should be used for training than for testing and validation. Hence, 70% of the data was used for training, 15% for testing and another 15% for validation. The correlation coefficient (R-value) between the outputs and targets is a goodness-of-fit measurement of the variance between the outcomes and targets. The R-values of the validation data set shown in Table 2 indicate a strong correlation between the experimental outputs and the network outputs of the designed architecture of the ANN Ra prediction.

Table 1 Verification of the developed model with the experimental data

Experiment No.	Cutting speed (m/min)	Depth of cut (mm)	Feed rate (mm/rev)	Material	Surface roughness (μm)		Relative Error (%)
					Experimental	ANN model	
1	150	1	0.1	Pure (unfilled) PTFE	2.45	2.4	2.0408
2	150	1	0.2		3.28	3.51	−7.0122
3	150	1	0.3		4.25	4.18	1.6471
4	150	2	0.1		3.31	3.06	7.5529
5	150	2	0.2		3.58	3.41	4.7486
6	150	2	0.3		3.78	3.55	6.0847
7	150	3	0.1		2.77	2.62	5.4152
8	150	3	0.2		2.88	2.71	5.9028
9	150	3	0.3		4.4	3.9	11.3636
10	200	1	0.1		3.88	3.82	1.5464
11	200	1	0.2		3.97	4.23	−6.5491
12	200	1	0.3		4.50	4.13	8.2222
13	200	2	0.1		2.85	2.90	−1.7544
14	200	2	0.2		2.96	3.13	−5.7432
15	200	2	0.3		3.65	3.99	−9.3151
16	200	3	0.1		2.12	2.22	−4.7169
17	200	3	0.2		3.64	3.80	−4.3956
18	200	3	0.3		4.23	4.27	−0.9456
19	150	1	0.1	25% carbon-filled PTFE	1.35	1.41	−4.4444
20	150	1	0.2		1.42	1.57	−10.5634
21	150	1	0.3		1.68	1.67	0.5952
22	150	2	0.1		1.69	1.81	−7.1006
23	150	2	0.2		1.87	1.83	2.1390
24	150	2	0.3		1.89	1.82	3.7037
25	150	3	0.1		1.55	1.58	−1.9355
26	150	3	0.2		1.56	1.57	−0.6410
27	150	3	0.3		1.65	1.70	−3.0303
28	200	1	0.1		1.62	1.64	−1.2346
29	200	1	0.2		1.77	1.74	1.6949
30	200	1	0.3		1.83	1.74	4.9180
31	200	2	0.1		1.64	1.65	−0.6098
32	200	2	0.2		1.71	1.81	−5.8480
33	200	2	0.3		1.81	1.87	−3.3149
34	200	3	0.1		1.61	1.75	−8.6957
35	200	3	0.2		1.89	1.84	2.6455
36	200	3	0.3		2.05	1.94	5.3659

(continued)

Table 1 (continued)

Experiment No.	Cutting speed (m/min)	Depth of cut (mm)	Feed rate (mm/rev)	Material	Surface roughness (μm)		Relative Error (%)
					Experimental	ANN model	
37	150	1	0.1	25% Glass fiber-FILLED PTFE	1.52	1.49	1.9737
38	150	1	0.2		2.00	1.90	5.0000
39	150	1	0.3		2.01	1.96	2.4875
40	150	2	0.1		1.80	1.79	0.5556
41	150	2	0.2		1.97	1.86	5.5838
42	150	2	0.3		1.99	1.83	8.0402
43	150	3	0.1		1.60	1.64	−2.5000
44	150	3	0.2		1.87	1.94	−3.7433
45	150	3	0.3		2.25	2.44	−8.4444
46	200	1	0.1		1.65	1.82	−10.3030
47	200	1	0.2		1.96	1.92	2.0408
48	200	1	0.3		2.21	2.14	3.1674
49	200	2	0.1		1.93	2.02	−4.6632
50	200	2	0.2		2.04	2.02	0.9804
51	200	2	0.3		2.59	2.29	11.5830
52	200	3	0.1		1.90	1.82	4.2105
53	200	3	0.2		2.05	1.89	7.8049
54	200	3	0.3		2.62	2.48	5.3435

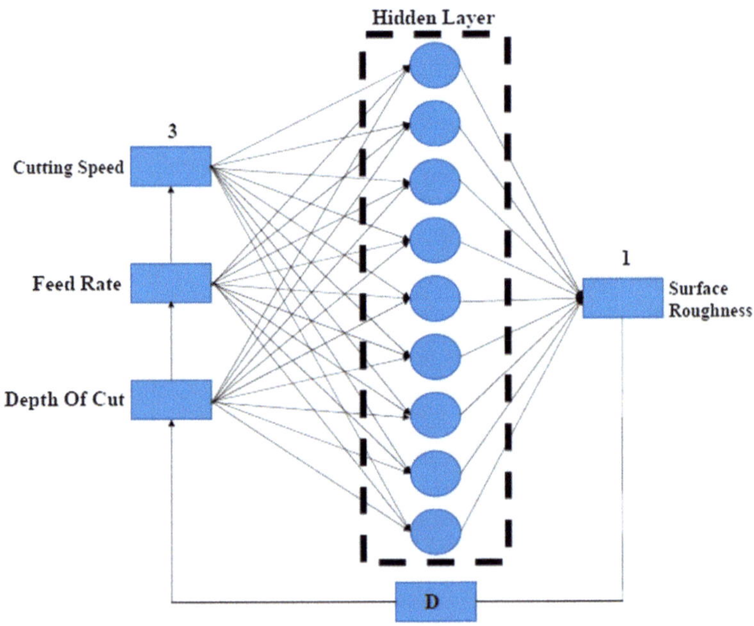

Fig. 6 Selected neural network architecture

Table 2 Correlation coefficients of validation datasets of the designed ANN

Cutting speed (m/min)	Pure (unfilled) PTFE	25% Carbon-filled PTFE	25% Glass fiber-filled PTFE
150	0.97	0.98	0.96
200	0.95	0.94	0.99

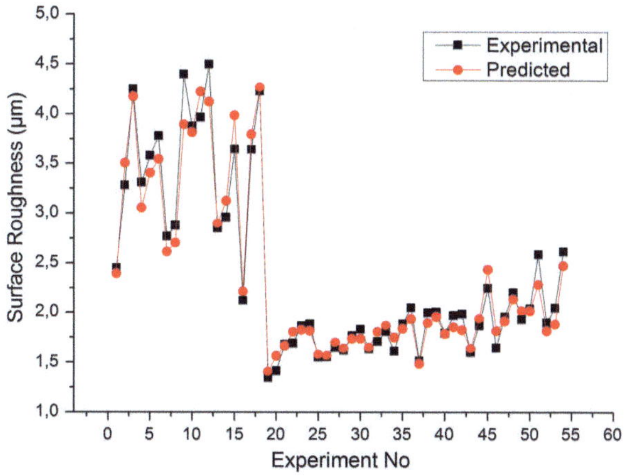

Fig. 7 Comparison of ANN results with experimental values

The predicted experimental data results are shown in Table 1. The average relative error between the experimental and predicted values was 4.66%. The average relative error between the experimental and predicted values together with the correlation coefficient of validation data (R-value) between the outputs showed that the well-trained network exhibited reliable accuracy in predicting the surface roughness values. Figure 7 shows the comparison of the ANN results with the experimental values. It can be seen that the neural network prediction results are very close to the experimental values.

4 Conclusion

An ANN-based prediction model was developed to determine the optimum cutting parameters (cutting speed, feed rate, and depth of cut) in terms of the surface roughness in the turning of 25% carbon-filled and 25% glass fiber-filled PTFE. In this study, in order to determine its efficiency, the performance of the ANN model was compared with the experimental results.

Differences between the average surface roughness values were observed after applying manufacturing parameters on pure (unfilled), 25% carbon-filled and 25% glass fiber-filled PTFE samples. It is generally recognized that feed rate is an important factor in the turning process of PTFE materials, and low cutting speed is believed to produce a better surface finish. Moreover, the effect of depth of cut on the surface roughness does not exhibit regular behavior. The lowest Ra value (1.35 μm) was obtained on the carbon-filled PTFE in turning at a cutting speed of 150 m/min, a feed rate of 0.1 mm/rev and a depth of cut of 1 mm, while the highest Ra value (4.4 μm) was observed on the pure PTFE in turning at a cutting speed of 150 m/min, a feed rate of 0.3 mm/rev and a depth of cut of 3 mm. Consequently, in the experiments, a better surface finish was obtained with the PTFE containing filler materials than with the unfilled PTFE.

When the ANN results were compared with the experimental values, the predictions of the neural network model were found to be accurate and reliable, with results very close to the experimental values. Thus, the proposed model can be used for prediction of the surface roughness in turning operations, with a promising potential for use in many other applications. The NARX model and Levenberg Marquardt (LM) algorithm were shown to be accurate for the optimization of the machining parameters.

References

1. Fetecau, C., Stan, F.: Study of cutting force and surface roughness in the turning of polytetrafluoroethylene composites with a polycrystalline diamond tool. Measurement **45**, 1367–1379 (2012)
2. Suresh, R., Basavarajappa, S., Gaitonde, V.N., Samuel, G.L.: Machinability investigations on hardened AISI 4340 steel using coated carbide insert. Int. J. Refract. Metal H. **33**, 75–86 (2012)
3. El-Gallab, M., Sklad, M.: Machining of Al/SiC particulate metal-matrix composites Part I: tool performance. J. Mater. Process. Technol. **83**, 151–158 (1998)
4. Kumar, R., Chauhan, S.: Study on surface roughness measurement for turning of Al 7075/10/SiCp and Al 7075 hybrid composites by using response surface methodology (RSM) and artificial neural networking (ANN). Measurement **65**, 166–180 (2015)
5. Sangwan, K.S., Saxena, S., Kant, G.: Optimization of machining parameters to minimize surface roughness using integrated ANN-GA approach. Proc. CIRP **29**, 305–310 (2015)
6. Akıncı, A., Akbulut, H., Yılmaz, F.: Floropolimer (Teflon) kaplamaların yapı ve özellikleri. UCTEA J. Chamb. Metall. Mater. Eng. **133**, 53–59 (2003)
7. Rooyen, L.J.V., Bissett, H., Khoathane, M.C., Kocsis, J.K.: J. Appl. Polym. Sci. (2016). doi:10.1002/app.43369
8. Li, F., Hu, K., Li, J., Zhao, B.: The friction and wear characteristics of nanometer ZnO filled polytetrafluoroethylene. Wear **249**, 877–882 (2001)
9. DuPont-Fluoroproducts: Teflon PTFE properties handbook: Tech. Rep. H-37051-3 (1996)
10. Karayel, D.: Prediction and control of surface roughness in CNC lathe using artificial neural network. J. Mater. Process. Technol. **209**, 3125–3137 (2009)
11. Jeyakumar, S., Marimuthu, K., Ramachandran, T.: Optimization of machining parameters of Al 6061 composite to minimize the surface roughness-modelling using RSM and ANN. Indian J. Eng. Mater. Sci. **22**, 29–37 (2015)
12. Leontaritis, I.J., Billings, S.A.: Input-output parametric models for nonlinear systems. Int. J. Control **41**, 303–344 (1985)

13. Ljung, L.: System Identification Theory for the User, 2nd edn. Prentice-Hall, Englewood Cliffs, New Jersey (1999)
14. Norgaard, M., Ravn, O., Poulsen, N.K., Hansen, L.K.: Neural Networks for Modelling and Control of Dynamic Systems. Springer, Berlin (2000)
15. Fausett, L.: Fundamentals of Neural Networks Architectures, Algorithms and Application. Prentice –Hall, New York (1994)

Investigation and Application of Fe–Co–Cu Based Diamond Cutting Tools with Different Bronze Content Used in Marble Production

Berrak Bulut, Onur Tazegul, Murat Baydogan and Eyup Sabri Kayali

1 Introduction

Marble is a metamorphic rock formed by the alteration of limestone or dolomite rock by heat and pressure. Marble is occurred more than 90% calcite [1]. Marble is a relatively soft rock, having a hardness degree of 3–4 in Mohs hardness scale and it is cut in the form of blocks and polished at the production process [1, 2]. Diamond Cutting Tools (DCTs) are a fast and modern cutting technique especially the diamond cutting wire method has become indispensable in today's marble production [3]. The diamond cutting wires consist of aligned beads on steel rope. The beads spacing distance is fixed and it is provided with springs or plastic parts [4, 5]. The beads are usually being produced by mixing diamond grit with the appropriate metal(s)/alloys powders via powder metallurgy processes [6, 7]. Processing of diamond in metal bonds often results in a reaction between the diamond surface and the surrounding metal matrix. The extent of this reaction depends on specific composition of metal powders, their particle size and distribution, processing temperature and time [8].

B. Bulut (✉) · O. Tazegul · M. Baydogan · E.S. Kayali
Faculty of Chemical and Metallurgy Engineering, Department of Metallurgical and Materials Engineering, Istanbul Technical University, Maslak, 34469 Istanbul, Turkey
e-mail: bebulut@itu.edu.tr

O. Tazegul
e-mail: tazegul@itu.edu.tr

M. Baydogan
e-mail: baydogan@itu.edu.tr

E.S. Kayali
e-mail: kayali@itu.edu.tr

L.F.M. da Silva (ed.), *Materials Design and Applications*,
Advanced Structured Materials 65, DOI 10.1007/978-3-319-50784-2_23

The main factor determining the performance of the tool during service is its wear resistance. A number of operating conditions such as feeding rate, depth of cut, peripheral speed, load, pressure, velocity, cutting mode, rock properties, working conditions etc. govern the wear rate of DCTs [9, 10]. Natural stones react differently and show different abrasiveness during cutting processes [11]. For this reason, it is very important to choose the right matrix material in the manufacturing process of DCTs. Depending on the properties of natural stone such as hardness and abrasivity, the metal matrix may contain some metals or alloys such as iron, cobalt, cobalt and tin, cobalt and bronze and tungsten carbide [12]. For a successful cutting operation, wear rates of the matrix and the diamond must be comparable. Otherwise, detachment of diamond from the matrix or embedding of diamond into the matrix may occur. In order to optimize the performance of the tool during cutting, type and amount of the additions are firstly determined by the hardness of natural stone to be cut [13, 14].

There are number of studies that have investigated Fe–Co–Cu–Sn based diamond cutting tools. Zeren et al. [15] studied sintering of polycrystalline diamond cutting tools. In that study, pressure-sintering experiments during direct sintering of diamond cutting tools were conducted at 730 °C for 3–15 min with a pressure of about 350 MPa. For the matrix, cobalt and nickel are used as the bonding agents and the bronze was used as the filling phase in various amounts. They concluded that the matrix firmly supported the diamonds preventing them from damaging during cutting and the bonding between the matrix and diamond must be strong enough for high performance. Aslantas et al. [16] investigated the performance of DCTs which are used in the marble cutting process. At this study, the computer controlled marble cutting machine was used and the 3-dimensional reactive force (F_x, F_y, F_z) occurring in the course of cutting were measured. They indicated that the most important factors causing lateral deviation of the DCTs are depth of cut and traverse speed. According the results, at smaller depths of cut the part of the DCTs that is inside the marble and the part that is outside were exposed to elastic deformation in different directions. Also at low peripheral speed, it was observed that the DCTs show sudden changes and was slogged at cutting. Rosa et al. [12] studied diamond-impregnated segmented discs used for hard stone cutting. Different types of matrices were used for manufacturing new segments for the cutting discs. Matrices containing Co + Bronze (50 wt%Co + 50 wt% of a 90Cu–10Sn bronze); Co + bronze + 10 wt% WC; and Co + bronze + 20 wt% WC, were produced and their behaviors were compared to that of a standard disc available in the market and commonly used by the stone industry for cutting granite. The commercial tool (Co + Bronze) showed the lowest values of resultant force and the highest values of tool consumption or tool wear rate.

The aim of this study is to evaluate the performance of the Fe–Co–Cu based matrix as a diamond bead and the effect of bronze additions on structural, physical and mechanical properties of the matrix.

2 Experimental Procedure

Commercial free sintering metal powder (W1) with a particle size of 13.62 μm was used as matrix material. 30% and 50% vol Bronze (90% Cu + 10% Sn) were added to W1 powder, which were designated as W1-30 and W1-50, respectively. Chemical composition of W1 powder is given in Table 1. The metal powders were mixed in 360° rotating chamber for 3 h. The powder mixtures were compacted in a uniaxial cold press by the application of 50 kN force. Green samples were prepared in 7 mm in diameter and 14 mm in length. Conventional Sintering was performed in a continuous tunnel furnace at 960 °C under hydrogen atmosphere for 1 h.

X-ray diffraction (XRD) analysis was used for quantitative identification of the phases by a Bruker D8-Advanced X-ray diffractometer using Cu Kα radiation. Microstructures of the samples were examined by a HITACHI TM-1000 Tabletop Scanning Electron Microscope (SEM) equipped with an Energy Dispersive Spectroscopy (EDS) unit. Density measurements, hardness and compression tests were performed for physical and mechanical characterization of the samples. Density of the samples was determined by the Archimedes method and divided by the theoretical density to calculate the relative density. Hardness measurements were performed in the Rockwell B scale by using Zwick/Roell ZHR hardness tester and the compression tests were carried out using Dartec universal testing machine at a crosshead speed of 1 mm/min. Abrasion wear tests were conducted in a DVT DA6 model abrasion tester. The samples were vertically installed in the rotary table of the tester which is covered with 80 grade sandpaper. The table was rotated with 40 rpm and 1 kg vertical load was applied on the processed composites. The composites were moved on the table for 20 m and samples were weighted before and after the wear test. Relative wear resistance was calculated as the ratio the maximum weight loss over the sample's weight loss. According to this definition, the relative wear resistance of the maximum weight loss specimen is 1.

When the examination of the pure and bronze doped matrixes has been completed, the beads of a cutting wire sintered at 960 °C via tunnel furnace. W1, W1-30 and W1-50 powders were used in the production of beads. The production of diamond cutting wires have been prepared in the form of a bead and a spring aligned respectively on steel rope as seen in Fig. 1.

Table 1 Chemical composition of W1 powder

	Chemical composition (wt%)					
	Fe	Co	Cu	Sn	Ni	W
W1	40	30	25	3.9	0.5	0.6

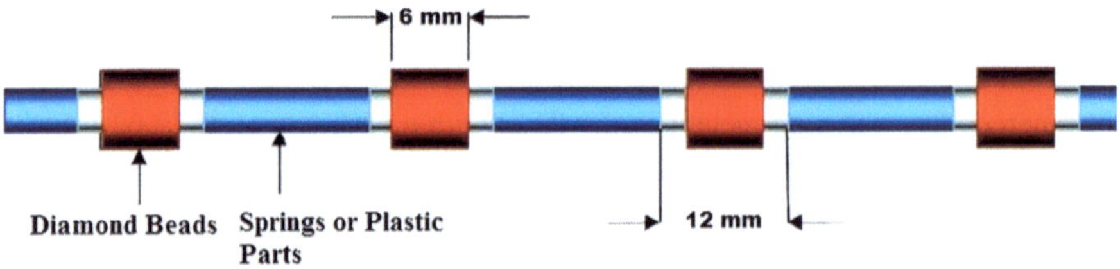

Fig. 1 Components of the marble cutting spring wire

3 Results and Discussion

Figure 2 shows XRD patterns of the samples investigated. Identified phases for W1 sample which were sintered at 960 °C for 1 h under hydrogen atmosphere are mainly Co_3Fe_7 and CuSn. By the addition of 30 and 50% of bronze into the matrix powder, the identified phases were essentially the same (Co_3Fe_7 and CuSn) except some variations in the relative intensities amongst the samples.

Figure 3 shows SEM micrographs of the samples investigated (Fig. 3a–c) and the corresponding EDS patterns of the samples (Fig. 3d–f). EDS analysis revealed that the microstructure of sample W1 mainly consists of Fe and Co coming from the Fe-Co intermetallic phase as shown in Fig. 2. Also, the fraction of the bright areas increased with increasing bronze addition to the matrix (Fig. 3a–c) and they are homogeneously distributed in the microstructure. According to the EDS analysis, the bright areas in the SEM micrographs of the samples corresponds copper- rich phases (Fig. 3d–f).

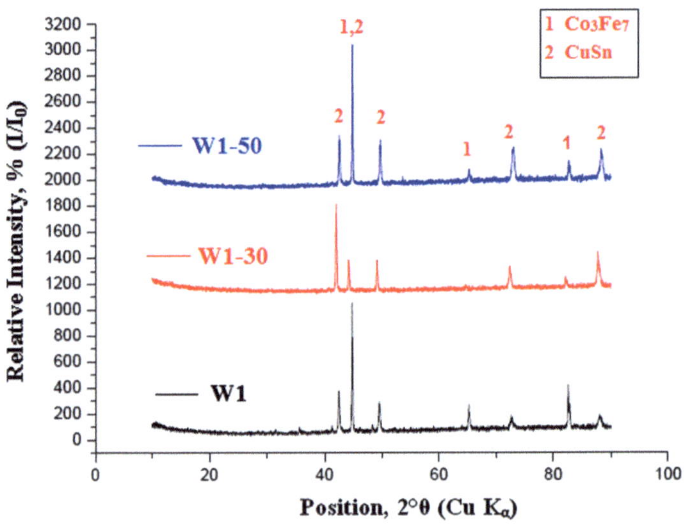

Fig. 2 XRD patterns of the samples

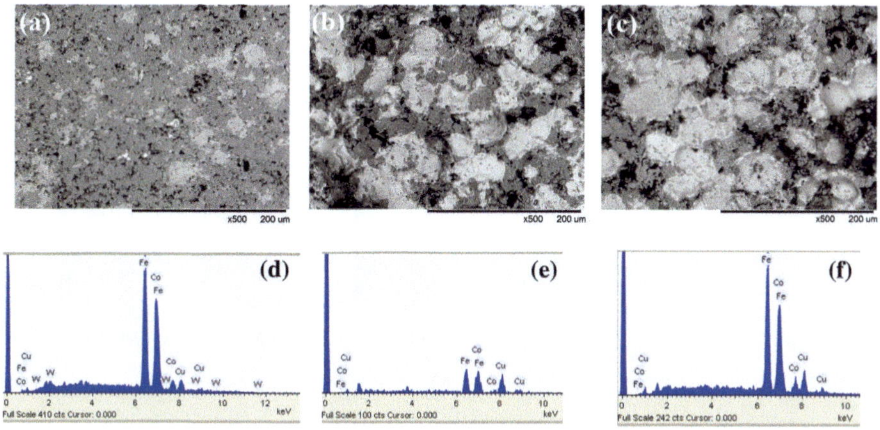

Fig. 3 SEM microstructures of **a** W1, **b** W1-30, **c** W1-50 and corresponding EDS analysis of the sintered samples **d** W1, **e** W1-30, **f** W1-50

Table 2 Relative density, hardness and compressive strength of the samples after sintering

Samples	Relative density (%)	Hardness (HRB)	Compressive strength (MPa)
W1	98.43	97 ± 0.6	1293 ± 3
W1-30	97.39	86 ± 1.2	1024 ± 3
W1-50	97.30	80 ± 1.7	976 ± 3

Table 2 lists relative density, hardness and compression strength of the samples investigated. The highest relative density was obtained in W1 samples. With the increasing amount of bronze addition into the powder composition, relative density slightly decreased. Beside density, hardness and compression strength were also decreased. The highest density, hardness and compression strength were obtained at W1 sample. Increasing amount of bronze addition led to more significant decrement in hardness and compression strength than that observed in the relative density. This result is in contrast to the previous study that investigated the diamond cutting tools with various bronze additions [17]. In that work, it was reported that bronze addition resulted in a decrement in mechanical properties due to accumulation of bronzes in certain regions. However, in the present work, bronze islands do not show any accumulation, but are uniformly distributed in the matrix, as shown in Fig. 3b, c. Finally it is reasonable to assume that uniformly distributed bronze islands as a softer component in the microstructure decreased hardness and compression strength of bronze containing samples.

The abrasion wear tests were performed with all the samples and the test results as relative wear resistances were given in Fig. 4. With increasing bronze content,

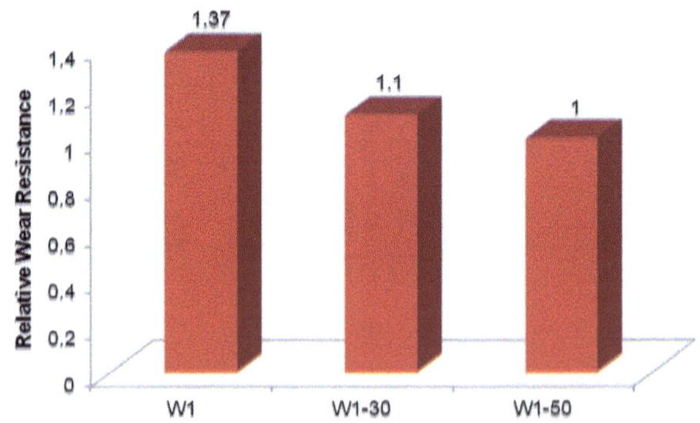

Fig. 4 The wear resistance of the all samples

the wear resistance of the samples were decreased. The relative wear resistance of the W1 sample is highest about 37% than that of W1-50 sample. It is known that optimization of wear rates of diamond and the matrix in a diamond containing cutting tool is necessary for a better cutting performance. Therefore it is expected that a softer matrix of W1-50 sample may result in better cutting performance due to reduced abrasion resistance.

Figure 5 shows an overview of the diamond beads and cutting wire processed in this work. In order to evaluate the cutting performance of diamond beads produced by using the matrix materials studied in this work, the cutting wires produced with these diamond beads are now under investigation by cutting marble in on-site field cutting operations. It is known that optimization of wear rates of diamond and the matrix in a diamond containing cutting tool is necessary for a better cutting performance, it is expected that a softer matrix produced by increased amount of bronze addition to W1 sample may result in better cutting performance due to reduced abrasion resistance of the matrix.

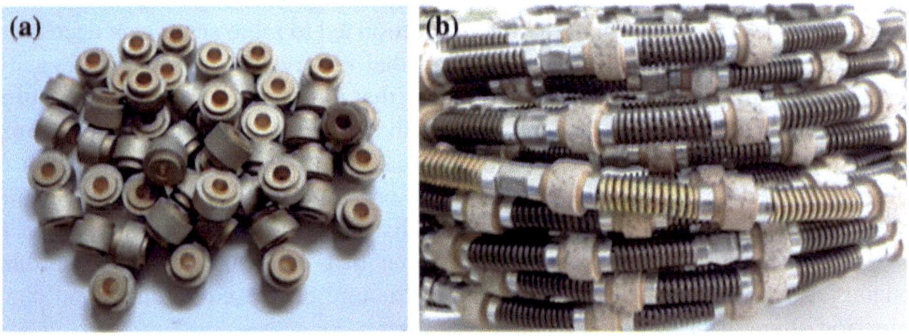

Fig. 5 **a** Diamond beads and **b** the cutting wire processed with W1-50 powders in this study

4 Conclusion

The Fe–Co–Cu based metal matrix was used to evaluate their performance as a diamond cutting tools used in marble production with the additions of two different quantities (30–50 vol. %) bronze. Microstructural and mechanical characterizations were performed by SEM, EDS, XRD, XRF, hardness, compression tests and wear test. Following results can be drawn from the present study:

1. The highest relative density, the highest compressive strength and the highest hardness values are obtained in the sample W1 with no bronze addition in its composition.
2. The additions of bronze to W1 sample decreases the theoretical density, hardness, compressive strength and wear resistance. Decrement in mechanical properties is higher than that in the theoretical density.
3. The presence of Co_3Fe_7 intermetallic phase gives rise to mechanical properties of W1 sample.
4. The cutting performance of the diamond beads produced by using the matrix materials studied are now under investigation by cutting marble in an on-site field cutting operations.

References

1. Çelik, M.Y., Sabah, E.: Geological and technical characterisation of Iscehisar (Afyon-Turkey) marble deposits and the impact of marble waste on environmental pollution. J. Environ. Manage. **87**, 106–116 (2008)
2. Hamza, R.A., El-Haggar, S., Khedr, S.: Marble and granite waste: characterization and utilization in concrete bricks. Int. J. Biosci. Biochem. Bioinform. **4**, 286–291 (2011)
3. Jain, S.C., Rathore, S.S.: Role of cut size area on the performance of diamond wire saw machine in quarrying of marble. Int. J. Min. Reclam. Environ. **23**, 79–91 (2009)
4. Hawkins, A.C., Antenen, A.P., Johnson, G.: The diamond wire saw in quarrying granite and marble. Dimens. Stone **8**, 44–50 (1990)
5. Tönshoff, H.K., Hilmann-Apmann, H.: Diamond tools for wire sawing metal components. Diam. Relat. Mater. **11**, 742–748 (2002)
6. Konstanty, J.: Powder Metallurgy Diamond Tools. Elsevier Science Technology, United Kingdom (2006)
7. Skury, A.D., Bobrovnitchii, G.S., Azevedo, M.G., Monteiro, S.N.: Obtention and characterization of the Sn-Ni-Fe-Cu alloys applied in diamond saw blade production. Mater. Sci. Forum **727–728**, 305–309 (2012)
8. Nitkiewicz, Z., Świerzy, M.: Tin influence on diamond–metal matrix hot pressed tools for stone cutting. J. Mater. Process. Technol. **175**, 306–315 (2006)
9. Luo, S.Y., Liao, Y.S.: Study of the behavior of diamond saw-blades in stone processing. J. Mater. Process. Technol. **51**, 296–308 (1995)
10. Ersoy, A., Buyuksagic, S., Atici, U.: Wear characteristics of circular diamond saws in the cutting of different hard abrasive rocks. Wear **258**, 1422–1436 (2005)
11. Tillmann, W., Kronholz, C., Ferreira, M., Knote, A., Theisen, W., Schütte, P.: Diamond-metal matrix interaction in tools fabricated by conventional and current-induced sintering. Int. J. Powder Metall. **47**, 29–36 (2011)

12. Rosa, L.G., Fernandes, J.C., Anjinho, C.A., Coelho, A., Amaral, P.M.: Long-term performance of stone-cutting tools. Int. J. Refract Metal Hard Mater. **49**, 276–282 (2015)
13. Oliveira, H.C.P., Cabral, S.C., GuimarÃ£es, R.S., Bobrovnitchii, G.S., Filgueira, M.: Processing and characterization of a cobalt based alloy for use in diamond cutting tools. Materwiss Werksttech. **40**, 907–909 (2009)
14. Kagnaya, T., Boher, C., Lambert, L., Lazard, M., Cutard, T.: Wear mechanisms of WC–Co cutting tools from high-speed tribological tests. Wear **267**, 890–897 (2009)
15. Zeren, M., Karagoz, S.: Sintering of polycrystalline diamond cutting tools. Mater. Des. **28**, 1055–1058 (2007)
16. Aslantas, K., Ozbek, O., Ucun, I., Buyuksagis, I.S.: Investigation of the effect of axial cutting force on circular diamond sawblade used in marble cutting process. Mater. Manuf. Process. **24**, 1423–1430 (2009)
17. Kahraman, Y.: Free sintering of matrix materials which are used for diamond cutting tools. M. Sc. Thesis. Istanbul Technical University. Istanbul, Turkey (2014)

Investigation of Surface Roughness and Tool Wear in End Milling of Al7075-SiC Co-continuous Composite

A.S. Prasanth and R. Ramesh

1 Introduction

A metal-ceramic composite in which both the phases are continuous and interpenetrating is termed as a co-continuous composite [1]. These composites exhibit superior mechanical properties and thermal shock resistance owing to their uniform three dimensionally interconnected microstructure [2, 3]. Further, the addition of a metal (ductile) phase to the ceramic, leads to a composite with enhanced toughness [4]. By varying the volume fraction of each phase, the desired hardness, strength and fracture toughness can be achieved [5]. Due to these advantages, such composites have been widely studied and applied in defence sector [6], aerospace [7] and automotive industries [8, 9]. Co-continuous composites were first developed using the reactive metal penetration (RMP) technique by Breslin [10]. Vecchia et al. [11] performed RMP of silica glass with 99.9% pure aluminium and Al–Si (10%) alloy as infiltrating alloys. It was found that the composites thus fabricated, had very low thermal expansion, high stiffness, improved toughness and impact strength, appreciable bending and compression strength compared with traditional ceramics. Co-continuous ceramic composites (C4) were manufactured by infiltration of silica glass preform with 1050 A aluminium alloy by Pavese et al. [1]. The fatigue limit was estimated to be 91 MPa at 10^7 cycles. Such composites were

A.S. Prasanth (✉)
Department of Mechanical Engineering, PSG College of Technology,
Coimbatore 641004, Tamil Nadu, India
e-mail: asp@mec.psgtech.ac.in

R. Ramesh
Department of Mechanical Engineering, PSG Institute of Technology
and Applied Research, Coimbatore 641062, Tamil Nadu, India
e-mail: ramesh@psgitech.ac.in

L.F.M. da Silva (ed.), *Materials Design and Applications*,
Advanced Structured Materials 65, DOI 10.1007/978-3-319-50784-2_24

suitable for applications involving high cycle fatigue. Banerjee and Roy [12] studied the displacement reaction between Al and SiO_2 which yielded an interpenetrating composite. The study reported that the composite possessed enhanced density and compressive strength. It was also amenable for near net-shape fabrication. Evarts [13] studied advanced processing techniques for co-continuous ceramic composites. The study reported that addition of copper to liquid aluminium bath reduced the characteristic microstructure length of the composite. This led to an increase in strength, toughness and hardness of the composite. Daehn and Breslin [9] attempted to create co-continuous composites for friction and braking applications. The study showed that the presence of SiC improved thermal conductivity of the composite. Since they were light weight with excellent frictional properties, these composites were proposed to be next-generation rotor materials. Clarke [14] elaborated on the various processing techniques to fabricate interpenetrating composite structures. Though numerous studies on processing, fabrication techniques and properties of C4 are present in literature, limited studies are available on machinability of co-continuous composites. Roy et al. [15] studied the factors influencing the behaviour of Al_2O_3/Al co-continuous composites during wirecut electrical discharge machining (WEDM). It was found that the EDM method created a heat-affected surface layer. This led to the formation of pores and cracks in the composite. Goswami et al. [16] evaluated MRR, cutting forces and surface finish during electrochemical grinding (ECG) of an Al_2O_3/Al interpenetrating phase composite through Taguchi-design-based experimental studies. It was reported that applied voltage, concentration and depth of cut were the three important process parameters to obtain the best surface finish. Despite their low cost and appealing properties, challenges in fabrication and machining of interpenetrating composites have limited their usage in industrial applications [16]. In particular, very few studies are available on end milling of Al7075-SiC co-continuous metal-ceramic composite. In the present research, such a composite was at first, manufactured by gravity infiltration method. Secondly, the effect of three end milling parameters namely, speed, feed and depth of cut on tool wear and surface roughness of the composite was studied. The final objective was to solve this multi-response optimization problem using Grey Relational Analysis (GRA).

2 Experiments

2.1 Selection of Metal and Ceramic Phase

A three-dimensionally porous SiC foam with 75% purity as shown in Fig. 1, was selected as the ceramic phase. The foam is a commercially graded ceramic filter used in foundries for entrapping harmful slags, sand and dross inclusions. The pore size of the foam was 10 ppi, with a porosity of 80%.

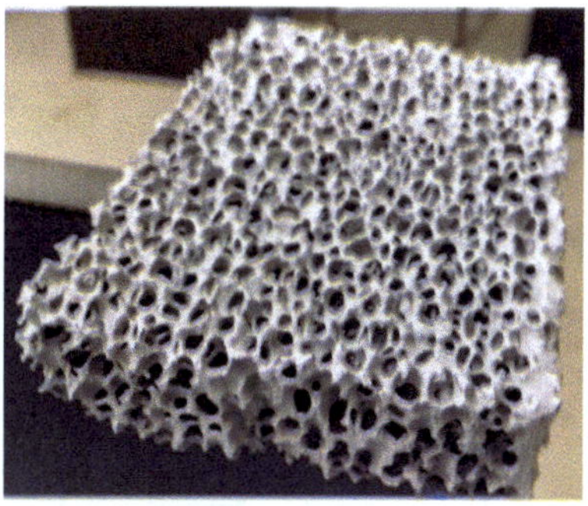

Fig. 1 SiC foam

Table 1 Spectroscopic analysis of Al7075

Metal	Al	Zn	Mg	Cu	Cr	Fe
% wt.	89.57	6.728	1.868	1.37	0.194	0.122

The base metal alloy used in the study was Al7075. Among aluminium wrought alloys, the 7-series alloys exhibit the highest strength. Hence, they are typically used in defence applications [17]. Spectroscopic analysis was performed to verify the chemical composition of the commercial grade Al7075 alloy. As can be seen in Table 1, the major constituent of this alloy is zinc, followed by magnesium. The analysis substantiated that the commercially acquired sample confirmed to ASM alloy specifications.

2.2 Manufacturing of the Composite

The composite was fabricated by gravity infiltration technique. Here, the molten metal alloy was infiltrated into the ceramic foam in a resistance furnace having a maximum temperature of 1100 °C and maximum power of 4500 W. The schematic of the experimental setup is depicted in Fig. 2.

The SiC foam was cut into a sample of size 50 × 25 × 25 mm. It was pre-heated in a graphite crucible to a temperature of 800 °C. Simultaneously, the Al alloy was melted in another crucible. The temperature was set to 150 °C in excess of its melting point in order to ensure better fluidity of the alloy. After 1.5 h, the molten Al 7075 is poured into the crucible containing preheated SiC. To arrest the possibility of the foam floating on to the surface due to buoyancy, a carbon rod was used to hold it in place. The setup was maintained at 800 °C for a period of one hour to allow the molten metal to infiltrate the pores of the SiC foam. The crucible was then taken out and cooled for one hour at room temperature. The co-continuous composite thus fabricated, was machined out with utmost care by a sequence of operations namely, shaping, milling, and grinding.

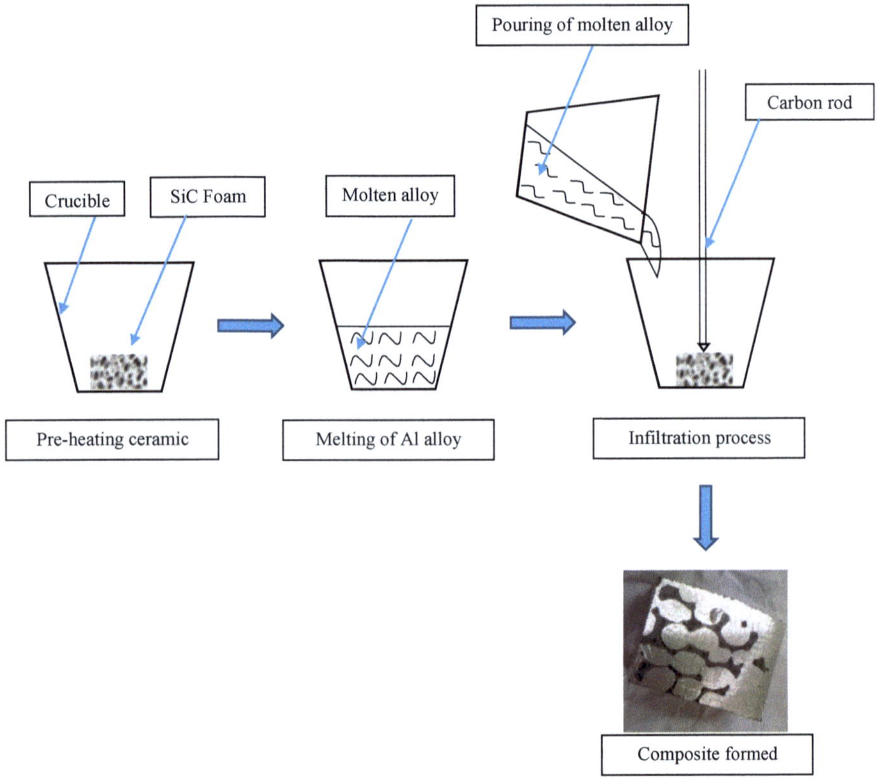

Fig. 2 Manufacturing process of co-continuous composite

3 End Milling of Al7075-SiC Composite

End milling was performed using a 5 mm solid carbide tool in uncoated condition as shown in Fig. 3. Generally, PCD tools are used in finish machining operations. However, the high cost of PCD tools increases the costs of machining composites. Hence, it is necessary to carry out machinability studies using carbide tools [18].

Slots were machined in the composite using a Makino VMC-S33 machine shown in Fig. 4. For aluminium alloys, the maximum rated speed and feed rate of the tool was 8960 rpm and 995 mm/min respectively. Speed, feed and depth of cut were the process parameters considered to investigate tool wear and surface roughness of the composite. The levels of the factors were fixed by conducting

Fig. 3 Solid carbide end mill cutter

Fig. 4 End milling of composite

Table 2 Parameters and their levels

Parameter	Unit	Level 1	Level 2	Level 3
Cutting speed	rpm	3000	5000	7000
Feed rate	mm/min	450	600	750
Depth of cut	mm	0.2	0.4	0.6

trial experiments and also, based on the rated speeds and feeds suggested by the tool manufacturer for aluminium alloys. Table 2 depicts the various levels of the parameters.

4 Taguchi Design

A Taguchi L9 orthogonal array [16] was chosen to analyse the results of machining. Two replications were carried out for each of the nine process designs and the averages were tabulated. The Taguchi array and response values for the experimental runs are shown in Table 3. The average surface roughness (R_a) in microns was measured using a KosakaLab Surfcorder and tool wear (wear-land wear) in mm was evaluated using DinoCapture software. The volume of material machined, material removal rate (MRR in mm^3/min) was also observed and tabulated in Table 3.

Table 3 L9 Orthogonal array for end milling of Al7075-SiC co-continuous composite

Exp. no.	Process parameter			Average response values		
	Speed (rpm)	Feed (mm/min)	Depth of cut (mm)	Surface roughness (μm)	Tool wear (mm)	MRR (mm^3/min)
1	1	1	1	2.95	0.027	45
2	1	2	2	2.87	0.026	1200
3	1	3	3	2.75	0.024	2250
4	2	1	2	1.92	0.023	900
5	2	2	3	2.65	0.022	1800
6	2	3	1	1.18	0.022	750
7	3	1	3	1.90	0.021	1350
8	3	2	1	1.85	0.020	600
9	3	3	2	1.08	0.018	1500

4.1 Signal to Noise (S/N) Ratio

Table 4 lists the signal to noise (S/N) ratios obtained by the analysis of Taguchi design shown in Table 3. In order to diminish the effect of noise factors, higher values of S/N ratios are preferred in experiments. In machining, lower surface roughness and tool wear are desired. Hence, smaller-is-better characteristic is chosen for calculating the S/N ratio. The parameters with the largest value of delta are considered to be most significant.

From Table 4, it can be inferred that, for lesser tool wear and surface roughness during end milling of the composite, cutting speed is the most significant parameter; followed by feed rate and depth of cut. The main effects plots depicted in Fig. 5 and Fig. 6 supplement the outcome that cutting speed is the significant parameter.

4.2 General Linear Model (GLM) ANOVA

Using the adjusted sum of squares technique, GLM ANOVA was computed to determine the percentage contribution of each of the end milling parameters. ANOVA for surface roughness and tool wear is shown in Tables 5 and 6.

Table 4 Response table for S/N Ratios of composite

Level	Surface roughness			Tool wear		
	Speed	Feed	Depth of cut	Speed	Feed	Depth of cut
1	−9.114	−6.879	−5.393	31.82	32.56	32.83
2	−5.190	−7.655	−5.164	33.02	32.94	33.12
3	−3.862	−3.631	−7.609	34.14	33.48	33.03
Delta	5.251	4.024	2.445	2.32	0.92	0.29
Rank	1	2	3	1	2	3

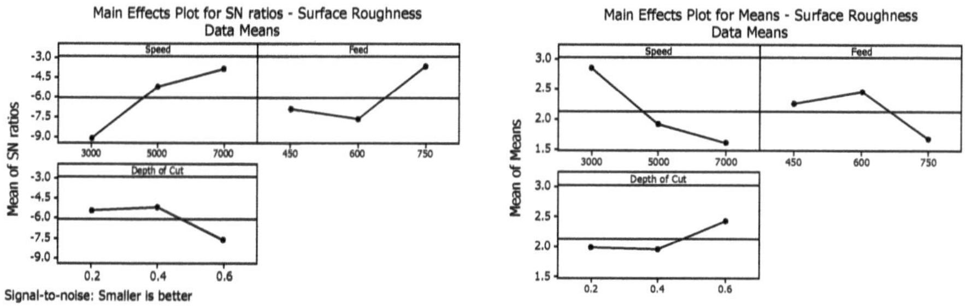

Fig. 5 Main effects plot for surface roughness of composite

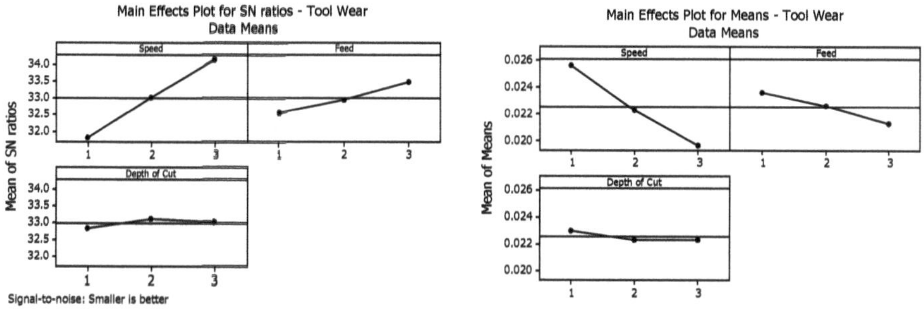

Fig. 6 Main effects plot for tool wear in composite

Table 5 Analysis of variance for surface roughness in composite

Source	DOF	Sum of squares	Adjusted sum of squares	Adjusted mean square	F-value	P-value
Speed	2	2.53182	2.53182	1.26591	25.79	0.037
Feed	2	1.00302	1.00302	0.50151	10.22	0.089
Depth of cut	2	0.42216	0.42216	0.21108	4.30	0.189
Error	2	0.09816	0.09816	0.04908		

Table 6 Analysis of variance for tool wear in composite

Source	DOF	Sum of squares	Adjusted sum of squares	Adjusted mean square	F-value	P-value
Speed	2	0.0000542	0.0000542	0.0000271	61.00	0.016
Feed	2	0.0000082	0.0000082	0.0000041	9.25	0.098
Depth of cut	2	0.0000009	0.0000009	0.0000004	1.00	0.500
Error	2	0.0000009	0.0000009	0.0000004		

In case of surface roughness, the R-square value obtained was 97.58%. Cutting speed was of high significance with highest F-value of 25.79 as can be seen in Table 5. Similarly, for tool wear, the R-square value obtained was 98.62% with cutting speed being of highest significance as can be seen in Table 6. Moreover, P-values being less than 0.05 affirmed the S/N ratio results for surface roughness and tool wear indicating cutting speed to be the dominant cutting parameter.

4.3 Percentage Contribution of Factors in Milling of Composite

The percentage contribution of factors towards the variation in responses namely, surface roughness and tool wear were estimated using statistical analysis of means. The expression for finding percentage contribution is given by Eq. (1).

$$\text{Percentage Contribution} = \frac{\text{Sum of squares}}{\text{Total of sum of squares}} \times 100 \tag{1}$$

The total sum of squares and individual sum of squares for surface roughness and tool wear were obtained from Table 5 and Table 6 respectively.

It is evident from Table 7, that cutting speed has a major impact on both surface roughness and tool wear. This confirmed the results from S/N ratio analysis in Table 4.

5 Grey Relational Analysis (GRA)

A part of grey system theory, GRA is performed in cases where information on the model is incomplete. GRA provides an efficient solution to a multiple input, multiple-output problem [19]. The relationship between cutting parameters and machining performance can be found using this technique. The sequence of steps in GRA is as follows.

Step 1: Data Pre-processing: Normalization of the Responses
The responses obtained from the experimental runs need to be normalized to ensure that the scatter range of the sequence is small. The expressions for two of the most common scenarios are shown in Eqs. (2) and (3).

Table 7 Percentage contribution of factors

Parameter	Surface roughness (%)	Tool wear (%)
Speed	62.44	84.43
Feed	24.73	12.77
Depth of cut	10.41	1.4
Error	2.42	1.4

For larger is better;

$$X_{ij} = \frac{Y_{ij} - \min(Y_{ij})}{\max(Y_{ij}) - \min(Y_{ij})} \quad (2)$$

For smaller is better;

$$X_{ij} = \frac{\max(Y_{ij}) - Y_{ij}}{\max(Y_{ij}) - \min(Y_{ij})} \quad (3)$$

where X_{ij} is the normalized response value and Y_{ij} is the response value obtained from experimental runs. During any machining process, larger values are desired for MRR and therefore Eq. (2) is utilized. However, for surface roughness and tool wear, smaller values are preferred. Hence, normalization is performed using Eq. (3). Table 8 shows the normalized results for the composite.

Step 2: Grey Relational Coefficient (ξ)
This value indicates the degree of relation between every sequence in the experimental runs. It is calculated as shown in Eq. (4)

$$\xi_{ij} = \frac{\Delta_{min} + \zeta\Delta_{max}}{\Delta_i + \zeta\Delta_{max}} \quad (4)$$

where 'ξ' is the grey relational coefficient, 'ζ' is the distinguishing coefficient. 'Δ' is treated as a quality loss from the target value; calculated as difference between '1' and normalized value. Table 8 displays the 'ξ' for the composite computed using Eq. (4).

Step 3: Overall Grey Relational Grade
On averaging the grey relational coefficients, the overall grey relational grade 'γ_i' is obtained according to Eq. (5).

Table 8 GRA for composite

Exp. no.	Normalized values (X_{ij})			Grey relational coefficients (ξ)			Grey relational grade (γ_i)	
	Surface roughness	Tool wear	MRR	Surface roughness	Tool wear	MRR	Grey grade	Rank
1	0	0	0	0.333333	0.333333	0.333333	0.333333	9
2	0.042781	0.111111	0.52381	0.343119	0.36	0.512195	0.405105	8
3	0.106952	0.333333	1	0.358925	0.428571	1	0.595832	3
4	0.550802	0.444444	0.387755	0.526761	0.473684	0.449541	0.483329	7
5	0.160428	0.555556	0.795918	0.373253	0.529412	0.710145	0.537603	6
6	0.946524	0.555556	0.319728	0.903382	0.529412	0.423631	0.618808	2
7	0.561497	0.666667	0.591837	0.532764	0.6	0.550562	0.561108	4
8	0.588235	0.777778	0.251701	0.548387	0.692308	0.400545	0.54708	5
9	1	1	0.659864	1	1	0.595142	0.865047	1

Table 9 Response table for GRG of composite

Process parameter	Grey relational grade			Max-Min	Rank
	Level 1	Level 2	Level 3		
Speed (A)	0.44476	0.54658	0.65775[a]	0.21299	2
Feed (B)	0.45926	0.4966	0.69323[a]	0.23397	1
DoC (C)	0.49974	0.58449[a]	0.56485	0.08475	3
Total mean value of the GRG = 0.54969					

[a]Optimum levels

$$\gamma_i = \frac{1}{n}\sum_{k=1}^{n} \xi_{ij} \qquad (5)$$

The grey relational grade (GRG) and corresponding rankings for composite is depicted in Table 8. A higher value of GRG represents better quality of the product.

The factor effects and optimal level of each factor can be determined on the basis of GRG. For each level of the machining parameters of the composite, the mean value is computed and displayed in Table 9. The highest value of GRG corresponds to the optimal levels of the machining parameters. Table 9 shows that the optimum levels are A3, B3 and C2 with feed being the most significant factor as evident from the rank.

In order to quantify the percentage contribution of the end milling parameters, an ANOVA was performed on the GRG values of the composite as shown in Table 10. The main effects plot for the same is displayed in Fig. 7. The percentage contribution shows that feed rate, followed by cutting speed are the major parameters, which significantly affect the responses. The R-squared value was close to 97%.

5.1 *Confirmation Tests*

The final step is to predict the grey relational grade and verify whether the performance characteristic has improved using the optimal level of cutting parameters. The estimated grey relational grade can be computed by the formula in Eq. (6).

Table 10 ANOVA for GRG of composite

Source	DOF	Sum of squares	Adjusted sum of squares	Adjusted mean square	F-value	Percentage contribution
Speed	2	0.068090	0.068090	0.034045	16.77	38.09
Feed	2	0.094802	0.094802	0.047401	23.36	53.04
DoC	2	0.011808	0.011808	0.005904	2.91	6.60
Error	2	0.004059	0.004059	0.002030		2.27

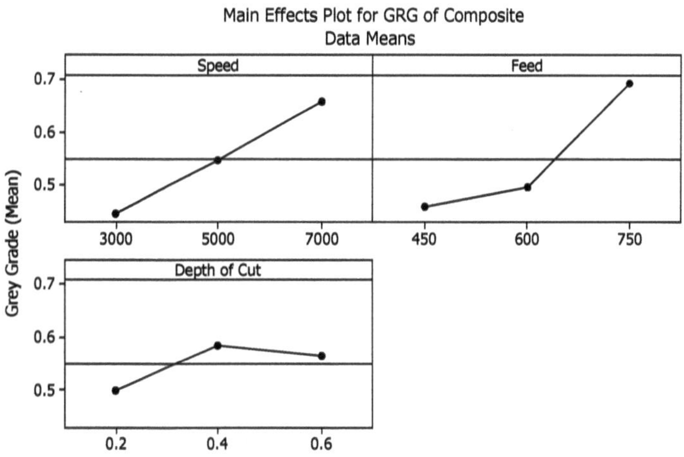

Fig. 7 Main effects plot for GRG

$$\hat{\gamma} = \gamma_m + \sum_{i=1}^{q} \bar{y}_i - \gamma_m \quad (6)$$

The predicted responses for surface roughness and tool wear were estimated by the formula in Eq. (7).

$$\text{Predicted Response} = \text{Avg(A3)} + \text{Avg(B3)} + \text{Avg(C2)} - 2 * (Y_{ij}) \quad (7)$$

Table 11 shows the correlation between predicted values and experimental observations for the optimum settings during end milling of the composite. It can be seen that, the experimental values correspond to the values predicted by GRA. Also, the surface roughness and tool wear after GRA analysis is found to be less when compared with the initial settings. This is characterized by an improvement in the grey relational grade.

Table 11 Improvement in machining performance of co-continuous composite

	Initial machining parameters	Optimal machining parameters	
		Prediction	Experiment
Setting level	A1B1C1	A3B3C2	A3B3C2
Surface roughness	2.95	0.98	1.08
Tool wear	0.027	0.018	0.018
GRG	0.33333	0.83608	0.86505
Improvement in GRG = 0.53172			

6 Conclusions

In this paper, the effect of varying three machining parameters namely speed, feed and depth of cut during end milling of a C4 consisting of Al7075-SiC was presented. Using GRA, from Table 9, it was found that the optimal setting for better surface finish and lesser tool wear was A3, B3 and C2. Furthermore, feed rate with 53.04% and cutting speed with 38.09% were found to be the significant factors contributing to the machining performance. Results from GRA analysis, presented in Table 11, have also shown that in case of the composite, the values of surface roughness have greatly improved by 63% and tool wear by 33% when compared with the initial machining settings. This paper has thus, applied Taguchi design and GRA to improve the multi-response characteristics during end milling of a co-continuous composite.

References

1. Pavese, M., Fino, P., Ugues, D., Badini, C.: High cycle fatigue study of metal-ceramic co-continuous composites. Scr. Mater. **55**, 1135–1138 (2006)
2. Lu, Y., Yang, J., Lu, W., Liu, R., Qiao, G., Bao, C.: The mechanical properties of co-continuous Si_3N_4/Al composites manufactured by squeeze casting. Mater. Sci. Eng. A **527**, 6289–6299 (2010)
3. Wang, L., Fan, Q., Li, G., Zhang, H., Wang, F.: Experimental observation and numerical simulation of SiC_3D/Al interpenetrating phase composite material subjected to a three-point bending load. Comput. Mater. Sci. **95**, 408–413 (2014)
4. Marchi, C.S., Kouzeli, M., Rao, R., Lewis, J.A., Dunand, D.C.: Alumina—aluminum interpenetrating-phase composites with three-dimensional periodic architecture. Scr. Mater. **49**, 861–866 (2003)
5. Li, G., Zhang, X., Fan, Q., Wang, L., Zhang, H., Wang, F., Wang, Y.: Simulation of damage and failure processes of interpenetrating SiC/Al composites subjected to dynamic compressive loading. Acta Mater. **78**, 190–202 (2014)
6. Chang, H., Binner, J., Higginson, R., Myers, P., Webb, P., King, G.: High strain rate characteristics of 3-3 metal-ceramic interpenetrating composites. Mater. Sci. Eng., A **528**, 2239–2245 (2011)
7. Han, J., Hong, C., Zhang, X., Wang, B.: Thermal shock resistance of TiB2-Cu interpenetrating phase composites. Compos. Sci. Technol. **65**, 1711–1718 (2005)
8. Etter, T., Kuebler, J., Frey, T., Schulz, P., Löffler, J.F., Uggowitzer, P.J.: Strength and fracture toughness of interpenetrating graphite/aluminium composites produced by the indirect squeeze casting process. Mater. Sci. Eng. A **386**, 61–67 (2004)
9. Daehn, G.S., Breslin, M.C.: Co-continuous composite materials for friction and braking applications. JOM **58**, 87–91 (2006)
10. Breslin, M.C.: United States Patent 5,214,011. 788, 783–788 (1993)
11. La Vecchia, G.M., Badini, C., Puppo, D., D'Errico, F.: Co-continuous Al/Al_2O_3 composite produced by liquid displacement reaction: relationship between microstructure and mechanical behavior. J. Mater. Sci. **38**, 3567–3577 (2003)
12. Banerjee, S., Roy, S.K.: Net-shape forming of bi-continuous Al_2O_3/Al composite by displacement reaction. Mater. Chem. Phys. **67**, 243–248 (2001)

13. Evarts, J.S.: Advanced processing techniques for co-continuous ceramic composites 102 (2008)
14. Clarke, D.R.: Interpenetrating phase composites. J. Am. Ceram. Soc. **75**, 739–758 (1992)
15. Roy, S.K., Pal, S., Ghosh, S.K., Dey, P.P.: Machining of interpenetrating phase alumina/aluminium composite by electrical discharge technique machining of interpenetrating phase alumina/aluminium. Trans. Indian Ceram. Soc. **67**, 75–80 (2008)
16. Goswami, R.N., Mitra, S., Sarkar, S.: Experimental investigation on electrochemical grinding (ECG) of alumina-aluminum interpenetrating phase composite. Int. J. Adv. Manuf. Technol. **40**, 729–741 (2009)
17. Cobden, R., Alcan, B.: Aluminium: Physical Properties, Characteristics and Alloys (1994)
18. Kök, M.: A study on the machinability of Al_2O_3 particle reinforced aluminium alloy composite. In: 11th International In-organic Bonded Fiber Composites Conference, pp. 272–281 (2008)
19. Kuo, Y., Yang, T., Huang, G.-W.: The use of grey relational analysis in solving multiple attribute decision-making problems. Comput. Ind. Eng. **55**, 80–93 (2008)

Machinability of an Aluminium Cast Alloy Using PCD Tools for Turning

R.B. Soares, A.M.P. de Jesus, R.J.L. Neto, P.A.R. Rosa, M. Machado and A. Reis

1 Introduction

Aluminium–Silicon alloys are the most common cast alloys for the production of automotive components, especially for parts where the alloy is a substitute for cast iron. The high acceptance of these alloys is explained by the necessity of reducing the vehicles weight, without compromising the components strength. Besides that, these alloys present an excellent castability, low coefficient of thermal expansion, excellent wear-resistance and hot tearing resistance [1, 2].

Pure aluminium has adverse machinability characteristics, because the material tends to adhere to the rake and flank faces of the cutting tool, forming the so called built-up edges (BUE) [3]. The BUE formation deteriorates the surface quality of the component, the cutting stability and causes an accelerated tool wear. However,

R.B. Soares (✉) · A.M.P. de Jesus · R.J.L. Neto · M. Machado · A. Reis
INEGI, Faculdade de Engenharia, Universidade do Porto, Rua Dr. Roberto Frias, 400, 4200-465 Porto, Portugal
e-mail: rsoares@inegi.up.pt

A.M.P. de Jesus
e-mail: ajesus@fe.up.pt

R.J.L. Neto
e-mail: rneto@inegi.up.pt

M. Machado
e-mail: mmachado@inegi.up.pt

A. Reis
e-mail: areis@inegi.up.pt

P.A.R. Rosa
IDMEC, Instituto Superior Técnico, Universidade de Lisboa, Av. Rovisco Pais, 1, 1049-001 Lisboa, Portugal
e-mail: pedro.rosa@tecnico.ulisboa.pt

L.F.M. da Silva (ed.), *Materials Design and Applications*,
Advanced Structured Materials 65, DOI 10.1007/978-3-319-50784-2_25

when cutting speed increases or the application of cutting fluid is performed, the amount of adhered material layer can be reduced [3]. Another problem is the formation of long chips which are detrimental to their extraction from the cutting machine, disturb the surface finish of the produced parts and generally reduces the productivity of the machining operation. Although, the cutting forces and specific cutting pressures related to the machining of aluminium alloys are usually much lower, when compared to steel, with about 70% lower values [4].

In particular case of the hypoeutectic AlSi9Cu3 alloy, the microstructure consists of a dendritic α-Al solid solution and Al–Si eutectic, where Si assumes an acicular shape [5]. Due to the relatively high silicon content, around 9%, pronounced wear on cutting tools becomes a more relevant factor in the process. In addition, the presence of various impurities such as Fe and the presence of another alloying elements such Cu, lead to the appearance of complex intermetallic phases Fe-rich and Cu-rich. The eutectic Al_2Cu improve the machinability of the alloy because this constituent acts as a natural chip breaker and improve the quality of the machined surfaces. In other hand, the Fe-rich phases are harmful to both machinability and mechanical properties. The Al_5FeSi which precipitates in the interdendritic and intergranular regions, in the form of long and sharp needles, are strongly detrimental to the alloy mechanical properties including fatigue, but the presence of Mn reduces the formation of this phase and more complex Fe-rich phases can appear. These phases present a more compact morphology, in "chinese-script" or in polyhedral crystals (sludge), hard inclusions and also affects the tool life [6, 7].

These factors associated to the concept of high speed machining applied for the industries in order to increase the productivity, challenge the conventional sintered carbide tools, because the tool wear are excessive. For this reason, the turning of aluminium alloys, particularly the alloys with high content of Silicon, the maximum efficiency can be achieved with the use of polycrystalline diamond tools (PCD) or with the use of diamond coatings, applied to conventional carbide tools, by chemical vapour deposition (CVD). These inserts show low friction coefficient, high thermal conductivity, high-temperature hardness, chemical stability and abrasive wear resistance. These properties explain the high applicability of these tools in high speed machining of aluminium alloys [8, 9].

However, in the modern machining processes developed in automated machines, especially in turning operations, where continuous chips are commonly produced, the effective control of the type (shape and size) of the formed chips is fundamental. The formation of long chips difficult their right extraction from the working area, leads to production equipment stops and consequent loss of productivity, increases the tool wear and contributes to poor surface quality of the components [4, 10].

For the reasons above referred there is an increasing need to develop favourable conditions for the chip breaking without energy of cutting process and tool life factors being affected. One-way to overcome this challenge consists of using chip breakers, which are complex three-dimensional geometries in the tool rake face. The use of chip breakers in PCD tools is not yet a very common practice due to the difficulty of its production, which is an unfavourable point comparatively to Tungsten carbide inserts.

More than 75% of carbide inserts currently used by the industry have chip breaker structures which are produced with a very controllable cost.

During the recent years a great effort has been placed on chip control and chip breaking analysis, analytical models or FEM based software being often used to achieve a design methodology of these chip breakers [10–13]. But the complexity of the cutting phenomena associated to the complexity of grooved tools puts several limitations on predictive models and according to the industrial point of view, the chip control and the design of new complex three-dimensional chip breakers are normally based on try and see methodologies, supported by the strong experience from tool producers.

Another challenging is associated with the manufacture of the chip breaker structure in the rack face in the PCD inserts due to the machining difficulty of this material. This material can be machined by Electrical Discharging Machining (EDM), with reduced free form, or by laser micromachining [14]. Nowadays, with the laser micromachining it is possible to shape smooth and three-dimensional complex chip breakers in PCD inserts, with the needed accuracy and surface roughness.

In this study, an experimental evaluation of the PCD inserts with and without chip breakers was carried out in a specific aluminium alloy, used for automotive parts. Machining tests were preformed in order to verify the performance of the cutting inserts and the influence of the chip breakers in the cutting forces, power consumption, specific cutting pressure, chip shape and size and surface roughness. For this propose, several cutting parameters are varied in external cylindrical turning operations.

2 Experimental Procedure

2.1 Material

The material selected for this study was the cast aluminium alloy specified as AlSi9Cu3, according to the NP EN 1706 standard. The test samples for the machinability turning tests were obtained by permanent mould casting process. This casting process was performed in a coreless induction furnace at 730 °C. Hydrogen degassing was carried out with the use of an appropriated component based on Nitrogen (Nitral C19) and the casting process was performed manually using a cooper permanent mould. This casting procedure aimed the provision of test cylinders of approximately 55 mm in diameter and 180 mm in length.

Table 1 compares the alloy chemical composition according to the NP EN 1706 standard and the composition of the alloy produced for this study, obtained by the spark spectrometry analysis. We realize that the chemical composition of the cast alloy is within the NP EN 1706 standard limits for the AlSi9Cu3 aluminium alloy.

Table 1 Chemical composition of AlSi9Cu3 used in this study (element wt. %)

	Si	Fe	Cu	Mn	Mg	Ni	Pb	Ti	Al
NP EN 1706 standard AlSi9Cu3	8.0–11.0	1.3 max.	2.0–4.0	0.55 max.	0.05–0.55	0.55 max.	0.35 max.	0.25 max.	Remaining
Alloy used in this study	9.88	0.159	3.35	0.012	0.139	0.005	0.018	0.062	

After casting process, a sample for microstructural analysis was cut from a machining test sample. This material sample was polished by a standard metallographic procedure. An Olympus PMG3 optical microscope was used for the metallographic examination across the transversal section of the machining test samples. After the microstructural analysis, the same sample was submitted to a Brinell hardness test, in three distinct zones, across the transversal section of the machining test specimens. For this purpose a hardness machine Emco M4U-075 was used. The hardness test results are represented in the Table 2.

Figures 1 and 2 show the microstructures of the alloy used in the experimental study. By comparison with other analyses in the literature [5, 15] , it can be verified

Table 2 Mechanical properties of AlSi9Cu3 used in this experimental study (average values of various tests)

Young's modulus (GPa)	66
Ultimate tensile strength (MPa)	196
Yield tensile strength (MPa)	91
Tensile elongation at failure (%)	3.2
Ultimate compressive strength (MPa)	524
Yield compressive strength (MPa)	199
Compressive elongation at failure (%)	30
Brinell hardness (HB)	87

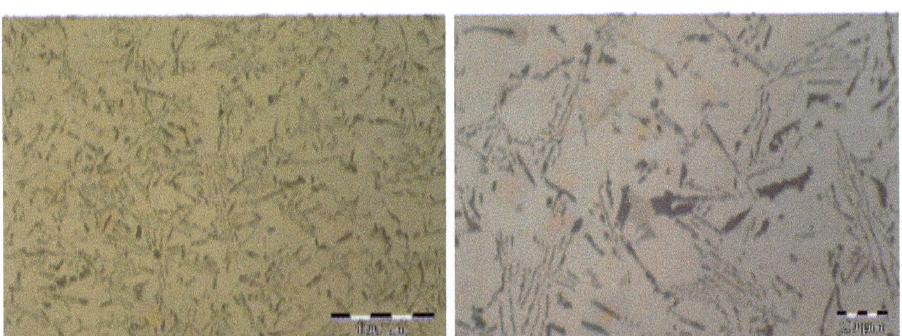

(a) (b)

Fig. 1 Microstructure of AlSi9Cu3 alloy used in this study: **a** α-Al solid solution and Al–Si eutectic, Si in acicular shape; **b** presence of Cu-rich and Fe-rich phases the dendritic α-Al solid solution and Al–Si eutectic, where Si assumes an acicular shape. Furthermore, it can be seen the presence of Fe-rich phase in polyhedral crystals and the presence of eutectic Al_2Cu (Fig. 1b). The presence of porosity (Fig. 2) confined to the central zone of the test samples, resultant of the casting and cooling process, did not influence the machining test results since a minimum diameter of 25 mm was not machined from the casted rods.

In addition to the previous material characterization, mechanical properties were evaluated from monotonic tensile tests in accordance with the EN 10002 standard and compression tests using cylindrical specimens, according to the ASTM E 9–89a standard. Figures 3 and 4 illustrate the resulting engineering stress–strain curves from both tensile and compressive tests respectively. Table 2 summarizes the mechanical properties extracted from the tensile tests. It can be observed that this material presents significantly different tensile and compressive strength behaviours. The tensile behaviour of this material is significantly less ductile than the

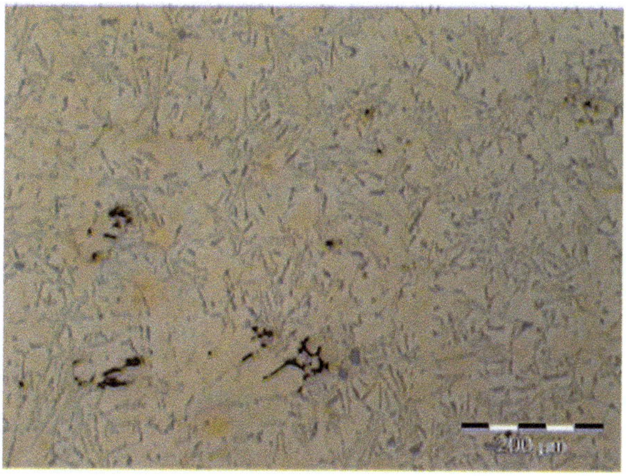

Fig. 2 Microestrutural analysis in central zone of machining test pieces with some porosity

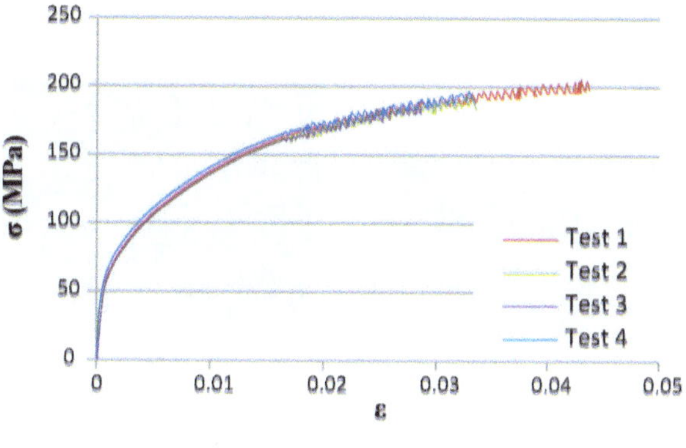

Fig. 3 Stress-strain curves from tensile tests

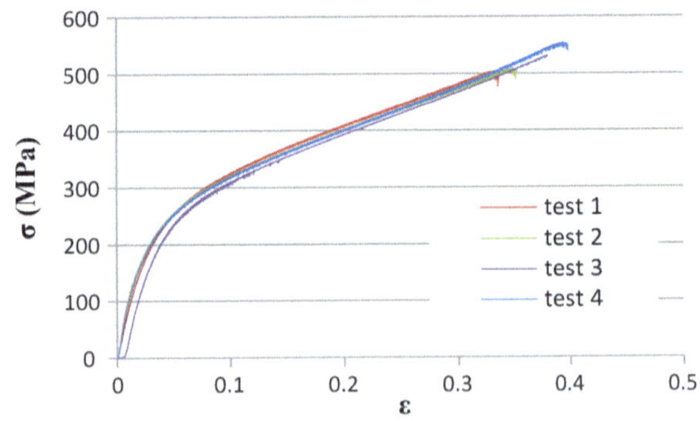

Fig. 4 Stress-strain curves from compressive tests

compressive behaviour as confirmed by the very distinct elongations. Also the strength values under tension are significantly lower than corresponding values under compression.

2.2 *Machinability Tests*

Machinability tests are evaluated using different solutions of commercially available cutting inserts proposed for the turning of aluminium alloys. Three different PCD tools were selected from Mapal tools manufacturer, one without chip breaker and two with different chip breaker geometries, CB1 and CB2 geometries. The CB1 geometry is indicated for finishing operations and the CB2 geometry is suggested for roughing operations. Figure 5 shows the PCD inserts and the designation of these inserts, according to the ISO 1832 standard. The same tool holder, with the specification SDJCL 2020 K11 was used in all cutting tests. The Table 3 presents the geometrical details of cutting inserts and tool holder.

A set of machining tests was performed in a EFI DU20 conventional horizontal lathe with 5.9 kW spindle power and a maximum spindle speed of 2500 rpm. These

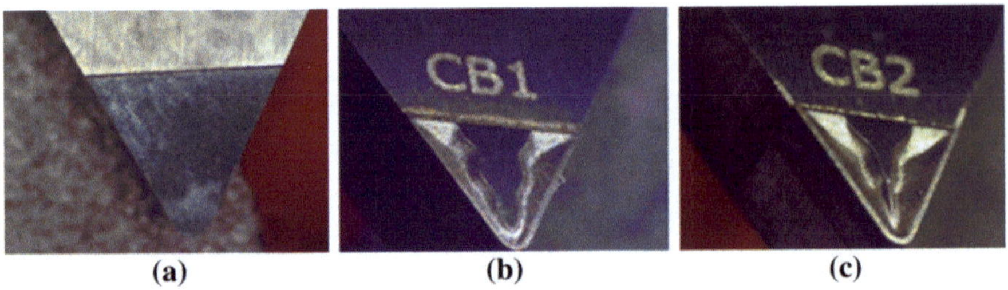

Fig. 5 PCD cutting inserts used in the machining tests: **a** flat rake face tool, DCGW 11T304 F01N-0AA; **b** CB1 geometry, DCGT 11T304 F01N-C1A; **c** CB2 geometry, DCGT 11T304 F01N-C2A

tests were instrumented in order to measure the cutting force components such as main cutting force (F_c), the feed force (F_f) and the passive force (F_p). A Kistler piezoelectric three-component dynamometer type 9257B, with a charge amplifier Kistler 5007 and a HBM Spider 8 acquisition system were used to acquire the signals of the cutting forces. The signals coming from dynamometer were evaluated using the *Catman* Software. Figure 6 illustrates the experimental setup concerning the turning tests on the conventional lathe.

The cutting parameters selected for the machining tests performed in the conventional lathe were defined based on the producer of the PCD inserts and also taking into account the limitations of the machine. Two spindle speeds were selected, namely 2500 and 900 rpm, with three cutting depths in the range of 0.25–1.5 mm and three feed rates in the 0.05–0.25 mm range. These cutting conditions are summarized in the Table 4.

Since the conventional lathe does not allow a constant cutting speed, the same machining sequence (a_p, f) was fully respected for each cutting insert, in order to allow a direct comparison of the result, which is generated for the same cutting velocities even they are varying with the diameter reduction. For the spindle speed of 2500 rpm resulted cutting speeds in the range of 350–432 m/min; for the spindle speed of 900 rpm resulted cutting speeds in the range of 88–117 m/min.

Table 3 Geometrical details of the inserts and tool holder used in machining tests

Insert shape (rhombic)	55°
Nose radius (r_ε)	0.4 mm
Clearance angle (α)	7°
Rake angle (λ_o)	0°
Inclination angle (λ_s)	0°
Position angle (χ)	93°

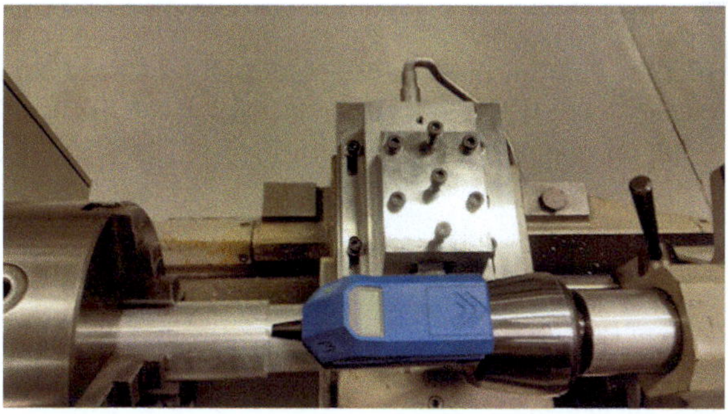

Fig. 6 Experimental setup for measurement of cutting forces and surface roughness

Table 4 Cutting parameters applied in cutting tests performed in the conventional lathe

Spindle speed (rpm)	2500; 900
Depth of cut (a_p) [mm]	0.25; 0.5; 1.5
Feed rate (f) [mm/rev]	0.05; 0.14; 0.25

Besides the cutting forces measurements, the chip typology and surface roughness were also evaluated during the machinability tests. For each turning test, chip samples are collected for the implementation of comparative tables. This procedure consists in organizing the chips corresponding to cutting parameters used in tests, in order to compare the different sizes and shapes and classify it according to the standard ISO 3685. Concerning the surface roughness measurements, three different zones of samples (cylinder), were assessed. This was performed according longitudinal lines equally spaced by 120°. This procedure allows the evaluation of the average roughness (R_a) and maximum height of roughness profile (R_{max}). The portable surface roughness tester Hommel Tester T500 was used for this analysis.

Once the conventional lathe machine used in this investigation only show a maximum modest spindle speed to the purpose of machinability tests with PCD inserts, additional tests were performed in a CNC GILDEMEISTER CXT400 lathe. This machine has a 36 kW spindle power and a maximum spindle speed of 5000 rpm, therefore, this machine permits the execution of machining tests with a maximum cutting speed of 690 m/min for the specimens diameters considered in the study. In addition to this cutting speed of 690 m/min, a smaller cutting speed of 300 m/min was used, and the same cutting depths and feed rates from Table 4 were applied again. These tests were performed under constant cutting speed. Only chip geometry and surface roughness's were evaluated in the tests performed in the CNC turning machine since the available load cells are not compatible with this machine.

3 Results and Discussion

3.1 Influence of Cutting Inserts and Cutting Parameters on Cutting Forces

Figures 7 and 8 illustrates typical cutting forces components, particularly the main cutting force (F_c), the feed force (F_f) and the passive force (F_p), along the machining time, for two different turning tests. In these specific cases, the insert used is the flat

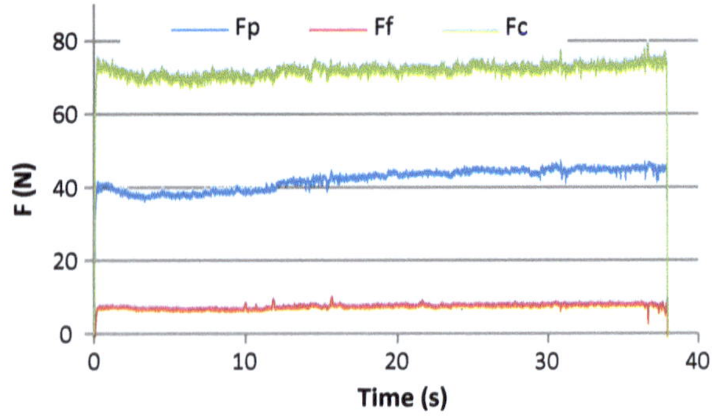

Fig. 7 Example of cutting forces components for insert without chip breaker. (v_c = 424 m/min, a_p = 1.5 m/min, f = 0.05 mm/rot)

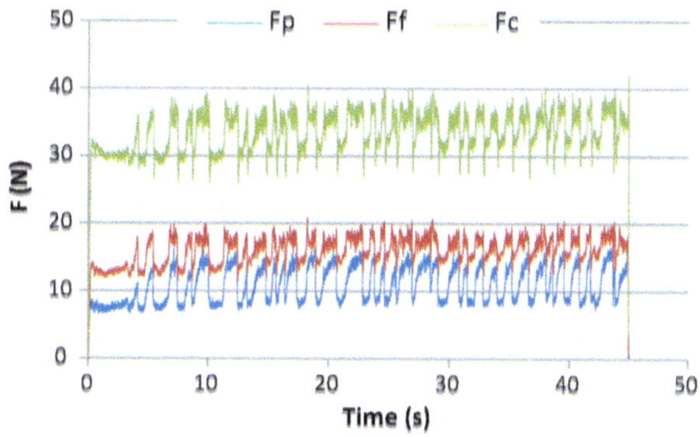

Fig. 8 Example of cutting forces components for insert without chip breaker (v_c = 116 m/min; a_p = 0.25 m/min; f = 0.14 mm/rot)—influence of the BUE in cutting forces

rake face tool and the major difference between these results is the cutting speed. For low cutting speeds (Fig. 8) the results show an oscillatory evolution of cutting force consistent with the BUE phenomena.

Figure 9 shows the results of the average of the main cutting force (F_c), obtained for the cutting parameters considered in the conventional lathe, for two spindles speeds applied in the tests. Results show that the main cutting forces increase with increasing cutting depth and feed rate. In fact, the increase of these parameters results in increasing chip section and consequently the shear force required to deform the chip also increases. On the other hand, the increase of the cutting speed produces an opposite effect, which can be justified with a temperature increase of the material, which leads to a softening of the material and consequently lower deformation resistance is observed. In addition, for lower cutting speeds the BUE phenomenon becomes more apparent and this fact leads to a greater friction between the tool and the workpiece, which justifies higher cutting forces.

Comparing the main cutting forces obtained for each tested cutting insert, it is possible to verify that the flat face tool typically has lower cutting forces

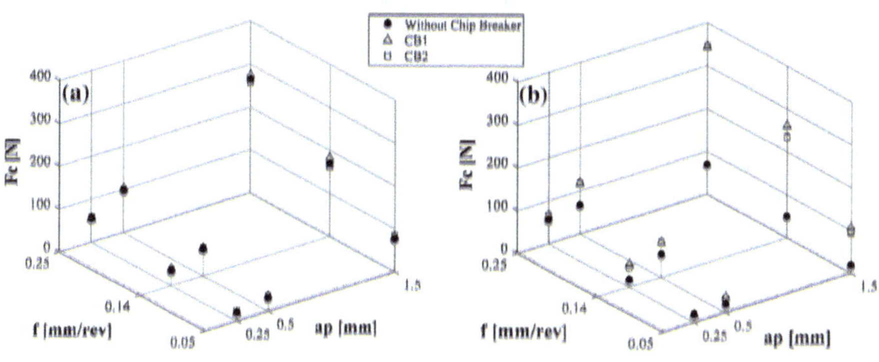

Fig. 9 Main cutting force variation with the feed rate and depth of cut for spindle speeds of 2500 rpm (**a**) and 900 rpm (**b**)

comparatively to inserts with chip breaker geometries present on the rake face. These results show that the presence of a groove in the rake face significantly increases the cutting forces which can be explained by the resistance to the natural flow of the chip generating higher cutting forces, as well as the increased frictional forces. Nevertheless, the forces generated in the cutting process are not dramatically influenced by the chip breaker geometries when high cutting speeds are considered.

3.2 *Influence of the Inserts and Cutting Parameters on Cutting Power*

Figure 10 shows the cutting power (P_c) as a function of the feed rate (f) and depth of cut (a_p), for each spindle speed used in the conventional lathe. The cutting power was calculated using the following equation, neglecting the cutting power generated by the feed movement:

$$P_c = v_c F_c \tag{1}$$

where F_c is the main cutting force, in N and v_c is the cutting speed in m/s.

The results concerning the cutting power show that the effect of different geometries of chip breakers do not have a significant influence in terms of energy required for the cutting process, in the case of the higher cutting speeds. Although the flat rake face insert is more advantageous from the energy consumption point of view, particularly for reduced f and a_p.

Another important observation is related to the cutting power evolution for the lowest range of tested cutting speeds, when the BUE phenomenon becomes quite evident. In this case, it appears that cutting power is much lower in flat rake face tool, revealing that the adhesion of the material is much lower in this type of tool,

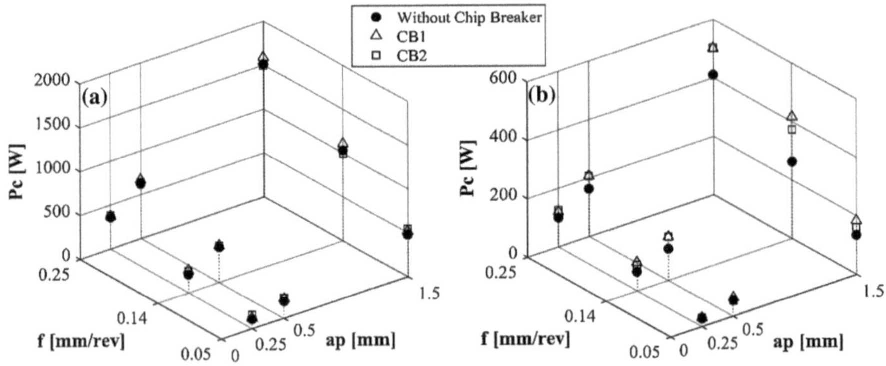

Fig. 10 Variation of cutting power with feed rate and depth of cut, for spindle speeds of 2500 rpm (**a**) and 900 rpm (**b**)

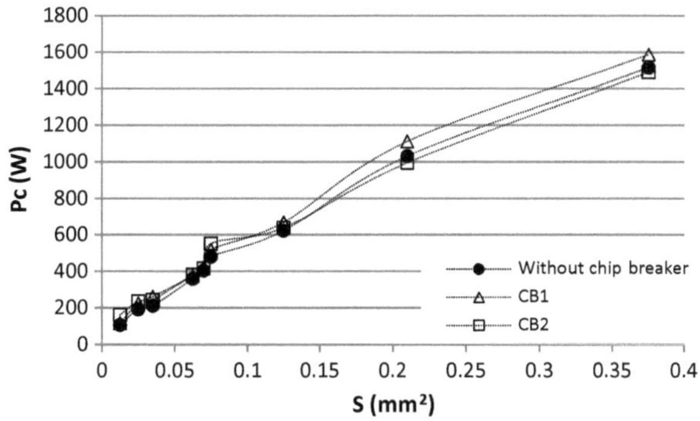

Fig. 11 Cutting power evolution as a function of the undeformed chip section (S) for 2500 rpm

and the groove in the rake face difficulties the evacuation of the adhered chips being a detrimental effect concerning the tool life.

Figure 11 represents the cutting power as a function of the undeformed chip section, for the spindle speed equal to 2500 rpm. In turning, the undeformed chip section (S), expressed in mm^2, is the product between the feed rate, in mm/rot and depth of cut, in mm. The analysis of the figure reveals the existence of some peaks or an oscillatory behaviour that are explained by the different contributions of the depth of cut and feed to the main cutting force values. In other words, for the increasing of the undeformed chip section the depth of cut and feed rate have different contributions in power consumption. For chip sections higher than 0.20 mm^2 can be observed that the use of the CB2 chip breaker geometry leads to substantially lower cutting power compared to the other inserts. This fact proves that obstacles to the chip flow can be produced still yielding an efficient chip control, with advantages in terms of energy consumption and also in terms of durability of the cutting tools.

3.3 Influence of the Inserts and Cutting Parameters on the Specific Cutting Pressure

The specific cutting pressure (K_s) was calculated using the following equation:

$$K_s = \frac{F_c}{S} \quad (2)$$

where Fc is the cutting force in N, and S the undeformed chip section in mm^2.

Figure 12 plots the specific cutting pressure K_s as a function of the feed and the cutting depth, for two different spindle speeds. It is clear the reduction of the specific cutting pressure when cutting depth or feed rate increases and the tendency to assume constant values for high values of these parameters. Another aspect is the influence of the cutting speed. For low cutting speeds it can be observed in some

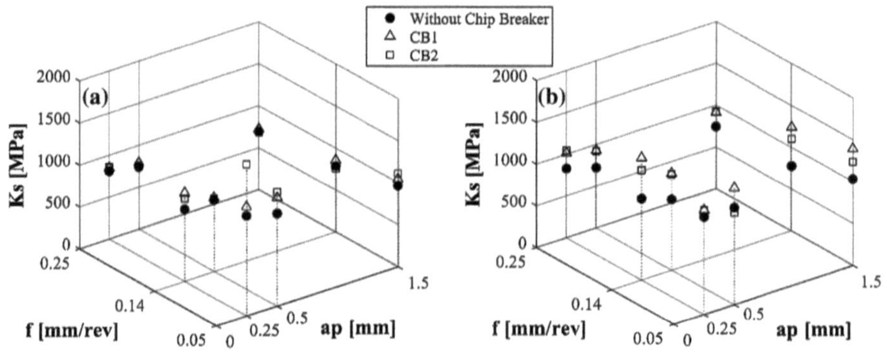

Fig. 12 Variation of specific cutting pressure (K_s) with feed rate and depth of cut for spindle speeds of 2500 rpm (**a**) and 900 rpm (**b**)

cases the increase of the Ks value. For these tested conditions it is notorious the effect of BUE formation, especially for inserts with chip breakers, where this phenomenon assumes a greater relevance and in some cases, produces a higher specific cutting pressure differently of what would be expected. Considering the higher cutting speeds, it can be concluded that all inserts presents similar Ks values under the same cutting conditions.

Figure 13 shows the evolution of the specific cutting pressure (K_s) with the undeformed chip section (S). From the figure it can be seen that the specific cutting pressure decreases with the increase of the chip section. In fact, it can be realized a size effect in the results obtained. For the lower chip sections, the friction effects between the tool and the chip, as well as the radius of the cutting tool tip can be an explanation for these results. For the higher tested undeformed chip section (0.38 mm^2), K_s values are in the range of 682–726 N/mm^2. These K_s values can be used as reference values for the industrial application, in particular for the estimation of the cutting forces.

Fig. 13 Comparison of specific cutting pressure with undeformed chip section (S) for spindle speeds of 2500 rpm

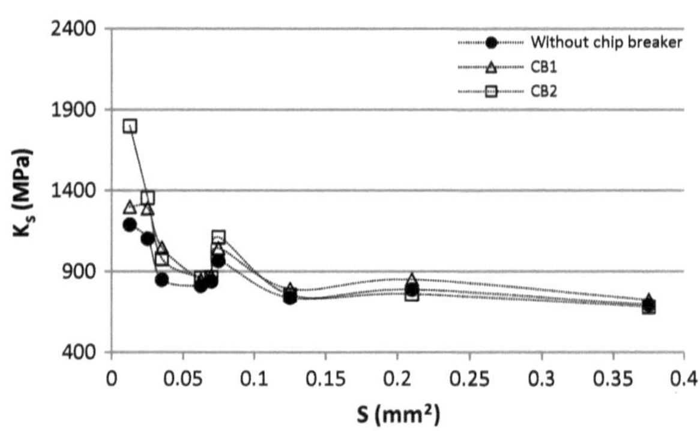

3.4 Influence of Cutting Inserts on Chip Morphology

Chip control (shape and size) is an aspect extremely important in turning operations, especially in case of ductile material, such aluminium alloys, machined with high cutting speed in automated machines. Effective control of the shape and size of the chips generated by the cutting process allows high levels of productivity and surface quality of the components.

Thus, for each machining test, chip samples were extracted, in order to allow a comparison between the inserts and to verify the effectiveness of the chip breakers for different cutting conditions. During some tests, especially for the flat face insert and the insert with CB2 chip breaker geometry applied in finishing conditions, long chips (e.g. ribbon chips) were obtained, which are harmful to the cutting process. The Tables 5 and 6 shows the comparison of the chips obtained for each insert. For the feed rate equal to 0.05 mm/rev we can observe that CB1 geometry is the only one that produced short chips (see Table 5). These results indicate that, this geometry is best suited for finishing operations, from the point of view of chip control. On the other hand, for the feed rate of 0.14 mm/rev the chip breaker CB2 presents a good chip control in these conditions while the flat face tool presents long chips and CB1 presents extremely short chips, producing higher cutting forces and power consumption, as previously seen. These results are represented in Table 6 and can be concluded that CB2 geometry works better for roughing operations.

For low cutting speed, in the range of 100–300 m/min, the execution of machining tests without cutting fluid causes the formation of BUE. The BUE formation results in short chips with a morphology revealing cutting instability (see

Table 5 Chip samples obtained for finishing operations ($f = 0.05$ mm/rev, $v_c = 690$ m/min)

a_p (mm)	Without Chip Breaker	CB1 Geometry	CB2 Geometry
0.25	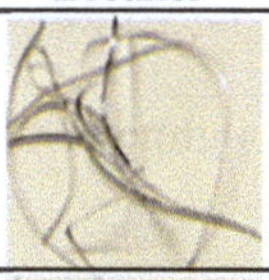	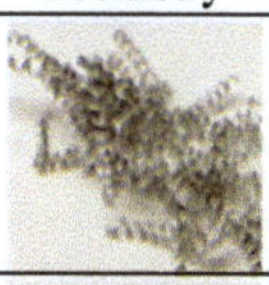	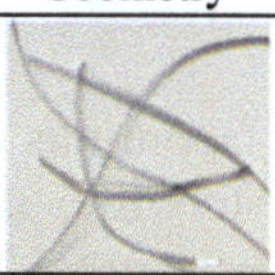
0.5			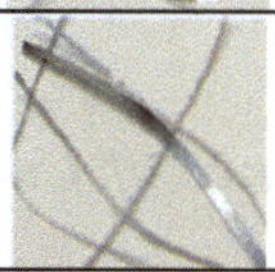
1.5	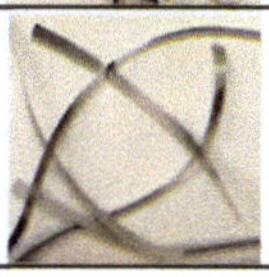		

Table 6 Chip samples obtained for semi-roughing operations ($f = 0.14$ mm/rev, $v_c = 690$ m/min)

a_p(mm)	Without Chip Breaker	CB1 Geometry	CB2 Geometry
0.25	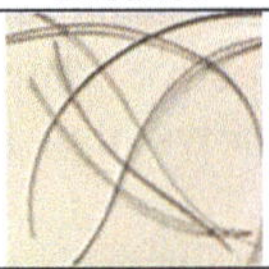		
0.5	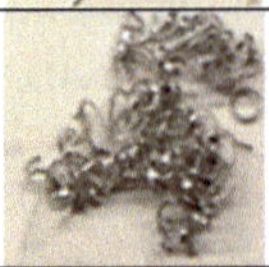		
1.5			

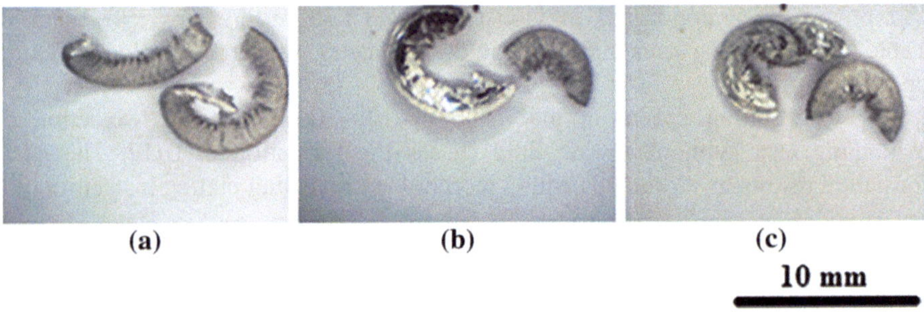

Fig. 14 Example of the effect of BUE formation in chip morphology for each insert used: **a** Without chip breaker; **b** CB1 geometry; **c** CB2 geometry

Fig. 14). This fact results in worse surface roughness, verified in the next section, and affects the tool life.

3.5 *Influence of Cutting Inserts on Surface Roughness*

Another important parameter in the evaluation of the machinability, depending of tool used, is related to the surface quality of the workpiece. Often, the surface quality is interpreted by the surface roughness of machined components. Therefore the average roughness or just roughness (R_a) and maximum height of roughness profile (R_{max}) were evaluated for each surface resulting from the machining tests.

For comparison purposes, experimental roughness results are compared with theoretical values represented by R_{at} and R_{maxt} and given by the following equations:

$$R_a = \frac{f^2}{32\, r_\varepsilon} \times 1000 \quad (3)$$

$$R_{tmax} = \frac{f^2}{8\, r_\varepsilon} \times 1000 \quad (4)$$

where f is the feed rate in mm/rev, and r_ε the nose radius of the insert in mm.

The quality of the machined surface is dependent of the cutting conditions selected for each test. In fact, the surface roughness it is more dependent of the cutting conditions and insert nose radius than the different type of inserts. The feed rate is the parameter that shows the higher influence in surface roughness, an increase in feed rate affecting negatively the surface quality (Fig. 15). Therefore, low feed rates are implemented in finishing machining.

Figures 16 and 17 show the influence of the cutting speed on the surface roughness. It can be concluded that the value of R_a decreases with the cutting speed. This aspect is directly associated to the BUE formation. For cutting speeds lower

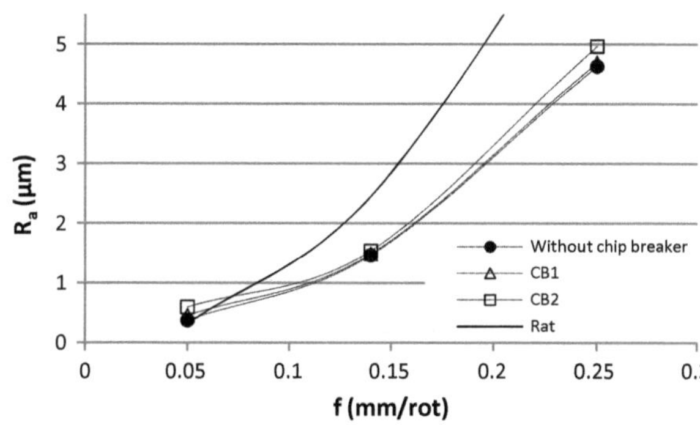

Fig. 15 Surface roughness (Ra) function of the feed rate (a_p = 0.25 mm; v_c = 690 m/min)

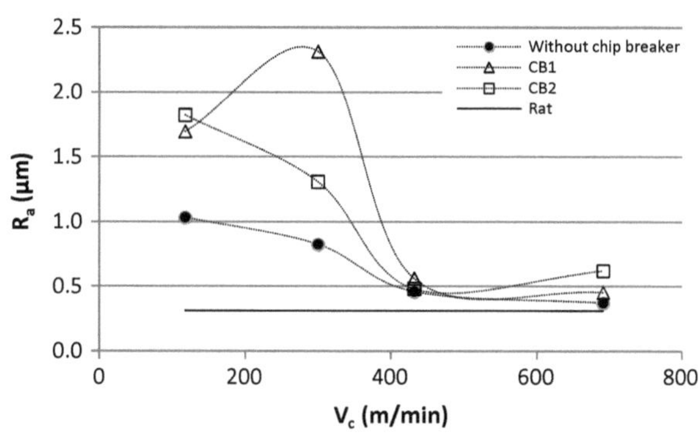

Fig. 16 Surface roughness (Ra) function of the cutting speed (a_p = 0.25 mm; f = 0.05 mm/rot)

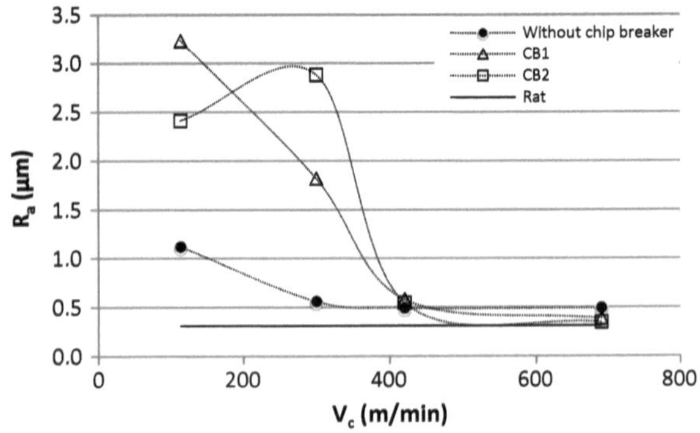

Fig. 17 Surface roughness (Ra) function of the cutting speed ($a_p = 0.5$ mm; $f = 0.05$ mm/rot)

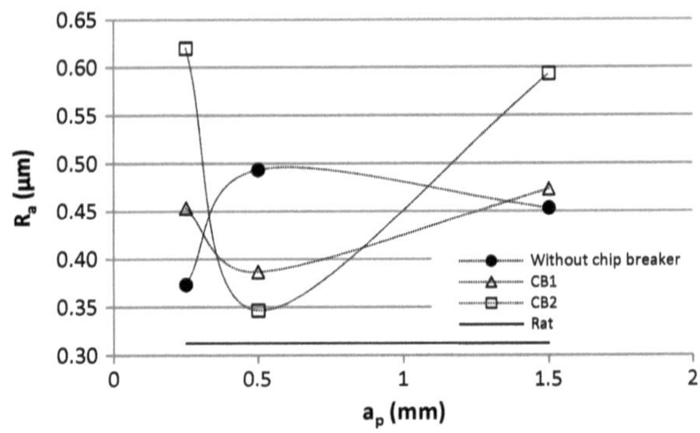

Fig. 18 Surface roughness (R_a) function of the cutting depth (a_p) ($f = 0.05$ mm/rot and $v_c = 690$ m/min)

than 300 m/min the surface roughness increases quickly. The adhesion of the material to the cutting tool edge as well as to new machined surfaces compromises the surface quality of the component. Although, in the flat rake face tool the adhesion of the material is less evident, consequently the influence in R_a value is less significant. Therefore, the flat face insert is the more indicated tool for low cutting speeds especially, in the case of dry machining.

The depth of cut also influences the surface roughness, this influence depending on cutting insert used. Figure 18 confirms that exist an optimal cutting depth for each insert to improve surface quality. The flat rank face tool presents the better results for R_a for depth of cut equal to 0.25 mm, whereas the inserts with chip breakers present the best results for depths of cut equal to 0.5 mm.

4 Conclusions

The influence of the chip breaker geometries and the cutting parameters (cutting speed, depth and feed rate) on machinability of the AlSi9Cu3 cast alloy were experimentally investigated and the following conclusions could be draw:

- The results showed a tendency to built-up edge formation for low cutting speeds for all of the inserts tested and this fact is responsible for higher cutting forces, power consumption and it also disturbs surface quality. However, this tendency of BUE formation can be reduced by the use of flat face tools;
- For chip sections higher than 0.20 mm^2 the use of the CB2 chip breaker geometry led to substantially lower cutting power, compared to the others inserts associated to efficient chip control, for this conditions;
- Regarding the specific cutting pressure values, it is notorious the size effect for lower undeformed chip sections, but in the case of higher undeformed chip sections the values the specific cutting pressure tends constant values. Considering the higher undeformed chip section tested (0.38 mm^2), K_s presents values in the range of 682–726 N/mm^2, for CB2 and CB1 chip breaker geometries respectively;
- The use of flat face inserts generates long chips for the majority of applied cutting conditions. On the other hand, the CB1 geometry presents an efficient chip control for low feed rates (0.05 mm/rev). This chip breaker is suitable for finishing operations, but this insert leads to higher power consumption. On the other side, the CB2 geometry presents an effective chip control for feed rates in the order of 0.14 mm/rev, and still presents, for these parameters, lower specific pressures values and lower cutting power;
- The surface roughness increases with the feed rate, but for feed rates equal to 0.05 mm/rev, there exists an optimum cutting depth. It can be possible to obtain surface roughness in the order of 0.35 μm for CB2 geometry, 0.37 μm for the flat face tool and 0.39 μm for the CB1 geometry.

Acknowledgements Authors acknowledge the funding of Project NORTE-01-0145-FEDER-000022—SciTech, co-financed by NORTE2020, through FEDER. Authors also acknowledge MAPAL Company which offered the cutting inserts.

References

1. Tanaka, T., Akasawa, T.: Machinability of hypereutectic silicon-aluminum alloys. J. Mater. Eng. Perform. **8**, 463–468 (1999)
2. Timelli, G., Fabrizi, A.: The effects of microstructure heterogeneities and casting defects on the mechanical properties of high-pressure die-cast AlSi9Cu3(Fe) alloys. Metall. Mater. Trans. A-Phys. Metall. Mater. Sci. **45**, 5486–5498 (2014)
3. Sood, P.K., Sehgal, R., Dwivedi, D.K.: Machinability study of stir cast hypoeutectic aluminum-silicon alloys during turning. J. Mater. Eng. Perform. **22**, 470–482 (2013)
4. Santos, M.C., Machado, A.R., Sales, W.F., Barrozo, M.A.S., Ezugwu, E.O.: Machining of aluminum alloys: a review. Int. J. Adv. Manuf. Technol. 1–14 (2016)
5. Puga, H., Costa, S., Barbosa, J., Ribeiro, S., Prokic, M.: Influence of ultrasonic melt treatment on microstructure and mechanical properties of AlSi9Cu3 alloy. J. Mater. Process. Technol. **211**, 1729–1735 (2011)

6. Ferraro, S., Timelli, G.: Influence of sludge particles on the tensile properties of die-cast secondary aluminum alloys. Metall. Mater. Trans. B-Proc. Metall. Mater. Proc. Sci. **46**, 1022–1034 (2015)
7. Davis, J.R.: Aluminum and aluminum alloys. ASM International, Materials Park, OH (1993)
8. Basavakumar, K.G., Mukunda, P.G., Chakraborty, M.: Influence of melt treatments and turning inserts on cutting force and surface integrity in turning of Al-12Si and Al-12Si-3Cu cast alloys. Surf. Coat. Technol. **201**, 4757–4766 (2007)
9. Kalyan, C., Samuel, G.L.: Cutting mode analysis in high speed finish turning of AlMgSi alloy using edge chamfered PCD tools. J. Mater. Process. Technol. **216**, 146–159 (2015)
10. Gonzalo, O., Quintana, I., Etxarri, J.: FEM based design of a chip breaker for the machining with PCD tools. Adv. Mater. Res. **223**, 133–141 (2011)
11. Shinozuka, J., Obikawa, T., Shirakashi, T.: Chip breaking analysis from the viewpoint of the optimum cutting tool geometry design. J. Mater. Process. Technol. **62**, 345–351 (1996)
12. Kim, H.G., Sim, J.H., Kweon, H.J.: Performance evaluation of chip breaker utilizing neural network. J. Mater. Process. Technol. **209**, 647–656 (2009)
13. Jawahir, I.S., Fang, X.D.: A knowledge-based approach for designing effective grooved chip breakers—2D and 3D chip flow, chip curl and chip breaking. Int. J. Adv. Manuf. Technol. **10**, 225–239 (1995)
14. Miyazawa, H., Takeuchi, S., Miyake, S., Murakawa, M.: Sintered diamond cutting inserts with chip breaker prepared by laser technique. Surf. Coat. Technol. **86–87**, 797–802 (1996)
15. Panušková, M., Tillová, E., Chalupová, M.: Relation between mechanical properties and microstructure of cast aluminum alloy AlSi9Cu3. Strength Mater-Engl. Tr. **40**, 98–101 (2008)

Optimization of Geometric Quality in a 5 Axis Machining of Curved Surfaces in a EN-AW-7075 Alloy by Taguchi Method

K. Castro, L.J. Segura, S.D. Castellanos and J. Lino Alves

1 Introduction

With the ever increasing demand of light materials within the industries such as automotive and aircraft, aluminium and its alloys have got much attention due to the advantages they possess. Strong, light, highly corrosion resistant, good heat and electricity conductor, and easy to form and process are some of the properties that makes aluminum one of the most utilized element in the market [1].

The ease of machinability of aluminum has considerably increased the importance of performing studies in this field. Additionally, the versatility that computer-aided design (CAD) and computer-aided manufacturing (CAM) systems along with new generation of computer numerical control (CNC) machines, allow to better plan and improve parameters to optimize finished characteristics on the machined parts. Considering that quality—defined as the degree of meeting

K. Castro · L.J. Segura (✉) · S.D. Castellanos
Departamento de Ingeniería Mecánica, Universidad de las Fuerzas Armadas-ESPE,
Av. General Rumiñahui, P.O.BOX 171-5-231B, Sangolqui, Ecuador
e-mail: ljsegura@espe.edu.ec

K. Castro
e-mail: leocastroskp18@gmail.com

S.D. Castellanos
e-mail: sdcastellanos@espe.edu.ec

S.D. Castellanos · J.L. Alves
INEGI, Faculty of Engineering, University of Porto,
Rua Dr. Roberto Frias 400, 4200-465 Porto, Portugal
e-mail: falves@fe.up.pt

K. Castro · L.J. Segura
Design and Manufacturing Department, Groupseg Cia. Ltda,
Quito, Ecuador

L.F.M. da Silva (ed.), *Materials Design and Applications*,
Advanced Structured Materials 65, DOI 10.1007/978-3-319-50784-2_26

customer requirements and specifications—has become the inflection point between choosing a supplier or not, companies and researchers have focused on developing studies regarding optimizing surface roughness (R_a), which is an important characteristic when assessing quality, for several machining processes as it is milling [2, 3].

Milling is one of the most used processes for producing parts in the automotive and aerospace sectors, where quality is strictly checked for the production of molds, pockets, and slots [4]. Thus, improving quality of finished products without making a substantial investment and in short period of time can be achieved by the use of Design of Experiments (DOE), whose main techniques include Taguchi Method, Factorial Design, and Response Surface Design Methodology (RSM) [5].

Wang et al. [6], presented a study on surface roughness in high-speed of AlMn 1 Cu using factorial design and partial least square regression. Analysis of variance (ANOVA) was used to determine the most influential cutting parameters-cutting speed, feed per tooth, and depth of cut. Additionally, partial least squares regression was used to determine the mathematical model of surface roughness and the biggest error obtained from the analysis was 14.64%.

Arokiadass et al. [7] made a study to improve surface roughness of Al/Si Cp MMC by utilizing RSM. The cutting parameters: spindle speed, feed rate, depth of cut and various weight of silicon carbide were considered to perform the experiments. They found that the effect of depth of cut on surface roughness was practically negligible.

Yang et al. [3] analyzed an approach to optimize surface roughness by using Taguchi Method, where ANOVA and Signal-to-Noise (S/N) ratio analyses are presented within the study. The controllable factors are depth of cut, cutting speed, feed rate, and tool diameter at three levels. The model predicted a mean of 0.56 μm and the experiments yield 0.58 μm, which is much better to the best response obtained for the L18 orthogonal array (0.69 μm).

Chen et al. [8] presented a different methodology—neural fuzzy system, to predict R_a. The parameters used to estimate the surface roughness are feed rate, spindle speed, depth of cut, vibration average per revolution, tool diameter, tool materials, and workpiece material, although the last three were not considered in the model since they were crisp parameters for the process. Toe system was developed in such a way that in-process real time measurements are taken, and the average deviation for the fuzzy net model is 10.5%.

As it can be seen, most of the available literature presents the optimization of R_a in the milling process by considering controllable parameters such as cutting speed, depth of cut and feed rate applied on flat surfaces; however, there is still little information regarding the optimization and influence of these and other parameters applied on the machining of convex and concave geometries, thus this being the paper's intention.

2 Setup Experiment

A 5-Axis CNC vertical machining center (DMU 60 eVo DECKEL MAHO) was used to conduct the experiments. This machine is equipped with a rotatory worktable that ranges from −5° to 110°, maximum load capacity of 600 kg, maximum spindle speed of 18,000 rpm, feed rate of 80 m/min and 36 kW drive motor (see Fig. 1). Ball nose end-mill cutter with diameter 5 mm was used. HYSOL XF water soluble coolant, 5% concentration was deployed during the tests.

EN-AW-7075 aluminium alloy specimens with dimensions 80 mm diameter and 40 mm thickness were clamped in the machine to run the experiments (see Fig. 2).

The characterization of the material was done in a Scanning Electron Microscope (SEM), Bruker Nano Quantax, Table 1 shows the distribution and percentage of the components in a sample of 25.16 mm width.

Since the tool path pattern can be easily chosen and modified by any CAM tool, the programming was done with MasterCAM X7. A Zig-Zag tool path pattern was picked, and the angle was chosen to be 0°, 45° and 90° far the different levels used in the orthogonal array of the experiment (see Fig. 3).

Fig. 1 Five axis vertical machining center employed in the experiment

Fig. 2 Machined specimens to be tested

Table 1 Result of the material characterization

Element	AN	wt%	Norm. wt%	Norm at.%
Aluminium	13	72.121	89.717	92.700
Zinc	30	4.928	6.130	2.613
Oxygen	8	1.759	2.188	3.813
Copper	29	1.433	1.783	0.782
Iron		0.146	0.182	0.091
		80.388	100	100

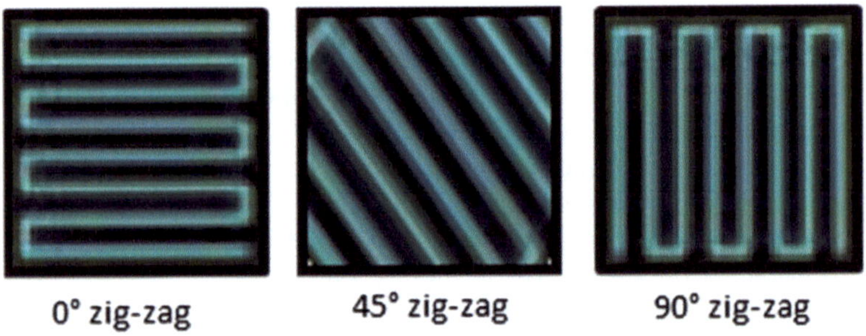

Fig. 3 Different toolpaths in the MasterCAM X7 software

The levels of the other parameters such as tilt angle (A), feed rate (B), and depth of cut (D) were established by using tool catalogue recommendations.

The roughness is defined as:

$$R_a = \frac{1}{L} * \int_0^L |y(x)|dx \qquad (1)$$

where R_a is the arithmetic average deviation from the center line, L is the sampling length and y(x) is the profile curve of the surface.

A Bruker NPFLEX optical interferometer (see Fig. 4) and a sampling area of 1.8 × 0.8 mm were utilized to measure the average surface roughness on each tested specimen.

3 Taguchi Method

Considering that customer's perception of quality is the most important factor when choosing a product, many companies look for quality through product and process optimization [9]. One of the most powerful and cost effective tools to achieve product and process optimization is the use of Taguchi Method. This methodology was proposed by Dr. Genichi Taguchi in 1960s [10]; thus, this study was developed following the structure shown in Fig. 5.

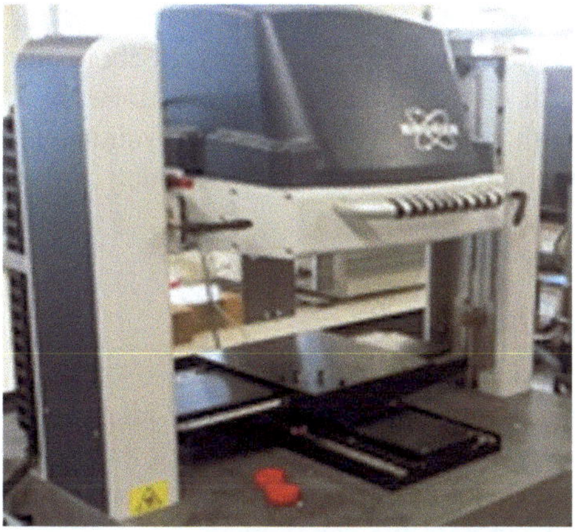

Fig. 4 Bruker NPFLEX equipment utilized to measure the roughness

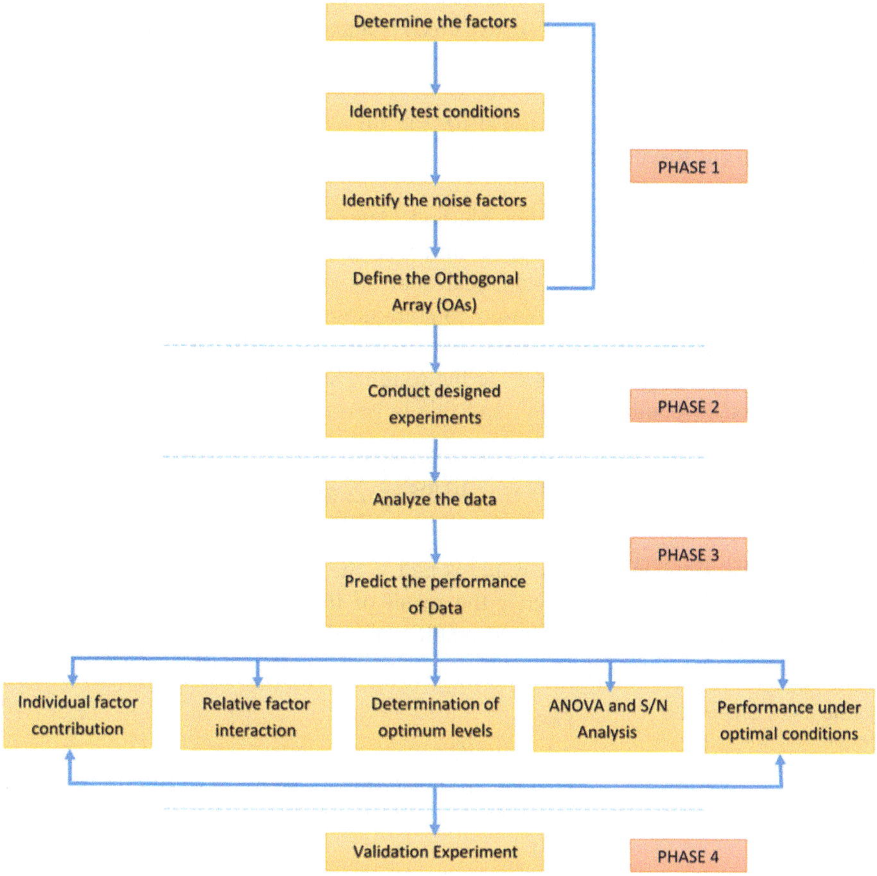

Fig. 5 Scheme of the Taguchi Method approach

The first phase is dedicated to defining the factors and choosing the adequate orthogonal array (OAs). Second phase is mainly all about performing experiments. Third phase deals with data analysis, which was done by the use of specialized software, and the fourth phase corresponds to the validation of the experiment by running and collecting data from the predicted model.

4 Experiments and Results

First of all, the four most influential factors were identified—tilt angle, feed rate, tool path pattern and depth of cut. Three levels for each factor was established as shown in Table 2. A deep knowledge of the process is indispensable when picking the factors to be analyzed with the experiments.

Additionally, noise factors—aspects that influence the response of a process but cannot be economically controlled [10]—such as weather conditions and machinery wear were considered when computing the Signal-to-Noise ratio (S/N) value. Defining the proper Orthogonal Array (OAs) is a key point to assure accuracy of the statistical model. For this reason a L_{27} OA is chosen, since it considers all the factors and their interactions (see Table 3). As it can be seen in Table 3, the main factors (A, B, C and D) are assigned for the tilt angle, feed rate, tool path pattern and depth of cut respectively. The other columns represent the interactions of the factors.

Following up with the Taguchi Method, data analysis has to be performed as required in Phase 2, shown in Fig. 5 and was conducted. A significance level of 0.05 was assigned for all the statistical analysis. Signal-to-noise (S/N) ratio was computed. There are several relations to compute (S/N) ratio, and this depends on the characteristic of the parameters that want to be optimized. Since the main objective of this study is to minimize the roughness, a Smaller-is-Better (SB) relation is utilized to obtain the ratio; and this is governed by the following equation:

$$S/N = -10\log\left[\frac{1}{n}\sum_{i=1}^{n} y_i^2\right] \quad (2)$$

where n is the number of measurements and y_i the response value. The bigger the value of the (S/N) ratio, the better the result of the response, which in this case means a minimum R_a. The obtained results can be seen in Tables 4 and 5.

Table 2 Details of the factors and levels for the experiment

Factors	Level 1	Level 2	Level 3
Tilt angle (A) [°]	10	12.5	15
Feed rate (B) [mm/min]	1591	1211	976
Tool path pattern Zig-Zag (C) [°]	0	45	90
Depth of cut (D) [mm]	0.2	0.4	0.7

Table 3 L_{27} Orthogonal array utilized for the Taguchi design

Trial no.	A	B	A × B	A × B	C	A × C	A × C	B × C	D	–	B × C	–	–
1	1	1	1	1	1	1	1	1	1	1	1	1	1
2	1	1	1	1	2	2	2	2	2	2	2	2	2
3	1	1	1	1	3	3	3	3	3	3	3	3	3
4	1	2	2	2	1	1	1	2	2	2	3	3	3
5	1	2	2	2	2	2	2	3	3	3	1	1	1
6	1	2	2	2	3	3	3	1	1	1	2	2	2
7	1	3	3	3	1	1	1	3	3	3	2	2	2
8	1	3	3	3	2	2	2	1	1	1	3	3	3
9	1	3	3	3	3	3	3	2	2	2	1	1	1
10	2	1	2	3	1	2	3	1	2	3	1	2	3
11	2	1	2	3	2	3	1	2	3	1	2	3	1
12	2	1	2	3	3	1	2	3	1	2	3	1	2
13	2	2	3	1	1	2	3	2	3	1	3	1	2
14	2	2	3	1	2	3	1	3	1	2	1	2	3
15	2	2	3	1	3	1	2	1	2	3	2	3	1
16	2	3	1	2	1	2	3	3	1	2	2	3	1
17	2	3	1	2	2	3	1	1	2	3	1	1	2
18	2	3	1	2	3	1	2	2	3	1	3	2	3
19	3	1	3	2	1	3	2	1	3	2	1	3	2
20	3	1	3	2	2	1	3	2	1	3	2	1	3
21	3	1	3	2	3	2	1	3	2	1	3	2	1
22	3	2	1	3	1	3	2	2	1	3	3	2	1
23	3	2	1	3	2	1	3	3	2	1	1	3	2
24	3	2	1	3	3	2	1	1	3	2	2	1	3
25	3	3	2	1	1	3	2	3	2	1	2	1	3
26	3	3	2	1	2	1	3	1	3	2	3	2	1
27	3	3	2	1	3	2	1	2	1	3	1	3	2

Table 4 Result of the experiment and its corresponding S/N value for convex machined specimens

Trial no.	Tilt angle (A)	Feed rate (B)	Tool path pattern (C)	Depth of cut (D)	R_a 1	R_a 2	R_a 3	R_a Mean	S/N
1	10	1591	0	0.2	0.385	0.155	0.258	0.266	10.991
2	10	1591	45	0.4	0.519	0.443	0.412	0.458	6.741
3	10	1591	90	0.7	0.446	0.480	0.430	0.452	6.888
4	10	1211	0	0.4	0.285	0.146	0.301	0.244	11.912
5	10	1211	45	0.7	0.570	0.493	0.488	0.517	5.707
6	10	1211	90	0.2	0.490	0.467	0.522	0.493	6.134

(continued)

Table 4 (continued)

Trial no.	Tilt angle (A)	Feed rate (B)	Tool path pattern (C)	Depth of cut (D)	R_a 1	R_a 2	R_a 3	R_a Mean	S/N
7	10	976	0	0.7	0.305	0.253	0.528	0.362	8.378
8	10	976	45	0.2	0.409	0.398	0.456	0.421	7.499
9	10	976	90	0.4	0.506	0.550	0.480	0.512	5.801
10	12.5	1591	0	0.4	0.393	0.513	0.654	0.520	5.501
11	12.5	1591	45	0.7	0.496	0.501	0.524	0.507	5.897
12	12.5	1591	90	0.2	0.723	0.668	0.691	0.694	3.168
13	12.5	1211	0	0.7	0.211	0.216	0.473	0.300	9.789
14	12.5	1211	45	0.24	0.588	0.998	0.791	0.789	1.876
15	12.5	1211	90	0.4	0.740	0.826	0.837	0.801	1.915
16	12.5	976	0	0.2	0.40	0.241	0.277	0.307	10.040
17	12.5	976	45	0.4	0.603	0.618	0.582	0.601	4.420
18	12.5	976	90	0.7	0.703	0.721	0.706	0.710	2.974
19	15	1591	0	0.7	0.5	0.511	0.565	0.543	5.296
20	15	1591	45	0.2	0.700	0.625	0.706	0.677	3.375
21	15	1591	90	0.4	0.797	0.857	0.836	0.830	1.615
22	15	1211	0	0.2	0.669	0.238	0.338	0.415	6.858
23	15	1211	45	0.4	0.759	0.692	0.766	0.739	2.618
24	15	1211	90	0.7	0.784	0.813	0.959	0.852	1.356
25	15	976	0	0.4	0.424	0.412	0.382	0.406	7.821
26	15	976	45	0.7	0.743	0.756	0.631	0.710	2.948
27	15	976	90	0.2	0.819	0.779	0.796	0.798	1.958

Table 5 Result of the experiment and its corresponding S/N value for the concave machined specimens

Trial no.	Tilt angle (A)	Feed rate (B)	Tool path pattern (C)	Depth of cut (D)	R_a 1	R_a 2	R_a 3	R_a Mean	S/N
1	10	1591	0	0.2	0.581	0.276	0.439	0.432	6.943
2	10	1591	45	0.4	0.837	0.789	0.843	0.823	1.688
3	10	1591	90	0.7	0.773	0.693	0.730	0.732	2.701
4	10	1211	0	0.4	0.325	0.600	0.515	0.480	6.133
5	10	1211	45	0.7	0.818	0.798	0.823	0.813	1.797
6	10	1211	90	0.2	0.739	0.684	0.650	0.691	3.198
7	10	976	0	0.7	0.420	0.530	0.520	0.490	6.152
8	10	976	45	0.2	0.625	0.632	0.573	0.610	4.285
9	10	976	90	0.4	0.892	0.712	0.745	0.783	2.082
10	12.5	1591	0	0.4	0.200	0.215	0.215	0.210	13.551

(continued)

Table 5 (continued)

Trial no.	Tilt angle (A)	Feed rate (B)	Tool path pattern (C)	Depth of cut (D)	R_a 1	R_a 2	R_a 3	R_a Mean	S/N
11	12.5	1591	45	0.7	0.500	0.340	0.303	0.381	8.168
12	12.5	1591	90	0.2	0.407	0.392	0.356	0.392	8.277
13	12.5	1211	0	0.7	0.382	0.375	0.398	0.385	8.288
14	12.5	1211	45	0.24	0.368	0.258	0.367	0.331	9.499
15	12.5	1211	90	0.4	0.078	0.680	0.597	0.431	6.053
16	12.5	976	0	0.2	0.402	0.310	0.320	0.344	9.207
17	12.5	976	45	0.4	0.089	0.643	0.621	0.451	5.702
18	12.5	976	90	0.7	0.408	0.398	0.430	0.412	7.697
19	15	1591	0	0.7	0.662	0.660	0.598	0.640	3.867
20	15	1591	45	0.2	0.850	0.693	0.662	0.735	2.620
21	15	1591	90	0.4	0.955	0.803	0.705	0.821	1.646
22	15	1211	0	0.2	0.241	0.259	0.520	0.340	8.799
23	15	1211	45	0.4	0.666	0.680	0.784	0.710	2.951
24	15	1211	90	0.7	0.896	0.882	0.850	0.876	1.148
25	15	976	0	0.4	0.163	0.273	0.329	0.255	11.563
26	15	976	45	0.7	0.836	0.749	0.758	0.781	2.136
27	15	976	90	0.2	0.851	0.742	0.693	0.762	2.328

Analysis of Variance (ANOVA) has also been done so that the significance of the factors and their effects can be quantified. Calculated F-value is compared to the standard value of $F_{0.05}$, which means, at the 5% of significance level. A factor or interaction is considering significant only if the calculated F-value is greater than $F_{0.05\text{–}2.12} = 3.89$.

For both cases-convex and concave machined specimens—the most influential factors towards a response (R_a) are the tilt angle (A) and tool path pattern (C). Based on the yield results, the other factors (B and D) and its interactions can be neglected, since they do not significantly affect the final response.

5 Prediction Model

In order to get the Prediction Model, an analysis of the factors was performed as shown in Fig. 6. As it can be seen for the convex-case, the factors A and C correlate positively with the roughness, meanwhile the factors B and D do not show much sensibility towards the finished surface; this was also evident according to the ANOVA in Table 6.

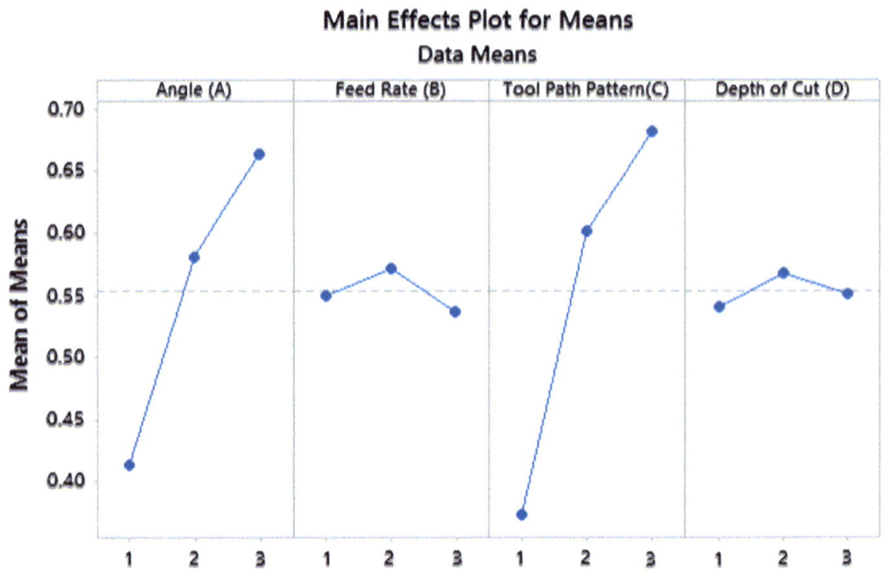

Fig. 6 Response of the main effect factors for the convex machined specimens

Table 6 ANOVA results of the raw data taken from the convex specimens

Source	DF	Adj SS	Adj MS	F-value	P
Tilt angle (A)	2	0.291	0.145	24.150	32.641
Feed rate (B)	2	0.006	0.003	0.490	0.665
Tool path pattern (C)	2	0.462	0.231	38.370	51.855
Depth of cut (D)	2	0.004	0.002	0.300	0.402
A × B	2	0.012	0.006	0.970	1.317
A × C	2	0.013	0.007	1.100	1.482
B × C	2	0.031	0.016	2.610	3.529
Error	12	0.072	0.006		8.108
Total	26	0.891			

According to the Taguchi model, the best performance for the convex-case is obtained with the following expression:

$$Ropt1_a = A_{(10^\circ)} + B_{(976)} + C_{(0^\circ)} + D_{(0.2)} - 3R_{MEAN} \quad (3)$$

where the values of the factors are taken from the Table 7 and it yields an optimized surface roughness of 0.2049 μm; this for the convex-case.

Similar analysis was done for the concave-case. Figure 7 shows that the response (R_a) is very sensitive to any change in factors A and C. On the other hand,

Table 7 Response table for the mean roughness (convex-specimens)

Level	Tilt angle (A)	Feed rate (B)	Tool path pattern (C)	Depth of cut (D)
1	0.4139	0.5497	0.3737	0.5400
2	0.5810	0.5722	0.6021	0.5679
3	0.6633	0.5363	0.6824	0.5503
Delta	0.2494	0.0359	0.3088	0.0279
Rank	2	3	1	4
R Mean	0.5530			

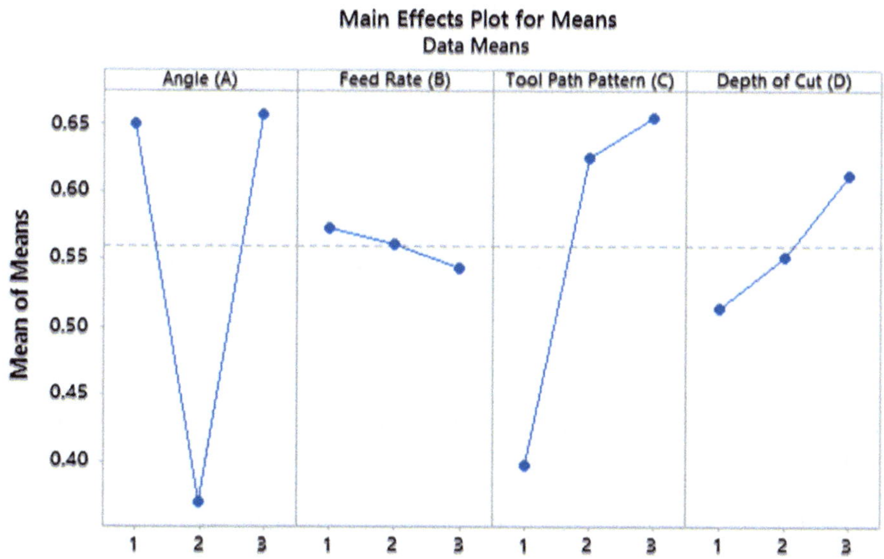

Fig. 7 Response of the main effect factors for the concave machined specimens

factors B and D do not affect considerably the final response as it can be seen in Table 8. The optimal roughness is determined by the equation:

$$Ropt2_a = A_{(12.5°)} + B_{(976)} + C_{(0°)} + D_{(0.2)} - 3R_{MEAN} \tag{4}$$

The values of the factors are taken from the Table 9 and an optimized surface roughness of 0.1448 μm is obtained. For the validation process, it is necessary to calculate a confidence interval (CI).

$$CI = \sqrt{F_{(1,DFe)} Ve \left[\frac{1}{n_{eff}} + \frac{1}{R}\right]} \tag{5}$$

Table 8 ANOVA results of the raw data taken from the concave specimens

Source	DF	Adj SS	Adj MS	F-value	P
Tilt angle (A)	2	0.485	0.242	28.630	44.713
Feed rate (B)	2	0.004	0.002	0.250	0.384
Tool path pattern (C)	2	0.358	0.179	21.170	33.064
Depth of cut (D)	2	0.044	0.022	2.590	4.046
A × B	2	0.022	0.011	1.290	2.020
A × C	2	0.065	0.033	3.860	6.036
B × C	2	0.004	0.002	0.230	0.365
Error	12	0.102	0.008		9.372
Total	26	1.084			

Table 9 Response table for the mean roughness (concave-specimens)

Level	Tilt angle (A)	Feed rate (B)	Tool path pattern (C)	Depth of cut (D)
1	0.6504	0.5732	0.3973	0.5144
2	0.3700	0.5619	0.6261	0.5516
3	0.6578	0.5431	0.6548	0.6122
Delta	0.2878	0.0301	0.2574	0.0978
Rank	1	4	2	3
R Mean	0.5600			

where $F_{(1,\ DFe)}$ is taken from tables at 95% confidence, and it is $F_{(1.12)} = 4.7472$ V_e refers to the variance of the error and it is $V_{e1} = 0.006$ and $V_{e2} = 0.008$ for the convex and concave case respectively [11]. R is the number of the test repetitions, which is 3 for both cases. n_{eff} is the effective number of replications and it is computed as follow:

$$n_{eff} = \frac{N}{1 + V_t} \tag{6}$$

where N is the total number of trials and it is 27 for this study. $V_t = 8$ is the total degrees of freedom related to the factors that influence significantly the response. CI's are ±0.1378 μm and ±0.1591 μm for convex and concave case respectively.

For both cases validation experiments were realized. The optimal combinations described in Eqs. 3 and 4 were used as set-up in the machine.

The experiment was run three times for each case yielding an average of 0.210 μm and 0.185 μm for convex and concave respectively. The obtained results fall between the values obtained in the optimization process, thus assuring the validity of the study.

Contrasting with other optimization studies that focus on developing optimization method in flat surfaces, pocket machining, for example, this study has focused on the quantification of the influence of the factors (A, B, C and D) in terms of surface roughness in convex and concave finished surfaces. The tool path pattern is crucial for this analysis, since it considerably influences towards finished surfaces as it can also be seen in [12]; tool path pattern do not behave the same way when machining flat surfaces with end-milling cutters and this can be neglected, as it can be inferred from the literature review [3–7].

6 Conclusions

According to this study, specifically from the ANOVA analysis, there are two significant factors—Tilt Angle and Tool Path Pattern, that influences the sensibility of the response (R_a). For the convex-case, these two factors influence by 84.50% the final response and the others factors and the interactions the rest to complete the 100%. Regarding the concave-case, A and C represent the 77.78% factor influence in the final response, the 22.22% corresponds to the influence in the response by the factors B and D and the interactions. Thus, it is concluded that factors B and D are not statistically significant in this optimization process.

For the convex-case, it is clear to see that factors A and C correlates positively with R_a, while factors B and D do not represent a big change in terms of surface roughness. From the concave-case analysis, it is not clear to see whether the factor A correlates either positively or negatively with the R_a; however, it is possible to see that the second level (12.5°) of factor A yields the best results when assessing the surface roughness. Factor C correlates positively with the response and R_a do not suffer a significant change if feed rate and depth of cut are changed from one level to another. Although factors B and D are not statistically significant in both cases, the analysis shows that levels three (976 mm/min) and one (0.2 mm) give better results regarding the optimization process.

After considering the ANOVA tables and the main effects plot for convex as well as concave case, the minimum surface roughness was found as, A = 10°, B = 976 mm/min, C = 0° and D = 0.2 mm for the first case and A = 12.5°, B = 976 mm/min, C = 0° and D = 0.2 mm for the concave-case. The validation experiments yielded averages of 0.210 μm and 0.185 μm for first and second cases.

The best response for the convex-case was achieved in the trial number 4, and an improvement of 13.93% was achieved for this case. Additionally, the best response for the concave-case was delivered by the trial number 10; thus getting a 11.90% reduction in the finished surface roughness.

Acknowledgements This study was developed in a collaboration between the Universidad de las Fuerzas Armadas-ESPE and University of Porto (U.PORTO) as part of an institutional research agreement. The authors would also like to have a special mention for the funding of Project NORTE-01-0145-FEDER-000022—SciTech—Science and Technology for Competitive and Sustainable Industries, cofinanced by Programa Operacional Regional do Norte (NORTE2020), through Fundo Europeu de Desenvolvimento Regional (PEDER) and Laboratorio de Procesos de Manufactura employees.

References

1. Report D: (2014) 2015 North American Light Vehicle Aluminum Content Study. drivealuminum.org. http://drivealuminum.wpengine.com/wp-content/uploads/2016/06/2015-Ducker-Study.pdf. Accessed Jan 2015
2. Radovanović, M., Petković, D., Janković, P., Madić, M.: Application of Taguchi method for determining optimum surface roughness in turning of medical stainless steel. Paper presented at the International Scientific Conference UNITECH 2014, Garrovo (2014)
3. Yang, J.L., Chen, J.C.: A systematic approach for identifying optimum surface roughness performance in end-milling operations. J. Ind. Technol. **17**(2), 2–8 (2001)
4. Lou, M.S., Chen, J.C., Li, C.M.: Surface roughness prediction technique for CNC end-milling. J. Industrial Technol. **15**(1), 2–6 (1998)
5. Mukherjee, I., Ray, P.K.: A review of optimization techniques in metal cutting processes. Comput. Ind. Eng. **50**(1–2), 15–34 (2006). doi:10.1016/j.cie.2005.10.001
6. Wang, Z.H., Yuan, J.T., Liu, T.T., Huang, J., Qiao, L.: Study on surface roughness in high-speed milling of AlMn1Cu using factorial design and partial least square regression. Int. J. Adv. Manuf. Technol. **76**(9–12), 1783–1792 (2014). doi:10.1007/s00170-014-6400-8
7. Arokiadass, R., Palaniradja, K., Alagumoorthi, N.: Surface roughness prediction model in end milling of Al/SiCpMMC by carbide tools. Int. J. Eng. Sci. Technol. **3**(6) (2012). doi:10.4314/ijest.v3i6.7
8. Chen, J.C., Savage, M.: A fuzzy-net-based multilevel in-process surface roughness recognition system in milling operations. Int. J. Adv. Manuf. Technol. **17**(9), 670–676 (2001). doi:10.1007/s001700170132
9. Roy, R.K.: A Primer on the Taguchi Method, 2nd edn. Society of Manufacturing Engineers (2010)
10. Pinar, A.M.: Optimization of process parameters with minimum surface roughness in the pocket machining of AA5083 aluminum alloy via Taguchi method. Arab. J. Sci. Eng. **38**(3), 705–714 (2012). doi:10.1007/s13369-012-0372-5
11. Kubiak, T.M., Benbow, D.W.: The Certified Six Sigma Black Belt HandBook, 2nd edn. ASQ Quality Press, Milwaukee (2009)
12. Kaymakci, M., Lazoglu, I.: Tool path selection strategies for complex sculptured surface machining. Mach. Sci. Technol. **12**(1), 119–132 (2008). doi:10.1080/10910340801913979

Part VIII
Joining

The Production-Related Influence of Iron Oxides on Steel Surfaces on the Adhesion of Fusion-Bonded Hybrid Structures

T. Reincke, S. Kreling and K. Dilger

1 Introduction and Motivation

Considering the required reduction of the exhaust emissions as well as the increasing market of electric vehicles within the automotive sector a major challenge consists in the realization of lightweight solutions. The high demands with regard to cycle times in the automotive serial production require innovative production and joining technologies for multi-material lightweight structures.

The Open Hybrid Lab Factory in Wolfsburg takes this innovative approach and examines the high-volume production-compatible manufacturing of multi-material structures of fiber reinforced thermoplastics and metals. Along with the research on injection molding and pressing of multi-material structures the roll forming process of hybrid parts manufactured in a continuous process offers a high development potential. With focus on the adhesion of polymer and metal within a multi-material structure the consequences of the heating process, integrated in a continuous process, on the metal surface have to be examined.

Beginning investigations focused on the first step in a roll formed production of multi-material composites which is the heating of the metal for the subsequent joining process of the metal and the carbon fiber reinforced thermoplastic to a multi-material part. Therefore a test setup was constructed which enable the induction heating of steel specimens as well as the hot gas heating of the carbon

T. Reincke (✉) · S. Kreling · K. Dilger
Institute of Joining and Welding, Technische Universität Braunschweig,
Langer Kamp 8, 38106 Brunswick, Germany
e-mail: t.reincke@tu-braunschweig.de

S. Kreling
e-mail: s.kreling@tu-braunschweig.de

K. Dilger
e-mail: k.dilger@tu-braunschweig.de

L.F.M. da Silva (ed.), *Materials Design and Applications*,
Advanced Structured Materials 65, DOI 10.1007/978-3-319-50784-2_27

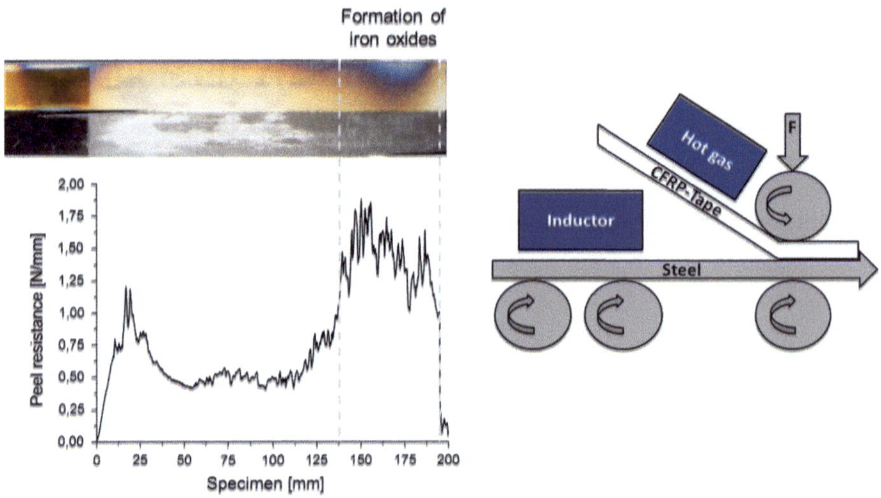

Fig. 1 Motivation: formation of iron oxides within a continuous fusion bonding process

fiber reinforced thermoplastics within a continuous process [1]. The examinations considered the influence of the steel temperature on the joint strength of a multi-material specimen. With regard to this, roller peel tests were carried out. The sample geometries and the measured area for the roller peel tests have been selected according to DIN EN 1464. The roller peel specimens consisted of a rigid adherend of uncoated DC 01 steel and a flexible adherend of unidirectional carbon fiber reinforced Polyamide 6. The results of a representative roller peel specimen as well as the schematic test setup are shown in Fig. 1.

Due to the induction heating of the metal in a continuous manufacturing process the heat is accumulated at the ends of the sample which results in heat colors respectively iron oxides on the uncoated metal [1]. Considering the areas of iron oxides on the steel surface the peel resistance significantly increases compared to the area without iron oxides.

Therefore this paper focuses on the correlation of iron oxides on the steel surface and the adhesion of carbon fiber reinforced thermoplastics towards these surfaces.

2 State of the Art

Based on the aim of the presented investigation fusion bonding as joining technology within sequential as well as continuous processes is described. In addition, the formation of iron oxides on steel surfaces due to tempering is considered.

2.1 Fusion Bonding

The fusion bonding process is a joining technology for thermoplastics or multi-material structures, i.e. metals and fiber reinforced thermoplastics. The thermoplastic matrix melts at a specific temperature with the consequence that the melted matrix of the fiber reinforced thermoplastic is able to wet the surface of the metal. As a result a multi-material joint by fusion bonding is created. A further advantage of fusion bonding is the possibility of its integration in a continuous roll forming process due to the non-required application of an adhesive. In addition no cure cycle occurs in contrast to an adhesively bonded multi-material structure. With focus on the recycling of the hybrid part debonding on demand can be realized by reheating of the joint to the melting temperature of the thermoplastic.

Concerning the fusion bonding of multi-material joints, various process technologies as well as pre-treatment methods have been investigated. Fusion bonding requires a defined heat input for the melting of the thermoplastic matrix. This can be achieved, among other things, by heating techniques such as induction heating, laser transmission heating, ultrasonic heating or resistance heating [2].

Most of the investigations in literature focus on multi-material joints of aluminum and glass or carbon fiber reinforced thermoplastics. Fusion bonded joints of aluminum and carbon fiber reinforced Polyamide 66 pre-treated by chemical or abrasive pre-treatment, carried out by induction heating, attained lap shear strengths in a range about 10 MPa [3]. Laser structured aluminum with glass fiber reinforced Polyamide 6 carried out by laser transmission heating attained lap shear strength of up to 25 MPa [4]. Amend achieved lap shear strengths up to 20 MPa at laser-heated aluminum, pre-treated by laser, and glass fiber reinforced Polyamide 66 [5]. In examinations concerning induction-heated multi-material joints of untreated DC 01 steel and glass fiber reinforced Polyamide 6 lap-shear strengths of up to 5 MPa were attained [6]. Examinations also focused on lap shear specimens with untreated steel adherends and a carbon fiber reinforced Polyamide 6 tape as 'adhesive' and attained lap shear strength up to 6 MPa [1].

The continuous fusion bonding process of overlapped Polypropylen and Aluminum was examined by Ramarathnam [7] whereas the investigations of Hümbert focused on overlapped Aluminum or DC 01 steel and glass fiber reinforced plastics [8]. The fusion bond of separately heated DC 01 steel and carbon fiber reinforced Polyamide 6 tape manufactured in a continuous process has been considered and evaluated by roller peel specimens [1].

In summary, state of the art examinations of fusion bonding joining technologies only focus on the non-continuous joining of multi-materials or continuous joining of mutual heated multi-materials. With focus on a continuous roll forming process and an improved joint strength the challenges are the reproducible heating of the metal as well as the fiber reinforced thermoplastic and the integration of a fusion bonding joining technology. Due to the separate heating of the untreated steel and the fiber reinforced plastics, iron oxides can occur on the steel surface prior to the subsequent fusion bonding process.

2.2 *Formation of Iron Oxides*

The heating of non-coated steel under atmospheric air, i.e. due to a tempering process, leads to the formation of iron oxides on the steel surface [9]. The tempering process also leads to different material properties in dependency of the tempering temperature and tempering time. As a consequence to the tempering at temperatures up to 200 °C heat colors occur. These heat colors are caused by the formation of thin iron oxide layers consisting of $Fe^{2+}(Fe^{3+})_2O_4$ which change the composition of the steel surface. The thickness of the oxide layer can be correlated with the characteristic temperature in the tempering process [9, 10].

With focus on the induction heating of steel edge effects which lead to increased heat generation at the edges have to be considered [11]. As a consequence heat colors on the steel surface, which have an influence on the adhesion of polymer and metal, can occur (see Fig. 1).

3 Experimental Set-up

3.1 *Materials*

The investigated materials were a DC 01 metal adherend with a thickness of 1.00 mm respectively 1.50 mm as well as an unidirectional carbon fiber reinforced Polyamide 6 tape. The Polyamide 6 has a melting temperature of 220 °C and a glass transition temperature of 47 °C. The fiber volume ratio is around 48% and the thickness of the material is 0.19 mm. The material properties of the two materials are shown in Table 1.

Table 1 Material properties

Metal adherend	DC 01
Yield strength (MPa)	Max. 280
Tensile strength (MPa)	270–410
Thickness (mm)	1.00/1.50
Coating	None
FRP-adherend	Carbon fiber reinforced thermoplastic
Matrix material	Polyamide 6
Melting temperature (°C)	220
Glass transition temperature (°C)	47
Thickness (mm)	0.19
Fiber volume ratio (%)	48

3.2 Lap Shear Test

The geometries of the specimens (100 mm × 25 mm) for the lap shear tests were selected according to DIN EN 1465. Both adherends consisted of steel and the carbon fiber reinforced Polyamide 6 tape was inserted as an 'adhesive' into the joining zone. The carbon fiber reinforced Polyamide 6 tape was not dried or conditioned previously. Compared to the non-tempered reference samples the steel samples for the determination of the influence of iron oxides were tempered for 15 min at the targeted tempering temperature in an oven. Therefore the oven was pre-heated on the targeted tempering temperature and the temperature was measured with thermocouples. The tempering time of the specimens started after achieving the targeted tempering temperature in the oven. As a consequence the specimens with higher tempering temperatures remained for a slightly longer time in the oven, due to the reduced temperature because of the opening and the insertion of the specimens. Afterwards the specimens were cleaned again with n-Heptane.

After that the two steel adherends and the carbon fiber reinforced thermoplastic were joined and fixed by clamps. All specimens were joined using the same joining pressure. The lap shear specimens were stored for 15 min at the predetermined joining temperature in an oven. The sample geometry of the lap shear specimens is shown in Fig. 2.

Considering the manufacturing of these specimens with only one layer of carbon fiber reinforced Polyamide 6 tape the resulting failure pattern can be divided into the types interfacial failure (AF), cohesive failure (CF) and substrate close cohesive failure (SCF). The cohesive failure means fiber breakage of the carbon fibers whereas the substrate close cohesive failure signifies a failure in the top layer of the matrix of the carbon fiber reinforced tape.

3.3 Test Parameter

The test parameters focused on the examination of the adhesion between the carbon fiber reinforced Polyamide 6 and the DC01 steel surface. As regards the influence of the joining temperature on the lap shear strength six different joining temperatures in the range of 220 °C to 300 °C were examined and considered as reference. This area was selected due to the melting temperature at 220 °C and the start of degradation of the Polyamide 6 at approximately 270 °C. Due to the manufacturing

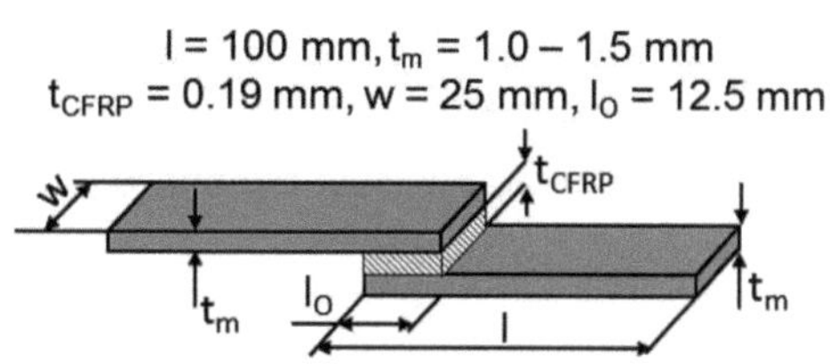

Fig. 2 Sample geometry lap shear test

Table 2 Test matrix lap shear tests

Aim of analyses	$T_{DC01\ tempered}$ (°C)	T_{join} (°C)	Comments
Reference	–	220, 240, 250, 260, 270, 300	Non-tempered, variation of T_{join}
Influence of the tempering temperature	220, 240, 250, 270, 300	250	Variation of $T_{DC01\ tempered}$, constant T_{join}
Influence of the joining temperature	220, 240, 250, 260, 270, 300	220, 240, 250, 260, 270, 300	$T_{DC01\ tempered} = T_{join}$

process the temperature of the DC 01 steel and the carbon fiber reinforced Polyamide 6 was equal during the joining process.

With reference to a comparison of different tempering temperatures and the related formation of iron oxides on the steel surface the tempering temperature was varied and the joining temperature was kept constant.

In addition, due to the temperature dependence of the viscosity of the Polyamide 6 matrix, the joining temperature as well as the tempering temperature of the DC 01 steel was kept constant. The test matrix for the lap shear specimens is shown in Table 2.

3.4 X-Ray Photoelectron Spectroscopy

The quantitative spectroscopic technique X-ray Photoelectron Spectroscopy (XPS) enables the determination of elemental composition, electronic as well chemical state within a material. The XPS measurements were carried out with an ESCALAB Mk. II using an MGK_α X X-ray source 1253.6 eV in the monochromatic mode at 200 W at a pressure below 5×10^{-7} Pa. The spot of the spectrometer was approximately 1000 µm. In addition, the pass energy for the survey scans was 180 eV with a step width of 1.0 eV and for the fine scans 50 eV with a step width of 0.1 eV. With the focus on preventing influences due to contaminations on the steel surface, i.e. because of oil, the surface was sputtered with Argon until the contaminations were removed.

3.5 Drop-Shape Analysis

The Drop-Shape Analysis was carried out with a Data Physics OCA 30 using Diiodomethane, Ethylene glycol and Water. The Drop-Shape Analysis measures the contact angle by using the image of a sessile drop and considered the points of intersection between the drop contour and the projection of the surface. In addition the surface energy with the polar and dispersive part was calculated. The polar and dispersive parts can be determined due to the different polar/dispersive behavior of the three test liquids.

4 Results

4.1 Surface Analyses

The surface analysis was carried out by X-ray Photoelectron Spectroscopy as well as Drop-Shape Analysis to determine the amount of oxygen and iron on the steel surface as well as to investigate the polarity of the steel surface.

4.1.1 X-Ray Photoelectron Spectroscopy

The XPS measurements focused on the examination of the steel surface considering especially the amounts of carbon (C1s), oxygen (O1s) and iron (Fe2p3). Therefore the survey scans show the total amounts of mainly C1s, O1s and Fe2p3 on the steel surface whereas the fine scans show the composition of the C1s, O1s and Fe2p3. Further elements were not considered so that the determined values were standardized to 100 atomic percentage (at %).

The results of the survey scans of the non-tempered and tempered DC 01 steel specimens are shown in Fig. 3.

Fig. 3 Results XPS survey scan

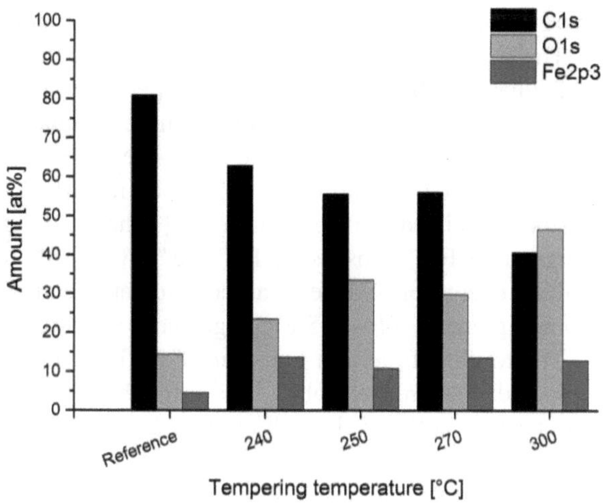

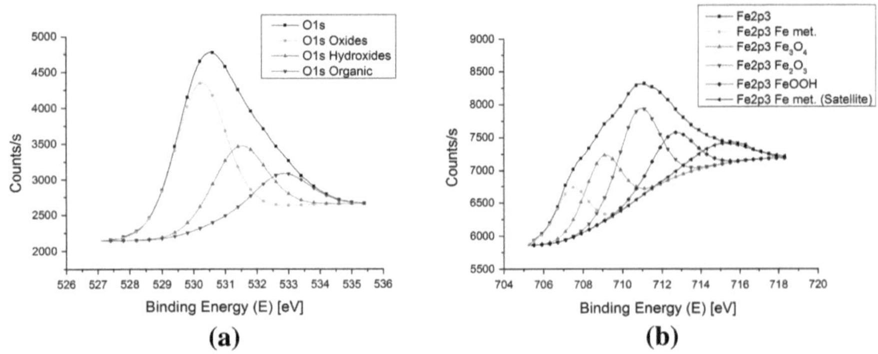

Fig. 4 Fine scans oxygen (**a**) and iron (**b**)

With regard to the survey scan the amount of carbon (C1s) is decreasing with increasing tempering temperature. In addition, the amount of oxygen (O1s) and iron (Fe2p3) is increasing with the highest value at the tempering temperature of 300 °C. The increase of iron and oxygen can be associated with the formation and the growth of the iron oxide layers on the steel surface.

In the following the approach for the evaluation of the fine scans is demonstrated for oxygen and iron. Therefore representative fine scans for oxygen and iron for a DC 01 steel sample tempered at 240 °C are shown in Fig. 4.

The O1 s fine scan demonstrated in Fig. 4a shows the composition of oxygen for a DC 01 steel sample tempered at 240 °C. The peak of the sample is located at a binding energy of ~530 eV with shoulders at binding energies of ~531 eV and ~533 eV. Regarding the measured peak binding energies of ~530 eV are assigned to oxides in metallic compounds [12]. Furthermore binding energies of ~531 eV correspond to oxygens in hydroxyl compounds [13] and binding energies of ~533 eV to oxygens in organic compounds [14].

The Fe2p3 fine scan in Fig. 4b shows the composition of iron for DC 01 steel which has also been tempered at a temperature of 240 °C. The peak is located at a binding energy of ~710 eV which can be assigned to Fe^{3+} ions as in Fe_2O_3 [15] and $Fe^{2+}(Fe^{3+})_2O_4$ [16]. On the right side of the peak two shoulders at a binding energy of ~707 eV and ~709 eV can be identified. Binding energies of ~707 eV can be associated with pure iron [17] whereas binding energies of ~709 eV are assigned to Fe^{2+} ions as in $Fe^{2+}(Fe^{3+})_2O_4$ [15]. In addition, the shoulder at a binding energy of ~712 eV can be associated with FeO(OH) [12]. The shoulder at a binding energy of ~715 eV might be a satellite.

The results of the iron fine scan (Fe2p3) of the non-tempered reference and the tempered steel are exemplary shown for the significant change in the iron (Fe2p3) composition in Fig. 5. In addition Fig. 5 only shows the change in the composition of pure metal and Fe_2O_3 due to its most significant importancy considering the influences of the tempering temperature.

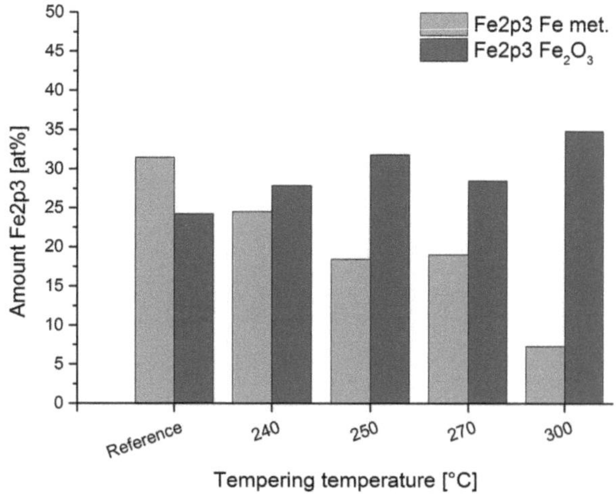

Fig. 5 Results XPS fine scan iron

The iron fine scan in Fig. 5 shows in dependency of the tempering temperature some differences in terms of the elemental composition. The amount of pure iron (~707 eV) decreases with higher tempering temperatures whereas the amount of Fe^{3+} ions as in Fe_2O_3 and $Fe^{2+}(Fe^{3+})_2O_4$ (~710 eV) increases. This can be explained with the formation of new oxide layers at higher tempering temperatures by a decrease in the amount of pure iron.

In conclusion of the XPS results the tempering of the DC 01 steel causes the formation of new oxide layers, due to the increasing amount of oxygen and iron as well as the conversion of pure iron to iron oxides.

4.1.2 Drop-Shape Analysis

The Drop-Shape Analysis has been carried out according to Sect. 3.5. The Drop-Shape Analysis only indicates the share of polar and dispersive parts due to its local measuring comparing with the XPS analysis and is also more effected by the roughness.

The ratio of polar to dispersive part of the surface energy of the measured, tempered specimens is shown in Fig. 6.

The amount of polar parts of the surface energy increases with higher tempering temperatures whereas the amount of the dispersive parts of the surface energy remains constant at a level of approximately 30 mN/m. Therefore the drop-shape analysis indicates that tempering of the steel specimens at a defined tempering temperature causes an increase of the amount of oxygen at the surface and subsequently a more polar surface.

Fig. 6 Drop-shape analysis

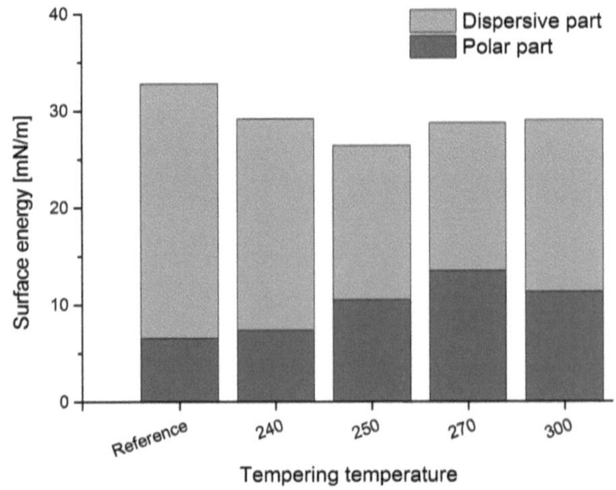

4.2 Influence of the Tempering Temperature on Lap Shear Strength

With focus on the determination of the influence of the tempering temperature of the DC 01 steel on the lap shear strength the specimens were tempered at various temperatures according to Sect. 3.3. The joining temperature of 250 °C and thus the viscosity of the matrix were kept constant (see Sect. 3.3). The results are shown in Fig. 7.

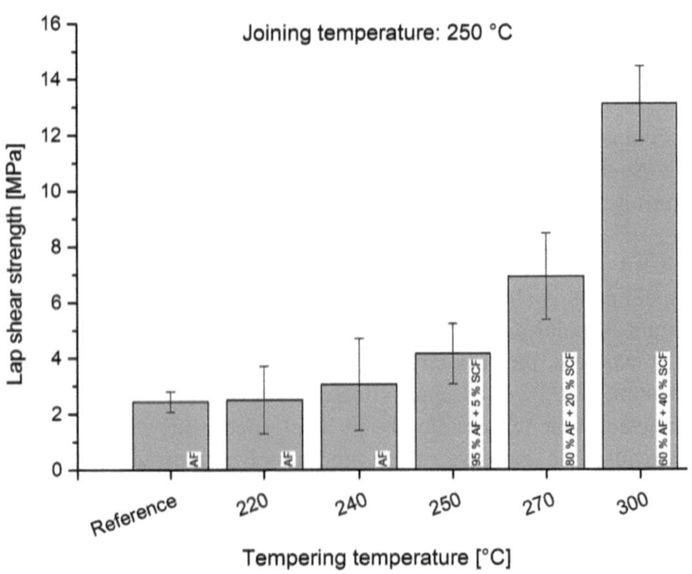

Fig. 7 Influence of tempering temperature on lap shear strength

The reference has not been tempered prior to joining. The reference shows an interfacial failure and has lap shear strength of 2.43 ± 0.36 MPa. Although the standard deviation of the tested specimens is relatively high, a clear tendency of the influence of the iron oxides on the lap shear strength is identifiable. Increasing the tempering temperature prior to joining relates to higher lap shear strengths. The highest lap shear strength of 13.12 ± 1.34 MPa is measured at a tempering temperature of 300 °C whereas the lowest lap shear strength with 2.49 ± 1.20 MPa belongs to a tempering temperature of 220 °C. The specimens tempered at 220 °C also show an interfacial failure whereas the specimens tempered at 300 °C have a mixed failure. The share of substrate cohesive failure (see Sect. 3.2) increases with higher tempering temperatures.

The formation of new oxide layers shown in Sect. 4.1.1 due to the tempering is a possible explanation for the increase of the lap shear strength. The increasing amount of the oxygen (O1s) which can interact with the Polyamide 6 matrix by hydrogen bonds supports this theory [18].

4.3 Influence of the Joining Temperature on Lap Shear Strength

To examine the interaction of the tempering temperature and the joining temperature on the lap shear strength, lap shear specimens were tempered at different temperatures between 220 and 300 °C and later on joined at different temperatures. These examinations focused on the targeted continuous manufacturing process described in Sect. 1. With regard to a continuous manufacturing process the joining temperature corresponds to the tempering temperature. In comparison with Sect. 4.2 the viscosity of the Polyamide 6 matrix varied and was not kept constant. The preparation of the specimens was carried out according to Sect. 3.2. The results and the comparison of the tempered samples to the reference specimens are demonstrated in Fig. 8.

In the range of 220–250 °C an increase of the lap shear strength of the tempered specimens from 4.77 ± 0.72 MPa up to 10.99 ± 0.80 MPa, due to the increasing joining temperature, is detectable. Regarding the failure pattern the share of an interfacial failure decreases whereas the share of substrate close cohesive failure increases. The joining above 250 °C does not cause a further significant growth of the lap shear strength. The maximum lap shear strength of 11.99 ± 0.92 MPa is achieved at a joining temperature of 300 °C.

Comparing the lap shear strength of the tempered with the non-tempered specimens, significantly higher values of the tempered samples are recognizable. The lap shear strength of the non-tempered specimen at joining temperatures between 240 and 270 °C is in the range of approximately 2–3 MPa and therefore roughly 70% lower compared to the tempered specimens. In addition, these specimens all show an interfacial failure. The joining temperature of 300 °C results in a

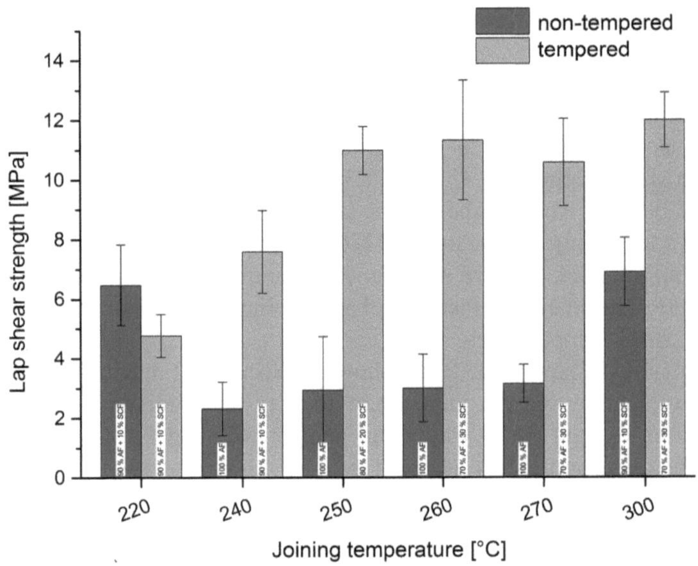

Fig. 8 Influence of joining and tempering temperature on lap shear strength

lap shear strength of 6.91 ± 1.15 MPa which can be explained with the start of the degradation of the Polyamide 6, beginning at temperatures around 270 °C. The start of degradation leads to degradation products enabling a better adhesion to the Polyamide 6. Regarding the increase of the lap shear strength of the non-tempered sample joined at a temperature of 220 °C the formation of oxides on the steel surface has to be considered. With regard to this the viscosity of the matrix is not high enough to wet the complete steel specimens. The specimens are not completely flat due to the manufacturing process of the samples. In addition possible deviations of the temperature because of the heating in an oven can occur. This leads, in combination with the melting temperature of the Polyamide 6 which is approximately 220 °C, to the formation of iron oxides on the edges of the steel surface. In consequence the heated air in the oven can affect the steel surface and create iron oxides which results in higher lap shear strengths. The shared interfacial and substrate close cohesive failure confirms this theory.

Comparing the results of the Sects. 4.2 and 4.3 the influence of resulting bending stresses during the lap shear test is visible. The investigations in Sect. 4.2 are carried out with DC 01 steel with a thickness of 1.0 mm whereas the examinations in this Chapter are carried out with DC 01 steel with a thickness of 1.5 mm.

Compared to the specimens in Sect. 4.2 which are all joined at a temperature of 250 °C the influence of the viscosity of the Polyamide 6 matrix can be identified. The viscosity increases with higher joining temperatures and relates to a significant increase of the lap shear strength.

5 Conclusions

In this paper, the production-related formation of iron oxides on uncoated steels is analyzed and as a consequence the improved adhesion of fusion-bonded hybrid structures consisting of steel and carbon fiber reinforced Polyamide 6 tape is justified.

The Drop-Shape Analysis indicates an increase of the polar amount of the surface energy with higher tempering temperature of the steel. The X-ray Photoelectron Spectroscopy validates the results of the Drop-Shape Analysis. Increasing tempering temperatures lead to a higher amount of oxygen on the surface. The amount of pure iron decreases with higher tempering temperatures whereas the amount of iron oxides increases. This can be explained with the formation of new oxide layers at higher tempering temperatures.

The lap shear examinations show the influence of the tempering temperature on the adhesion of the steel and the carbon fiber reinforced Polyamide 6. Referring to the lap shear strength, the tempering of the steel increases the joint strength of the multi-material bond. The iron oxides on the surface prove advantageous for the adhesion of steel and carbon fiber reinforced Polyamide 6. The viscosity of the matrix has to be considered for the joint strength of the hybrid bond, as well.

Further investigations have to examine the influence of ageing processes on the adhesion of the carbon fiber reinforced Polyamide 6 and the iron oxides on the steel surface. Considering the induction heating of steel samples in a continuous process different processing speeds and temperatures of the thermoplastic tape have to been investigated. In addition, the influence of induction heating on various steel coatings, i.e. zinc-coated steel with regard to corrosion protection, has to be examined.

Acknowledgements This research and development project TRoPHy2 (02PQ5135) is funded by the German Federal Ministry of Education and Research (BMBF) within the Forschungscampus "Open Hybrid LabFactory" and managed by the Project Management Agency Karlsruhe (PTKA). The authors would like to thank the partners of the participating research institute of the German Aerospace Center as well as the project partner data M Sheet Metal Solutions GmbH, Dreistern GmbH, EDAG GmbH & Co. KGaA, Salzgitter Mannesmann Forschung GmbH and Volkswagen AG Nutzfahrzeuge. The author is responsible for the contents of this publication.

References

1. Reincke, T., Kreling, S., Dilger, K.: Roll forming of multi-material composites. In: SAMPE Conference Proceedings, Long Beach, USA (2016)
2. Ageorges, C., Ye, L., Hou, M.: Advances in fusion bonding techniques for joining thermoplastic matrix composites: a review. Compos. Part A **32**, 839–857 (2001)
3. Mitschang, P., Velthuis, R., Emrich, S., Kopnarski, M.: Induction heated joining of aluminum and carbon fiber reinforced nylon 66. J. Thermoplast. Compos. Mater. **22**, 767–801 (2009)
4. Heckert, A., Zaeh, M.F.: Laser surface pre-treatment of aluminum for hybrid joints with glass fiber reinforced thermoplastics. J. Laser Appl. 27S29005 (2015)

5. Amend, P., Pfindel, S., Schmidt, M.: Thermal joining of thermoplastic metal hybrids by means of mono- and polychromatic radiation. Phys. Proc. **41**, 98–105 (2013)
6. Lippky, K., Löchte, C., Schnurr, R., Fischer, F., Raatz, A., Dietrich, F., Dilger, K., Dröder, K.: Production technology for the pre-assembly of multi-material-preforms—process design, handling- and joining technologies. In: 14th SAMPE Europe Technical Conference, Tampere, Finland (2014)
7. Ramarathnam, M., Libertucci, M., Sadowski, M.M., North, T.H.: Joining of polymers to metal. WJ Res. Supplements 483–490 (1992)
8. Hümbert, M., Mitschang, P.: Interdependency of welding speed and distance between coil and roller during induction welding of metal and glass fiber reinforced Polyamide 6. In: 20th International Conference on Composite Materials, Copenhagen, Denmark (2015)
9. Waseda, Y., Suzuki, S.: Characterization of Corrosion Products on Steel Surfaces. Springer, Berlin Heidelberg New York (2006)
10. Feßmann, J., Orth, H.: Angewandte Chemie und Umwelttechnik für Ingenieure—Handbuch für Studium und betriebliche Praxis. ecomed Verlagsgesellschaft AG & Co. KG, Augsburg (2002)
11. Rudnev, V., Loveless, D., Cook, R.L., Black, M.: Handbook of Induction Heating. Marcel Dekker, New York (2002)
12. Barr, T.L.: An ESCA study of the termination of the passivation of elemental metals. J. Phys. Chem. **82**(16), 1801–1810 (1978)
13. Jin, S., Atrens, A.: ESCA-studies of the structure and composition of the passive film formed on stainless steels by various immersion times in 0.1 M NaCl solution. Appl. Phys. A **42**(2), 149–165 (1987)
14. Clark, D. T., Thomas, H. R.: Applications of ESCA to polymer chemistry. XVII. Systematic investigation of the core levels of simple homopolymers. J. Polym. Sci. Polym. Chem. Ed. **16** (4), 791–820 (1978)
15. Clayton, C.R.: A bipolar model of the passivity of stainless steel: the role of mo addition. J. Electrochem. Soc. **133**(12), 2465 (1986)
16. Marcus, P., Grimal, J.M.: The anodic dissolution and passivation of NiCrFe alloys studied by ESCA. Corros. Sci. **33**(5), 805–814 (1992)
17. Di Castro, V., Ciampi, S.: XPS study of the growth and reactivity of thin films. Surf. Sci. **331–333**, 294–299 (1995)
18. Üzum, M.: Metal/polymer hybrids: multiscale adhesion behaviour and polymer dynamics. Technische Universität Berlin, Fakultät II—Mathematik und Naturwissenschaften, Berlin (2015)

Comparison of Stepped, Curved, and S-Type Lap Joints Under Tensile Loading

Şerif Çitil

1 Introduction

In load-bearing structures, bonding joints are used in many areas of the aviation industry. These joints are low in labor, time, and cost; they have high corrosion and fatigue resistance, crack delay, and good damping. They are versatile and can bond different types of material. These outstanding properties of bonding joints, in addition to lightness and easy applicability, have been the focus of many studies.

Adin [1] investigated the effect of bonding joint angle in Scarf Lap Joints. Soykök [2] drilled small holes in sheets and joined the sheets by filling the small holes with adhesive and compared joint types with different geometries. Jen [3] investigated the fatigue of Scarf bonding joints at different angles. Arenas et al. [4] bonded different types of materials. Grant et al. [5] subjected single lap joints to three—and four-point bending. Dasilva and Lopes [6] used mixed flexible and rigid adhesives in bonding joints and proposed the use of more flexible adhesives near the end of joints. Sancaktar and Karmarkar [7] examined the effects of the surface in multi-stepped double Scarf bonding joints. Çitil et al. [8] demonstrated that the intermediate element had no effect on double-strap bonding joints. Ayaz et al. [9] proposed a rapid solution to repair damaged pipes.

Single—and double-effect joint types are the most common of bonding joints. Da Silva and Adams [10] analyzed stress on double-strap joints in a given temperature range using different types of materials and adhesives. Çitil et al. [11] embedded reinforced covers instead of using classic double-strap joints. Apalak and

Ş. Çitil (✉)
Faculty of Engineering, Department of Mechanical Engineering,
Adıyaman University, 02040 Adıyaman, Turkey
e-mail: scitil@adiyaman.edu.tr

L.F.M. da Silva (ed.), *Materials Design and Applications*,
Advanced Structured Materials 65, DOI 10.1007/978-3-319-50784-2_28

Engin [12] analyzed the elasticity and plasticity of single—and double-strap joints bonded with adhesive. They investigated the formation and development of damage loads under different loading conditions. Temiz [13] performed finite element analysis of four-point bending loading with AA2024-T3 aluminum sheets bonded with the double-strap method.

Decreasing stresses from the damage load and joint type and creating a safe and durable bonding joint depend on the design of the model [14]. The model has great importance for bonding joints [15–21]. The creation of a durable bonding joint depends on the type of the adhesive, the joined parts, the surface on which it is applied, overlap length, thickness, temperature, and the type of joint model [22].

Undesirable peeling stresses in bonding joints depend on the direction of the load. These stresses make the joint insecure and indurable, which has motivated many scientists to design appropriate joint models to eliminate undesirable stresses [23, 24]. Marques and Da Silva [25] proposed giving certain angles to the inner side of the patch and end points of the adhesive to reduce the stresses that occur at the ends of double-strap bonding joints.

The joint model has a great importance in bonding joints. It is possible to make the joint model more durable and safer by modeling the connecting surfaces of the parts. This study proposes two different models to make bonding joints safer. Different geometric surfaces were formed with the same overlap length and thickness. The most common type of lap joint in the literature (stepped) was compared with two new joint models (curved, S-type). Finite element analysis was performed by creating three different bonding joints with the same overlap length and thickness. Experimental tests were performed to verify the models.

2 Materials and Methods

AA 2024-T3 aluminum alloy sheets, which are very commonly used in aviation, were used in the experimental models. Sheets with h = 6.5 mm thickness were processed in the vertical processing center (CNC) and cut to w = 25 mm in width and L = 100 mm in length. Type I (stepped), Type II (curved), and Type III (S-type) samples were created by processing the end portions of the samples, as shown in Fig. 1 for L_1 = 30 mm overlap length.

Samples were placed into a mold (Figs. 2 and 3) and bonded with the structural adhesive produced by 3 M Scotch-Weld. This adhesive is a DP810 bicomponent acrylic, resistant to humid environments, and cures at room temperature (24 °C). A thickness of 0.2 mm was applied and let dry for 24 h at room temperature. Before bonding, surfaces were washed with acetone and completely dried.

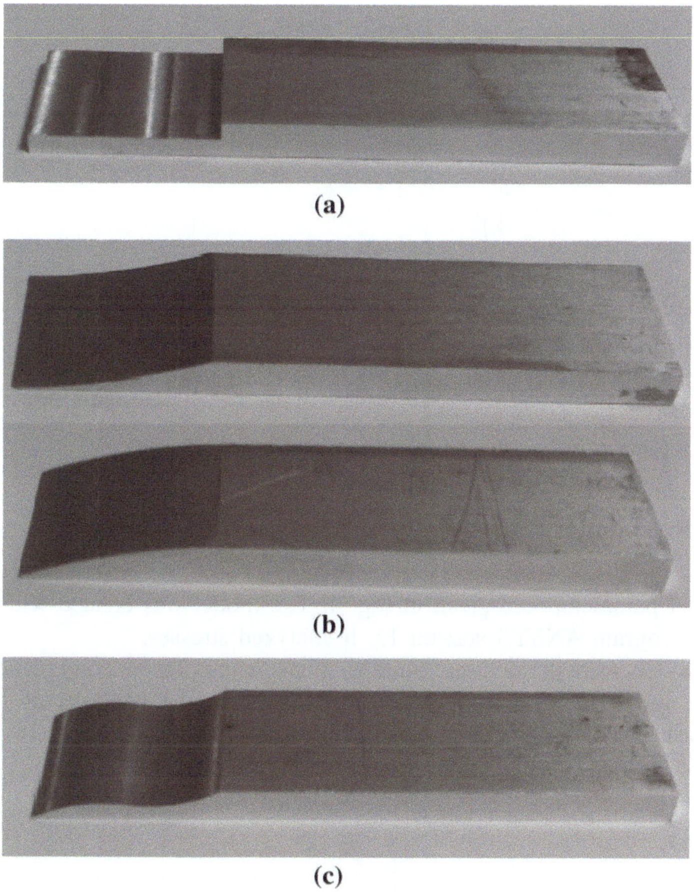

Fig. 1 Test samples: **a** Type I (stepped), **b** Type II (curved), **c** Type III (S-type)

Fig. 2 The mold used to prepare the samples

Fig. 3 Test samples: **a** step lap joint, **b** curved lap joint, **c** S-lap joint

3 Finite Element Model

The model parameters are given in Fig. 4. The model was created with the finite elements program ANSYS version 15. It analyzed stresses.

The models were divided into finite elements using plane 82 and plane 183 with 8 and 6 nodal points and a rectangular prism. In the analyses of lap joints, elasto-plastic analysis was done on the adherent and adhesive (Fig. 5). Stress and strain distributions were obtained using the Von Mises yield criterion.

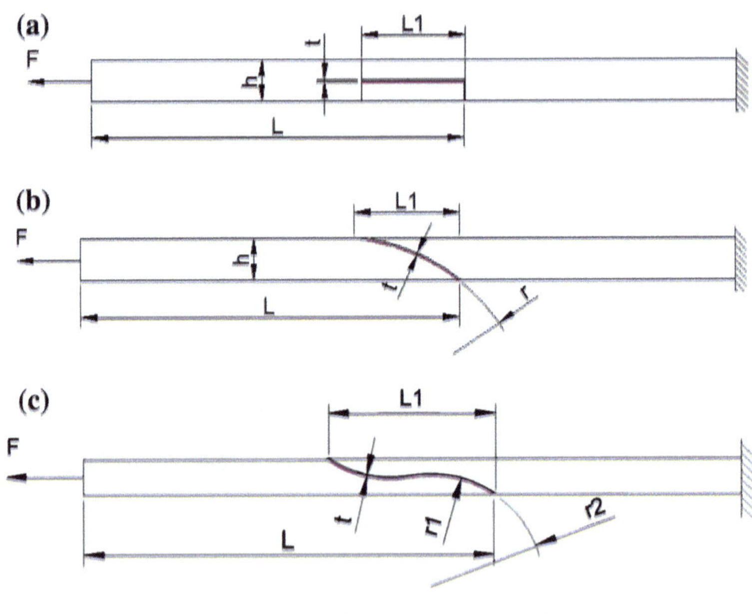

Fig. 4 Lap joint types **a** Type-I (stepped), **b** Type-II (curved), **c** Type-III (S-type)

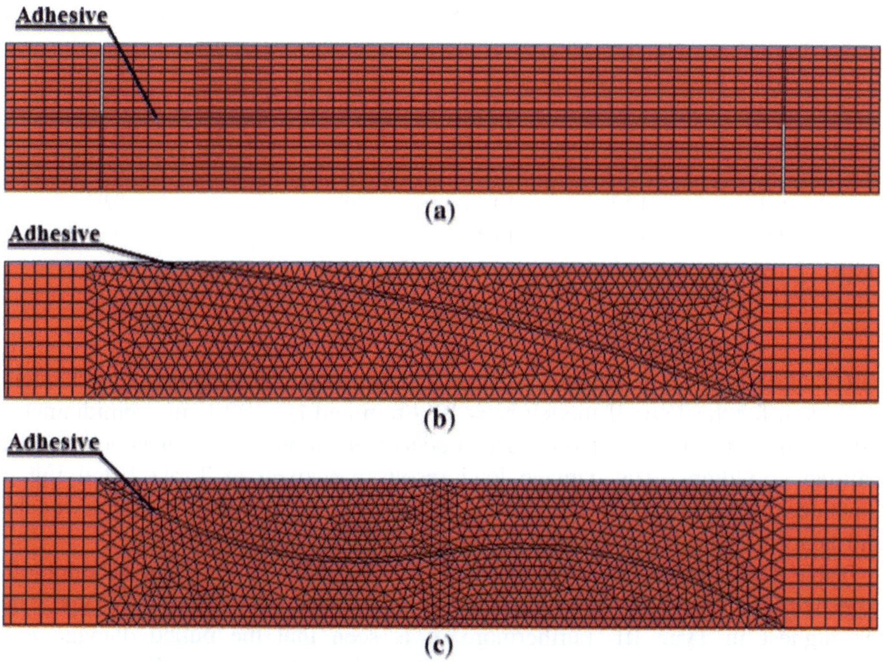

Fig. 5 The finite element model of lap joints: **a** Type I, **b** Type II, and **c** Type III

Table 1 Physical and mechanical properties of the adherent (AA 2024-T3) and adhesive (DP810)

	Aluminum alloy (AA2024T3)	Adhesive (DP810)
E_a (MPa)	71.875	425.082
ν_e	0.33	0.35
σ_y (MPa)	430	11.9023
σ_t (MPa)	482	24.2385
ε_t (mm/mm)	0.16	0.115

Table 1 lists the physical and mechanical data of the adhesive and aluminum A2024-T3 material used in the non-linear finite element models.

4 Results and Discussion

4.1 Numerical and Experimental Results

The results from finite element models have a 94% similarity to experimental models. In the finite element analyses, the values taken for three different models were L = 70 mm, overlap length L_1 = 30 mm, adhesive thickness t = 0.2 mm and

Table 2 Numerical tensile damage loads of joint types for $L_1 = 30$ mm overlap length, t = 0.2 mm adhesive thickness and h = 6.5 mm adherent thickness (Newton)

	Num. damage load results (kN)			Exp. damage load results (kN)	P_{FEM}/P_{EXP}
Joint type	h = 6.5 mm	h = 10 mm	h = 13.5 mm	h = 6.5 mm	For h = 6.5 mm
Type I	8.4	10.3	10.46	8.9	0.94
Type II	11.2	13.2	13.8	12.8	0.87
Type III	13.9	16.8	18.2	14	0.99

part thickness h = 6.5, 10, and 13.5 mm. In addition, r = 99.9 mm semidiameter was taken for the Type II model; $r_1 = 19.9$ mm and $r_2 = 20.1$ mm semidiameters were used for the Type III model. The models were subjected to tensile loads under limiting conditions. The damage load results are given in Table 2 and Fig. 5. Considering that the numerical solution is approximate, this approach is reasonable and hence the results from finite element models are acceptable.

Table 1 shows that damage loads significantly changed in different bonding joint types with the same overlap length. The lowest damage load occurred in Type I and the highest in Type III. Furthermore, it is seen that the pulled damage load decreased as the part thickness increased depending on the part thickness.

4.2 Stress Distribution Results

The stress distribution results of the stepped, curved, and S-type lap joints subjected under tensile loads are examined in the following graphics. The stresses in adhesive–sheet interface located in the x-directional overlap length area are given for Type I, Type II, and Type III bonding models. These lines are shown with the A–B line for the x directional interface (Fig. 6).

The lowest damage load pressure was taken into account to obtain an equal pressure force in bonding joints. The lowest damage load was in the Type I model with 6.5 mm thickness, which was applied to all models.

4.3 The Stress Distribution Along the A–B Line for Type I

The Type II and III models were produced as alternatives to the Type I joint and have the same overlap length as Type I. However, the surface area the adhesive was applied to was increased by creating different surface geometries. As a result of the

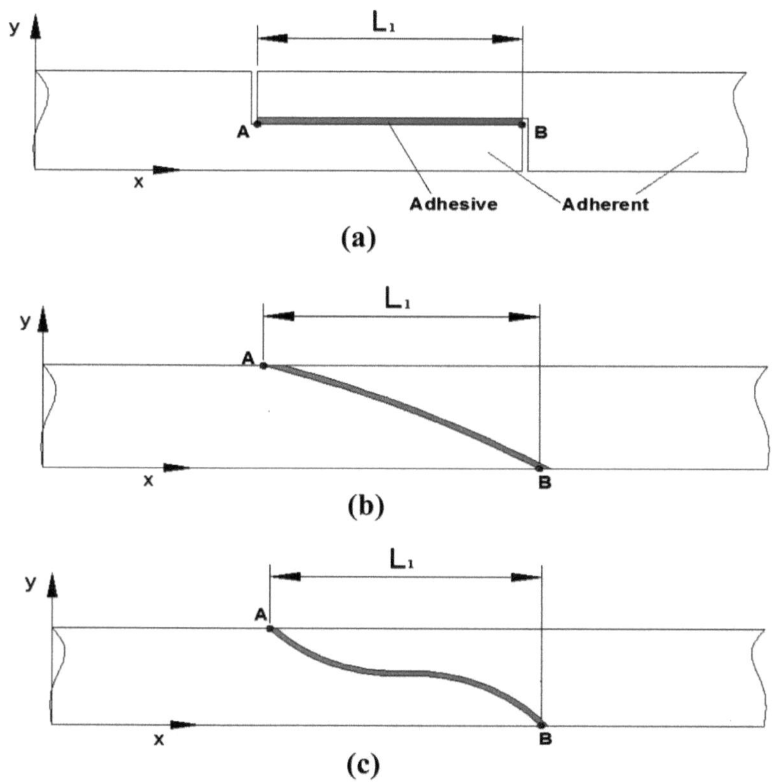

Fig. 6 Stress distribution lines for bonding models (A–B line): **a** Type I, **b** Type II, and **c** Type III

calculations, the surface areas of Type I, Type II, and Type III models were approximately 750, 763, and 787 mm^2, respectively. In addition, the adhesive is forced to shear in the Type I model. However, in the Type II and Type III models, the load affected the adhesive horizontally and vertically and was subjected to both shear and normal tensile loads. Thus, both normal stresses and shear stresses occurred on the normal surface of the adhesive (Figs. 7, 8 and 9).

When the stress distributions along the A–B line for Type I were analyzed, it was observed that x and y directional maximum normal stresses ($\boldsymbol{\sigma_x}$ and $\boldsymbol{\sigma_y}$) occurred in the area close to point A, due to peel stresses, and decreased toward B point. In shear and equivalent ($\boldsymbol{\tau_{xy}}$, $\boldsymbol{\sigma_{eqv}}$) stresses, it began to increase from point A, became constant toward point B and reached a maximum around point B. As can be seen in the figures, the adhesive was forced to shear in particular in the Type I model. In Type II model, normal, shear, and equivalent ($\boldsymbol{\sigma_x}$, $\boldsymbol{\sigma_y}$, $\boldsymbol{\tau_{xy}}$, $\boldsymbol{\sigma_{eqv}}$) stress distributions occurred minimally around point A and reached a maximum around

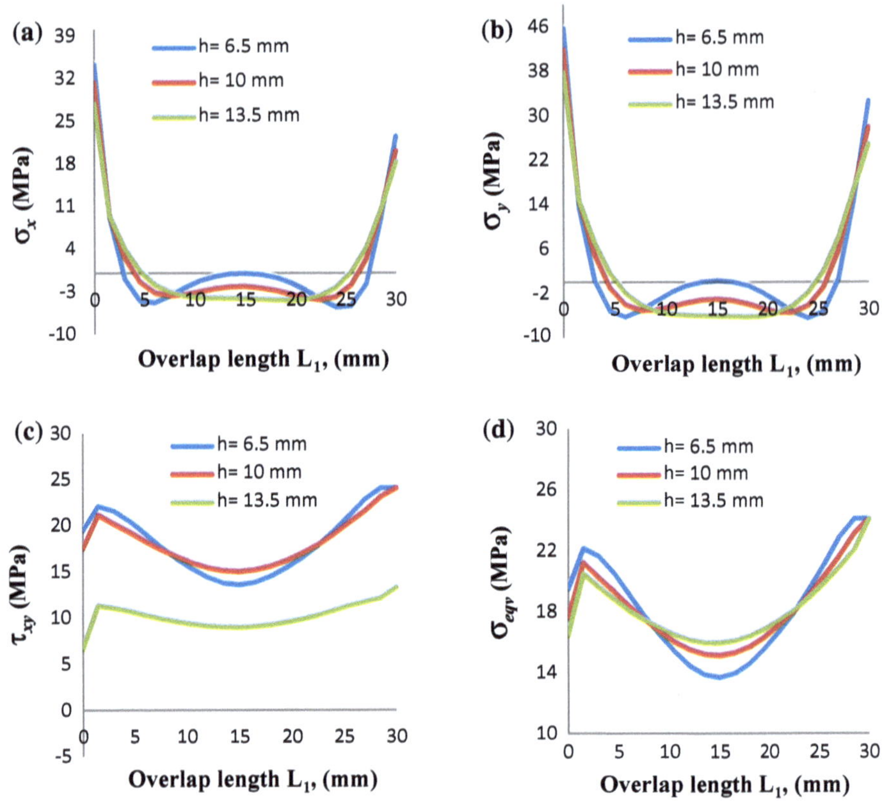

Fig. 7 Stress distributions along the A–B line for Type I: **a** σ_x, **b** σ_y, **c** τ_{xy}, and **d** σ_{eqv} in the model, adherent thickness (h)

point B. The adhesive was subjected to shear and tensile forces in the Type II model. In the Type III model, normal and equivalent stresses (**σ_x, σ_{eqv}**) occurred at a maximum at points A and B. It was at a minimum between the two points. Shear stresses (**σ_y, τ_{xy}**) were at a minimum at the end portions and a maximum in the central part of the A–B line. In the Type III model, the adhesive was subjected to tensile forces, especially at end the points, and to shear forces in the central part of the A–B line.

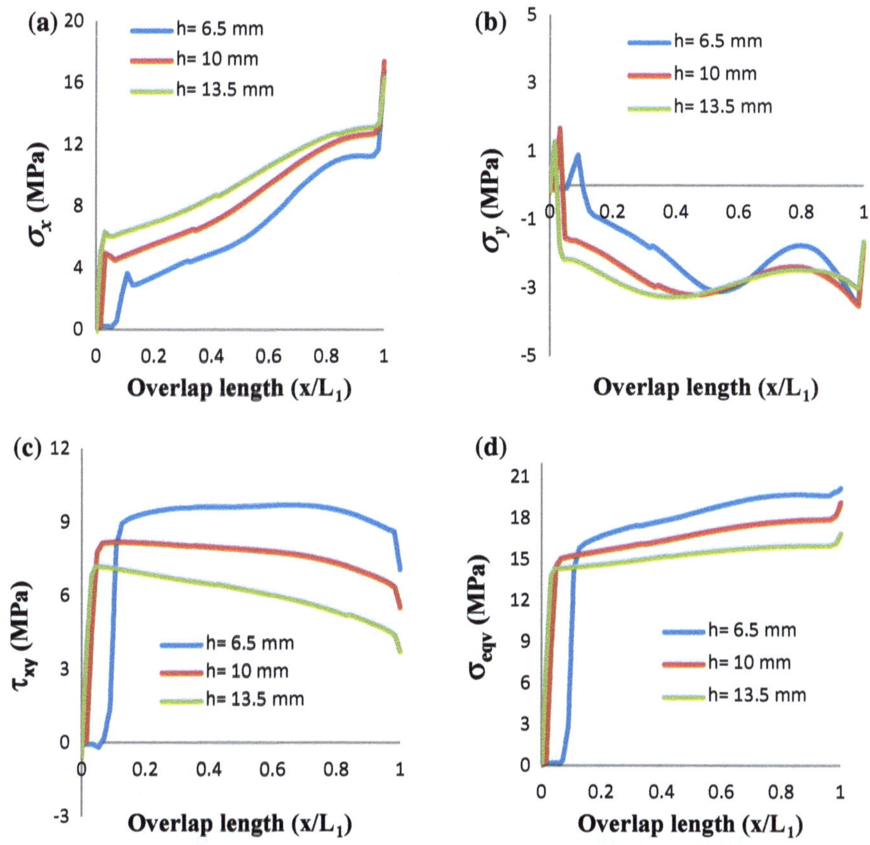

Fig. 8 Stress distributions along the line A–B for Type II: **a** σ_x, **b** σ_y **c** τ_{xy}, and **d** σ_{eqv} in the model, adherent thickness (h)

Stress-strain graphics of the joined models show that stresses decreased as part thickness increased along the A–B line in all models and the damage loads they withstood significantly increased. The reason for this is that the surface area of the material increased as the part thickness increased.

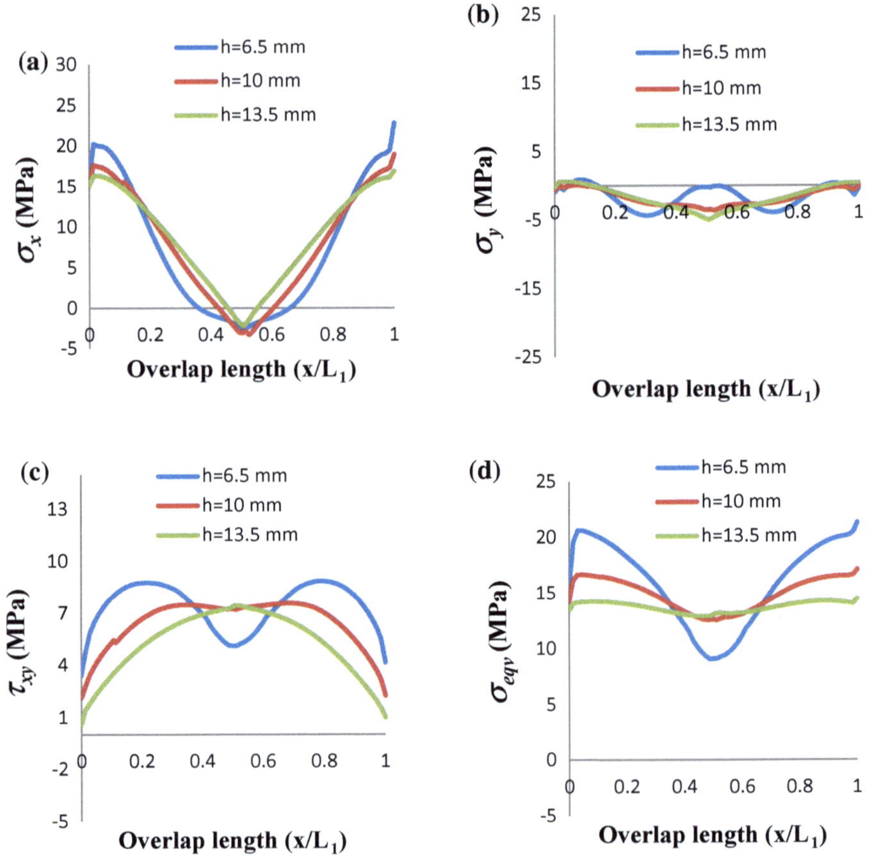

Fig. 9 Stress distributions along the A–B line for Type III: **a** σ_x, **b** σ_y **c** τ_{xy}, and **d** σ_{eqv} in the model, adherent thickness (h)

5 Conclusions

In this study, an attempt was made to develop two joint models (Type II and III) as alternatives to conventional stepped joints (Type I). Type II and III models were experimentally and numerically compared to the conventional joint model (Type I). We conclude the following.

- The experimental and numerical data were in 94% agreement.
- Improving surface geometry increased the damage load that lap joints withstood when subjected to tensile loads.
- Type II and III joint models withstood more damage load than Type I because the surface area on which the adhesive was applied changed the surface geometry.

- The damage loads they withstood significantly increased because the load that was applied was distributed depending on the surface geometry and the types of loads applied to the adhesive (normal and shear loads).
- Thicker sheets led to decreased stress and greater damage loads could be withstood.

Acknowledgements The author is thankful to Adıyaman University scientific research projects for financial support (MÜFBAP/2014-0004).

References

1. Adin, H.: The investigation of the effect of angle on the failure load and strength of scarf lap joints. Int. J. Mech. Sci. **61**, 24–31 (2012)
2. Soykok, I.F.: End geometry and pin-hole effects on axially loaded adhesively bonded composite joints. Compos. Part B **77**, 129–138 (2015)
3. Jen, Y.M.: Fatigue life evaluation of adhesively bonded scarf joints. Int. J. Fatigue **36**, 30–39 (2012)
4. Arenas, J.M., Cristina Alía, C., Narbón, J.J., Ocaña, R., González, C.: Considerations for the industrial application of structural adhesive joints in the aluminium—composite material bonding. Compos. Part B. **44**, 417–423 (2013)
5. Grant, L.D.R., Adams, R.D., da Silva, L.F.M.: Experimental and numerical analysis of single-lap joints for the automotive industry. Int. J. Adhes. Adhes. **29**, 405–413 (2009)
6. Da Silva, L.F.M., Lopes, M.C.Q.: Joint strength optimization by the mixed-adhesive technique. Int. J. Adhes. Adhes. **29**, 509–514 (2009)
7. Sancaktar, E., Karmarkar, U.: Mechanical behavior of interlocking multi-stepped double scarf adhesive joints including void and disbond effects. Int. J. Adhes. Adhes. **53**, 44–56 (2014)
8. Çitil, Ş., Ayaz, Y., Temiz, Ş.: Stress analysis of adhesively bonded double strap joints with or without intermediate part subjected to tensile loading. J. Adhes. (2015). doi:10.1080/00218464.2015.1075885
9. Ayaz, Y., Çitil, Ş., Şahan, M.F.: Repair of small damages in steel pipes with composite patches. Mater Sci. Eng. Technol. (2016). doi:10.1002/mawe.201600526
10. Da Silva, L.F.M., Adams, R.D.: Adhesive joints at high and low temperatures using similar and dissimilar adherends and dual adhesives. Int. J. Adhes. Adhes. **27**, 216–226 (2007)
11. Çitil, Ş., Temiz, Ş., Altun, H., Özel, A.: Determination of mechanical properties of double-strap adhesive joints with an embedded patch. J. Adhes. Sci. Technol. 25(18), 2555–2567 (2011)
12. Apalak, M.K., Engin, A.: Geometrically non-linear analysis of adhesively bonded double containment cantilever joint. J. Adhes. Sci. Technol. **11**(9), 1153–1195 (1997)
13. Temiz, Ş.: Application of bi-adhesive in double-strap joints subjected to bending moment. J. Adhes. Sci. Technol. **20**, 1547–1560 (2006)
14. Edwards, K.L.: A brief insight into the selection and use of engineering adhesives for preliminary joint design. Mater. Des. **19**, 121–123 (1998)
15. Goglio, L., Rossetto, M., Dragoni, E.: Design of adhesive joints based on peak elastic stresses. Int. J. Adhes. Adhes. **28**, 427–435 (2008)
16. Castagnetti, D., Dragoni, E., Spaggiari, A.: Failure analysis of bonded T-peel joints: efficient modelling by standard finite elements with experimental validation. Int. J. Adhes. Adhes. **30**, 306–312 (2010)
17. Bendemra, H., Compston, P., Crothers, P.J.: Optimisation study of tapered scarf and stepped-lap joints in composite repair patches. Compos. Struct. **130**, 1–8 (2015)

18. Kimiaeifar, A., Lund, E., Thomsen, O.T., Sørensen, J.D.: Asymptotic sampling for reliability analysis of adhesive bonded stepped lap composite joints. Eng. Struct. **49**, 655–663 (2013)
19. Engerer, J.D., Sancaktar, E.: The effects of partial bonding in load carrying capacity of single lap joints. Int. J. Adhes. Adhes. **31**, 373–379 (2011)
20. Pinto, A.M.G., Campilho, R.D.S.G., Mendes, I.R., Aires, S.M., Baptista, A.P.M.: Effect of hole drilling at the overlap on the strength of single-lap joints. Int. J. Adhes. Adhes. **31**, 380–387 (2011)
21. Haghpanah, B., Chiu, S., Vaziri, A.: Adhesively bonded lap joints with extreme interface geometry. Int. J. Adhes. Adhes. **48**, 130–138 (2014)
22. Dasilva, L.F.M., Carbas, R.J.C., Critchlowb, G.W., Figueiredo, M.A.V., Brown, K.: Effect of material, geometry, surface treatment and environment on the shear strength of single lap joints. Int. J. Adhes. Adhes. **29**, 621–632 (2009)
23. Adams, R.D., Pepiatt, N.A.: Stress analysis of adhesive-bonded lap joints. J. Strain Anal. **9**, 185–196 (1974)
24. Adams, R.D., Harris, J.A.: The influence of local geometry on the strength of adhesive joints. Int. J. Adhes. Adhes. **7**, 69–80 (1987)
25. Da Silva, L.F.M., Marques, E.A.S.: Joint strength optimization of adhesively bonded patches. J. Adhes. **84**, 915–934 (2008)

Bonding Strength of Hot-Formed Steel with an AlSi Coating and Approaches to Improve It by Laser Surface Engineering

Alexander Wieczorek, Matthias Graul and Klaus Dilger

A pre-treatment process by laser surface melting for 22MnB5+AS150 is presented, which is able to improve mechanical properties of the coating for adhesive bonding. The aim of the process is to improve the coating by keeping the properties of the base material to the greatest possible extent. It is shown, that the bonding properties of the coating can be improved, however a heat affected zone arises.

1 Introduction

For lightweight purpose hot-formed high strength steels (also known as advanced high strength steels, AHSS) like 22MnB5+AS150 are widely used in nowadays cars [1]. Steel of 22MnB5+AS150 is characterized by a martensitic microstructure, high tensile strength (up to 1500 MPa) and an aluminum-silicon coating to prevent the steel from oxidation and decarburization [2]. Hot formed steels are mainly joined by spot welding and adhesives. Due to the fact that welding has a negative influence on the martensitic structure, bonding has an advantage in joining martensitic steels, because it does not influence the base material with heat. Adhesives for body-in-white application dispose over 35 MPa in lap shear strength

A. Wieczorek (✉) · M. Graul
Department of Advanced Development, Volkswagen AG, Berliner Ring 2, 38440 Wolfsburg, Germany
e-mail: alexander.wieczorek@volkswagen.de

M. Graul
e-mail: matthias.graul@volkswagen.de

K. Dilger
Institute of Technology, Institute for Welding and Joining (Ifs), University of Brunswick, Langer Kamp 8, 38106 Brunswick, Germany
e-mail: k.dilger@tu-braunschweig.de

L.F.M. da Silva (ed.), *Materials Design and Applications*,
Advanced Structured Materials 65, DOI 10.1007/978-3-319-50784-2_29

[3]. Previously investigations of bonded joints showed high bonding strength, but a coating ablation as fracture pattern [4]. Obviously, the coating is the weakest part in the compound. However, neither the reason of coating ablation, nor the influence of high dynamic loading on the bonding strength is fully understood. Furthermore, there is no technique known to improve the coatings properties to prevent the coating ablation in this particular case of testing.

This work deals with the achievable bonding strength of aluminized high strength steels in lap shear test. Specimens in initial state have an intermetallic AlSi-coating due to hot forming process. Bonding strength of three different coating conditions are examined in this work. Specimen in initial coating state are compared to specimen, without an AlSi-coating. The AlSi-coating of these specimen was removed by shot blasting. In addition, laser surface treated specimen will be compared. These specimen are initial state specimen which are treated by a surface laser process. The lap shear tests are carried out at two different loading conditions and the influence on the bonding strength in relation to the coating condition are evaluted.

2 State of the Art

The steel as delivered is 22MnB5 with an AlSi10Fe3-coating with a thickness of approximately 30 μm. This coating is applied by hot-dipping process. In the delivered condition, the yield strength is about 600 MPa at an elongation of 20%. In the hot formed condition, the tensile strength is up to 1500 MPa at an elongation of 5%. To achieve this special condition, a heat treatment process is necessary [2]. In a first step, the steel is heated up to approximately 900–950 °C for at least 4 min in a roller heath furnace to ensure a complete austenitic microstructure. Because eutectic phases in the AlSi10Fe3-coating have a melting point of about 575 °C, the coating becomes partly liquid during the process [5]. During the heating process, iron diffuses from the substrate into the coating and aluminum vice versa. As a result, intermetallic compounds like Al_5Fe_2 and AlFe with dissolved Si content form and lead to a solidification of liquid phases [6].

In austenitic state, the workpiece is transferred directly from the roller heat furnace into a cooled forming die. It is then formed and quenched simultaneously at a cooling rate of more than 30 K/s to ensure a complete martensitic structure. After this hot-forming process the part can be trimmed by laser [7].

The evolving of the coating during the soak time is detailed described in [6]. A suggested initial state is shown in Fig. 1. The approximately 40–50 μm thick coating can be characterized by five points. First micro cracks (1) arise in the coating, due to differences of thermal expansions coefficients (α) of the base material (iron, $\alpha = 11.8 \times 10^{-6}\ K^{-1}$) and coating (mainly aluminum, $\alpha = 23.1 \times 10^{-6}\ K^{-1}$) [8]. Second, there are Kirkendall voids (2), which result from different diffusion coefficients of iron and aluminum. Third, the main phase of the coating is Al_5Fe_2 (3). It has a fracture toughness of $K_{IC} = 1\ MPa\sqrt{m}$ and a

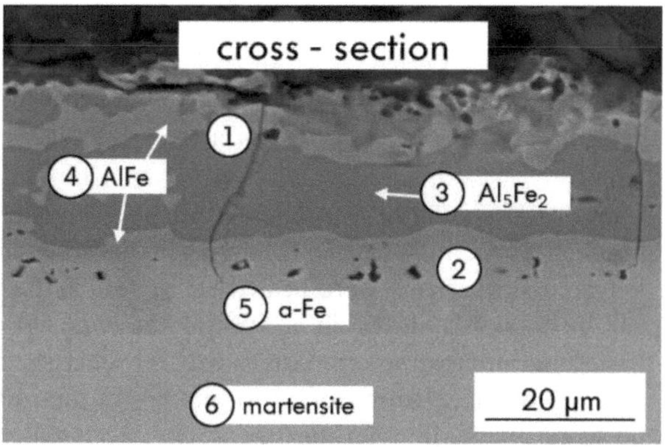

Fig. 1 Cross-section of 22MnB5+AS150, 6 min. soak time

hardness of about 900–1050 HV [9]. Fourth, layers of AlFe (4) range through the coating. AlFe is more ductile (300–650 HV) and has a higher fracture toughness ($K_{IC} = 26\ MPa\sqrt{m}$) [10]. Fifth, there is an α-Fe-layer (5) with dissolved aluminum. This is due to the fact of diffused aluminum, which stabilized the body-centred-cubic lattice of iron in austenitic state [11]. This layer thickness is increasing by dwell time due diffusion of aluminum. Last, there is martensitic base material of 22MnB5.

Dilger et al. investigated the bonding strength of this type of 22MnB5+AS150 steel and found, that in lap shear testing at 2 m/s, the bonding strength decreased of about 30% compared to samples with down grounded coating. On a random basis it was tried to ablate the brittle coating with a pulsed laser but no satisfying parameters were reported [12].

Bergweiler used a laser process to anneal areas of hardened 22MnB5. However, it was noticed that the coating was not affected by this laser annealing process [13].

Windmann et al. removed the brittle aluminum-silicon coating by shot-blasting with corundum to improve the wetting behaviour of laser brazing consumables [14]. The bonding strength of specimen with a removed coating by shot blasting has not been reported so far.

Therefore, the aim of this research is to investigate the bonding strength of zinc coated steel (HX420 LAD+Z, 1.0935) and different conditioned substrates of 22MnB5+AS150 (1.5528) by quasi static and fast performed lap shear tests. Furthermore a laser pre-treatment process will be introduced, which is able to improve the mechanical properties of the coating of 22MB5+AS150 by surface melting.

3 Experimental Setup

To investigate the influence of substrates or their coating to the joint strength, two different substrates in different conditions are used: Galvanized steel HX420LAD +Z and 22MnB5+AS150 in hot formed condition. Specimens of 22MnB5+AS150

Table 1 Experimental material

Steel	Coating	Material no.	$R_{p0,2}$ in MPa	R_m in MPa	A_{80} in %
HX 420 LAD	Zinc	1.0935	420–520	470–590	≥ 17%
22MnB5	AlSi10Fe3	1.5528	1100	1500	6%

in produced condition as reference (a) are compared to shot blasted (b) and laser pre-treated specimen (c). To achieve a hardened condition of 22MnB5+AS150, the delivered steel sheets were heated up to 930 °C for 6 min in a roller heat furnace and then cooled down in a special die with more than 30 K/s to achieve a martensitic microstructure. Afterwards the steel sheets were cut by a laser into lap shear geometry form of 150 mm × 25 mm. Galvanized specimens were mechanically cut into lap shear geometry. Both materials had a thickness of 1.5 mm (Table 1).

The coating of shot blasted specimen was removed by corundum with a grain diameter of 0.2 mm to 0.3 mm. Laser surface melting was performed with a fiber laser (Power $P_L = 400$ W and a wavelength of about 1070 nm) on samples in reference condition (a).

All specimens were degreased with acetone with a flash-off time of 5 min. The single lap shear test specimens (according to DIN EN 1465 [15]) were joined with an overlap length of 12.5 mm with an epoxy based adhesive, which had an E-modulus of 3000 MPa. An homogenous adhesive thickness between adherends was ensured by glass beads with a diameter of 0.2 mm. The adhesive of the bonded joints was cured at 180 °C for 30 min in a furnace and conditioned at room temperature for 24 h. Each test series consisted of five specimens. Lap shear test were performed with a ZWICK100. The time, force and elongation signals of the test machine were measured and analysed with the software TestXpert2. Two different speeds of the testing machine were set, first 10 mm/min (quasi static loading) and second 200 mm/min (dynamic loading). The testing machine accelerated up to the set value in 10 s, which means an acceleration of 6 mm/min^2 for quasi static testing and 1200 mm/min^2 for dynamic loading respectively.

Metallographic examinations were carried out with cross–sections in microscopy and SEM investigations. Cross-sections were grinded and if needed, etched with nital. SEM investigations were performed by a Tescan *Vega TS 5136 XM* scanning electron microscope at 20.0 kV and a BSE Detector.

4 Results

As shown in Fig. 2 the test samples of HX420LAD+Z had an average bonding strength of about 36.83 ± 0.52 MPa in quasi static testing. The fracture pattern exposed a cohesive adhesive failure. For 22MnB5+AS150 (b), which was shot blasted before, an average bonding strength of 36.69 ± 0.54 MPa was recorded. The bonding strength of a combination of HX420LAD+Z and 22MnB5+AS150

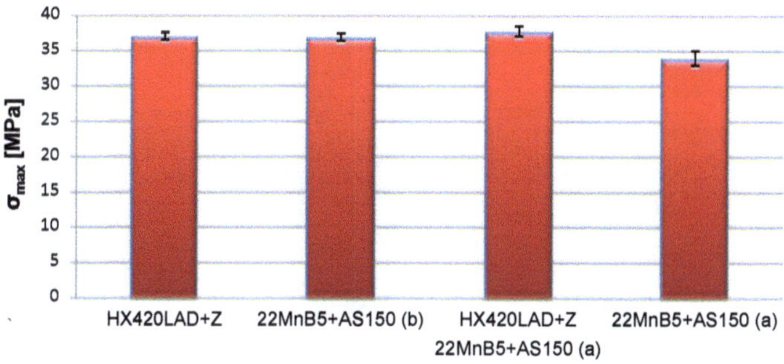

Fig. 2 Bonding strength of different substrates in quasi static lap shear tests with 6 mm/min^2

(a) was at the same value 37.54 ± 0.71 MPa, however a complete coating ablation of the 22MnB5+AS150 (a) adherend was present. The lap shear test of bonded 22MnB5+AS150 (a) with 22MnB5+AS150 (a) showed an average bonding strength of about 33.74 ± 1.04 MPa and the fracture pattern was complete coating ablation. As long as there is cohesive failure of the adhesive, there is no significant change in bonding strength related to the substrate. In this particular case there is a change of about 3 MPa in bonding strength related to coating ablation (Fig. 3).

At higher shear accelerations (1200 mm/min^2) the bonding strength differs from the quasi static results. Specimen of HX420LAD+Z reached a maximum bonding strength of 35.77 ± 0.45 MPa with a cohesive adhesive failure. Specimen of type 22MnB5+AS150 (b) reached a similar bonding strength of 35.82 ± 0.79 MPa with a cohesive adhesive failure, too. Bonded 22MnB5+AS150 (a) resulted in an average bonding strength of 25.28 ± 1.42 MPa (Fig. 4) with a nearly complete

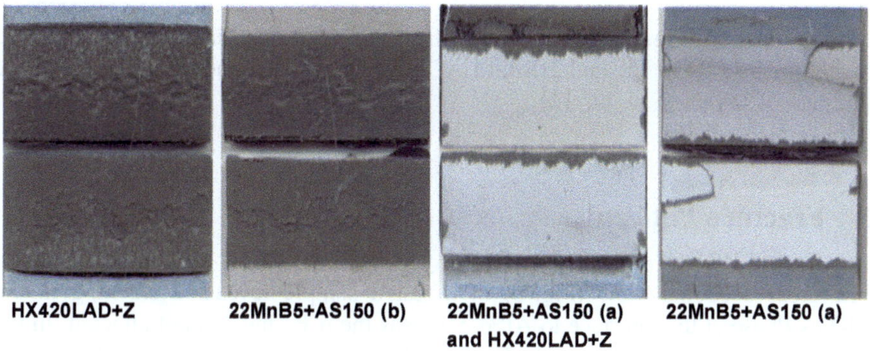

Fig. 3 Fracture patterns of quasi static lap shear samples

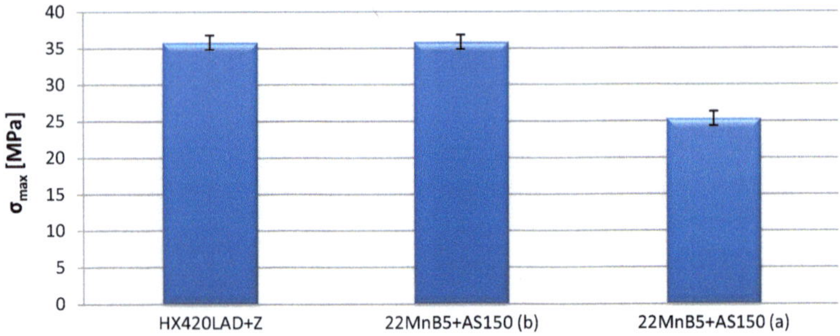

Fig. 4 Bonding strength of different substrate combinations at 1200 mm/min^2

Fig. 5 Fracture patterns of different substrates at 200 mm/min

coating ablation (Fig. 5). This is about 30% less of the bonding strength compared to 22MnB5+AS150 (b) and HX420LAD+Z, which is in good correlation to the findings of Dilger et al. [4, 12].

5 Fracture Pattern

The cross sectional view (Fig. 6) of a tested sample reveals a cracking in the Al_5Fe_2-Phase at the interface to AlFe, whereas the α-Fe-layer is not affected. In the area of the crack, residues of Al_5Fe_2 and AlFe are still upon the α-Fe-layer. It is suggested, that the interfacial boundary layer of Al_5Fe_2 and AlFe is the weakest part in the system, which is due to the fact, that hard and soft phases with different mechanical properties are close to each other.

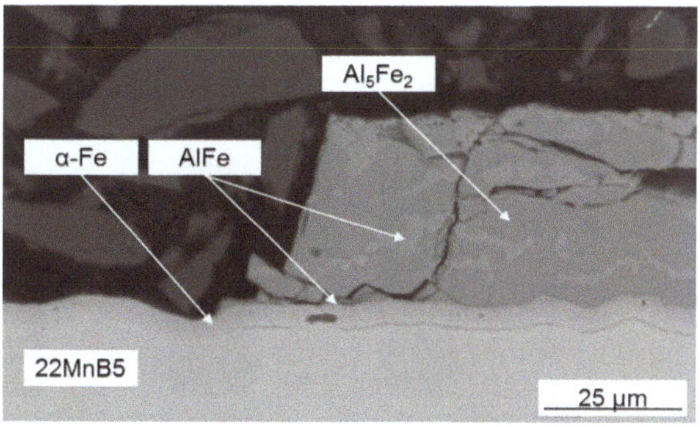

Fig. 6 Cross-sectional view of a tested specimen with ablated coating in SEM

6 Laser Pre-treatment

The coating-fracture pattern leads to the conclusion, that the aim of a laser surface treatment should affect the interface of Al_5Fe_2 and AlFe, so that at least more Fe diffuses into the Al_5Fe_2 which will lead to a transformation to the more ductile AlFe-phase [6]. Therefore lap shear specimens of 22MnB5+AS150 (a) were radiated by laser in an area of 25 × 25 mm as shown in Fig. 7 by a focused spot size of 150 µm. The influence of the beam guidance was investigated by two different configurations, which were selected on a random basis: First a crosswise pattern with a track pitch of 200 µm and second a parallel pattern (rectangular to the length of the specimen) with a track pitch of about 150 µm.

As seen in a representative cross-sectional view (Fig. 8), the laser spot melted an area (1) of approximately 150 µm width and a depth of approximately 80–100 µm.

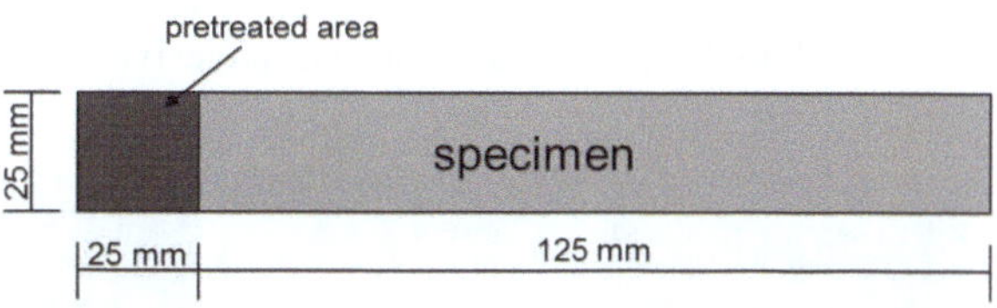

Fig. 7 Area of the laser pre-treatment

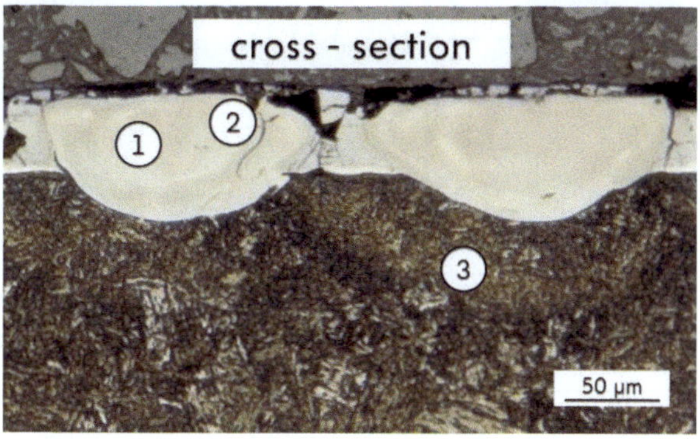

Fig. 8 Cross–section of laser pre-treated 22MnB5+AS150

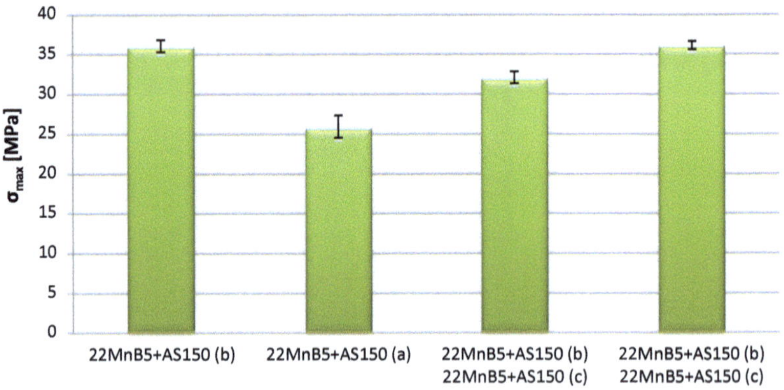

Fig. 9 Bonding strength of different substrates and pre-treatments at 1200 mm/min^2. Condition of the specimen from left to right: blasted, untreated, blasted/crosswise pattern, blasted/parallel pattern

Micro cracks are present in the solidification (2). Furthermore a heat affected zone (3) of about 50 μm depth was revealed.

For the experiments in bonding strength, five specimen of type 22MnB5+AS150 (a) per test series were pre-treated and bonded with blasted 22MnB5+AS150 (b). The testing was performed with a high acceleration as this is the critical condition for the brittle AlSi-coating. The resulting bonding strengths are presented in Fig. 9.

As reference, the bonding strength of specimen of 22MnB5+AS150 (b) and 22MnB5+AS150 (a) are presented. In contrast, specimens with the crosswise pattern showed an average bonding strength of 31.84 ± 0.76 MPa which is an increase of about 6.56 MPa. The fracture pattern, as shown in Fig. 10, exposed a coating ablation. The parallel beam guidance with a track pitch of 150 μm performed as well as the blasted specimen with an average bonding strength of about 35.93 ± 0.46 MPa and a cohesive adhesive failure. It can be stated, that a

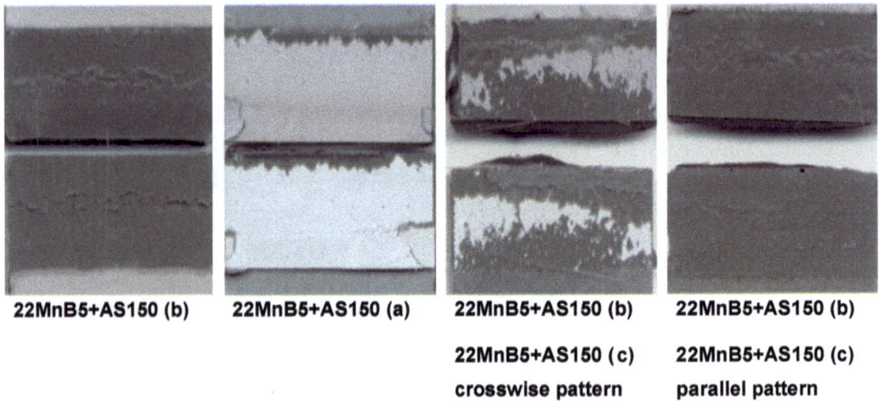

Fig. 10 Fracture pattern of different substrates tested with 1200 mm/min^2

sophisticated melting of the coating took place, where the weak interfacial boundary layer of Al_5Fe_2 and AlFe was removed.

7 Discussion

The experiments showed an influence of the testing acceleration to the bonding strength. As long as the adhesive is broken in a cohesive manner, it can be stated that the maximum bonding strength is achieved; only restricted by the properties of the adhesive itself. It is suggested, that lap shear samples of 22MnB5 with an aluminium-silicon coating will show a coating ablation if tested dynamically and if the adhesive shear strength is more than 25 MPa.

The coating is sensitive to sudden load. As shown, lower bonding strength is related to the effect of coating ablation in case of abrupt loading situations. The crack evolves and grows in the Al_5Fe_2-phase close to the interface to the AlFe-phase. Obviously, this configuration is the weakest link in the system of steel, coating and adhesive. It is suggested, that a heat influence by a laser would at least transfer brittle Al_5Fe_2 into more ductile AlFe by diffusion due to heat absorbing. It is shown, that a suitable laser parameter was found to melt the coating and to strengthen it hereby. As presented, the crack exposure interface is vanished. The bonding strength increases respectively to the beam guidance strategy. A crosswise pattern with a track pitch of 200 µm increases the bonding strength not to the highest possible amount. Whereas the parallel beam guidance strategy with a track pitch of 150 µm raises the achievable bonding strength to the maximum. This improved coating is not as brittle as before, which is stated due the fact that the coating is not ablating in dynamic lap shear testing. Furthermore, the coating is still in place and can protect the steel against oxidation at least in a passive way. However, there is a heat influence zone which may cause a reduction of the yield strength of the base material. Future investigations need to show the influence of the coating melting to the base material as well as to corrosion resistance.

8 Conclusions

To sum up, the mechanical property of the aluminum-silicon coating leads to lower bonding strength under dynamic loading in lap shear conditions. Mainly, the brittle phase of Al_5Fe_2 in combination with its boundary layer to more ductile AlFe is responsible for a failure within the coating. Though, a surface laser melting improves the properties by melting the brittle phase of Al_5Fe_2 into a more ductile condition. Thereby the crack exposure interface disappears. Due to this effect, it is possible to reach the maximum shear strength under dynamic loading conditions without coating ablation. However, a heat affected zone occurs close to the melting pool.

References

1. Karbasian, H., Tekkaya, A.E.: A review on hot stamping. J. Mater. Process. Technol. **210**, 2103–2118 (2010)
2. ArcelorMittal: product information. (2016)
3. da Silva, L.F.M., Öchsner, A., Adams, R.D.: Introduction to adhesive bonding technology. In: Handbook of Adhesion Technology, pp. 1–10 (2011)
4. Dilger, K., Kreling, S.: Adhesive bonding techniques for advanced high-strength steels (AHSS). In: Welding and Joining of Advanced High Strength Steels (AHSS), pp. 167–179 (2015)
5. Zolotorevsky, V.S., Belov, N.A., Glazoff, M.V.: Casting Aluminum Alloys (2007)
6. Windmann, M., Röttger, A., Theisen, W.: Phase formation at the interface between a boron alloyed steel substrate and an Al-rich coating. Surf. Coat. Technol. **226**, 130–139 (2013)
7. Wagner, A., Lütke, M., Wetzig, A., Eng, L.M.: Laser remote-fusion cutting with solid-state lasers. J. Laser Appl. **25**, 052004 (2013)
8. Haynes, W.M.: CRC Handbook of Chemistry and Physics (2011)
9. Köster, U., Liu, W., Liebertz, H., Michel, M.: Mechanical properties of quasicrystalline and crystalline phases in Al-Cu. J. Non-Cryst. Solids **153–154**, 446–452 (1993)
10. Kubošová, A., Karlík, M., Haušild, P., Prahl, J.: Fracture behaviour of Fe3Al and FeAl type iron aluminides. Mater. Sci. Forum **567–568**, 349–352 (2008)
11. Mein, A., Fourlaris, G., Crowther, D., Evans, P.J.: The influence of aluminium on the ferrite formation and microstructural development in hot rolled dual-phase steel. Mater. Charact. **64**, 69–78 (2012)
12. Kreling, S., Bischof, M., Frauenhofer, M., Dilger, K.: AiF FOSTA UsiBond—Adhesive Bonding of Press Hardened Steels. In: Joining in Car Body Engineering, 12–14 April 2011, Bad Nauheim (2011)
13. Bergweiler, G.: Lokale Wärmebehandlung mit Laserstrahlung zur Verbesserung der Umform- und Funktionseigenschaften von hochfesten Stählen, Ph.D. Thesis, RWTH Aachen (2013)
14. Windmann, M., Röttger, A., Kügler, H., Theisen, W., Vollertsen, F.: Laser beam welding of aluminum to Al-base coated high-strength steel 22MnB5. J. Mater. Process. Technol. **217**, 88–95 (2015)
15. DIN EN 1465: Adhesives—Determination of tensile lap-shear strength of bonded assemblies (2009)

Micro Cork Particles as Adhesive Reinforcement Material for Brittle Resins

A.Q. Barbosa, L.F.M. da Silva, A. Öchsner, E.A.S. Marques and J. Abenojar

1 Introduction

Adhesive bonding is one of the most commonly applied techniques for joining materials, mainly due to its adaptability, being able to reliably join a large range of materials [1, 2]. Several industrial sectors have been increasingly applying adhesives, from cutting edge applications with high mechanical stress and requirements (i.e. aeronautical, aerospace, automotive and electronic), to more conventional uses (i.e. packaging and construction) [3–7]. Epoxy resins are the most commonly used structural adhesives, due to their good mechanical, thermal and chemical properties; offering greater strength, rigidity, toughness, durability and chemical resistance than polyesters, but at a higher price and with more difficulties in processing [8]. Epoxy resins are monomeric or oligomeric compounds containing two or more epoxy rings that may be opened catalytically, or stoichiometrically, by reaction with multi-functional amines or carboxylic acids, to give a cross-linked network. When epoxies are in the polymerized state, they are amorphous and highly reticulated due to their thermoset nature [8]. Properties useful for applications of structural engineering result from this microstructure, such as high strength and modulus of

A.Q. Barbosa · E.A.S. Marques
INEGI, Rua Dr. Roberto Frias, 400, 4200-465 Porto, Portugal

L.F.M. da Silva (✉)
Faculty of Engineering, Department of Mechanical Engineering,
University of Porto, Rua Dr. Roberto Frias, 4200-465 Porto, Portugal
e-mail: lucas@fe.up.pt

A. Öchsner
Griffith School of Engineering, Griffith University (Gold Coast Campus),
Building G39 Room 2.22, Parklands Drive, Southport Queensland 4214, Australia

J. Abenojar
Materials Performance Group, Materials Science and Engineering Department,
Universidad Carlos III de Madrid, Leganés, Spain

L.F.M. da Silva (ed.), *Materials Design and Applications*,
Advanced Structured Materials 65, DOI 10.1007/978-3-319-50784-2_30

elasticity, low creep and good thermal strength [1, 9]. The epoxy microstructure also present some disadvantages, causing brittleness, with a low resistance to the initiation of cracks and their propagation [3, 4]. Consequently, in the past forty years there has been considerable research focusing on the improvement of the toughness of these thermosets [5, 6, 10–12]. Several published works describe some solutions to improve the toughness of brittle adhesives, such as: addition of a second polymer with a good toughness [13], use of woven or knitted reinforcement [14], inclusion of micro or nano inorganic (silicates, glass, alumina, etc.) [15] or organic particles [16]. In addition, natural materials can be used as reinforcement materials (i.e. cork or wood fibres) [10, 17–20].

This chapter provides an overview of the recent developments in the use of cork as reinforcement particles in the adhesive technology [10, 18–22]. The main parameters that influence the cork particle performance (size, particle/matrix interaction and volume fraction) and in what way they influence the toughness of the final composite/adhesive are discussed.

2 Cork as Reinforcement Material

Cork, constituting the external covering of cork oak (Quercus suber L.) stem and branches, is a biological material known to have several unique properties [23–26]. This specie stands out from the oaks family mainly for the bulky peel of woven bark (cork). Cork is the main product of cork oak exploration and is considered as a truly sustainable product, because it is renewable and biodegradable. Cork harvesting is an environmentally friendly process, during which not a single tree is cut down as, after being removed, the bark renews itself before the next harvesting [10].

To properly characterize cork one should be aware of its macro and microscopic properties, since both are quite interesting for engineering applications. Macroscopically, cork is light, elastic and considerably impermeable to liquid and gas, a thermal and electric insulator and acoustic and vibration insulation absorber; innocuous and also unaffected by microbial activity, with the ability to be compressed without lateral expansion. Microscopically, cork is constituted by layers of dead cells arranged in a honeycomb shape. The cell membranes have a certain degree of sealing, and are filled with a gas close to air, which occupies about 90% by volume. The combination of the honeycomb shape and the closed cell membranes is what provides the cork with its characteristics [27, 28]. Cork wall cells are essentially formed by suberin, lignin, and cellulose.

Although initially limited, interest in the understating of the properties of cork increased significantly with the industrialization of the bottle stopper production process. While stoppers are the most widespread and perhaps the oldest application of cork, nowadays there is significant interest in expanding the use of this material. During its production of stoppers, a large amount of cork is not used because it is unsuitable or is discarded during production. This material is often granulated and used in a variety of applications, mainly for production of cork/rubber composite

[24]. It can be used in the automobile industry, aeronautical industry, manufacture of joints for electric and gas equipment, acoustic and vibration insulation, oil spills, footwear, among several others uses [10, 29, 30].

Cork is a closed cell material with prismatic cells that work together and elastically orthotropic axisymmetric relative to the direction of the tree. This type of structure is of particular interest, as it can be used to reinforce a brittle adhesive, particularly to improve its toughness. In addition, the closed cells can also efficiently absorb impacts. However, the properties of an adhesive/cork composite are not only dependent on the material properties, but also on the interfacial adhesion properties between the cork and the resin, size and amount of cork particles and mixing conditions [31].

3 Parameters that Influence the Cork Particle Performance as Reinforcement Material

Cork particles may act as a reinforcing material and improve the fracture toughness of adhesives. However, this is not the only change they introduce, as several other mechanical properties are also inherently changed. Among them are the impact toughness, Young's modulus, maximum strain and stress and glass transition temperature.

It is also important to remember that the properties of a reinforced adhesive are not solely governed by the properties of the adhesive matrix or the reinforcing material. In fact, there are several other parameters involved in the determination of the toughness of the material, which largely influences the performance of the composite material. In the following section, the role of the volume fraction, particle size and the particle/matrix interface, are discussed.

3.1 Size

The size of reinforcing particles is one of the most important parameters and has been the focus of significant research. Several authors have studied the effect of particle size on the mechanical performance of thermosetting polymers [32–34]. The influence of particle size on the fracture toughness is not consensual, but some studies [35, 36] indicate that, for composites with micro scale particles, fracture toughness generally increases with particle size. The stiffness of the particles must also be taken into account, as the addition of rigid micro-scale particles to polymers often increases their strength while decreasing their toughness. In some cases, these particles may also induce stress concentrations, and act as crack initiation points [37]. In fact, this is one of the main reasons behind the favoured use of nano sized particles in detriment of micro-sized particles [10]. Additionally, particle size is also a material parameter that affects not just the fracture toughness but also the toughening mechanisms in reinforced adhesives [38]. It is quite simple to control the particle size and its importance is felt at all stages of production. Particle size can be measured using a variety of techniques including sieving, sedimentation, optical scattering and diffraction from particle suspensions. Nonetheless, the most suitable technique must be chosen according to the particle shape and dimension [39]. For cork particles, sieving is a particularly suitable technique, allowing easy separation of different ranges of sizes. Particle size and particle properties can be adjusted to improve toughness [36, 40]. The influence of size on fracture toughness was experimentally investigated by Barbosa et al. [18]. In that study, two different range sizes were used, bigger particles (125–250 μm) and smaller particles (38–53 μm), as seen in Fig. 1. Smaller particles present a damaged honeycomb cell structure, with open and single cells. In contrast, bigger particles present a honeycomb cell structure with multiple cells, which are open on the edge of the particle and closed on the particle core. These features are crucial for determining the mechanical properties of cork/resin composite.

Araldite 2020, from Huntsman Advanced Materials (Pamplona, Spain), was the selected adhesive because it is quite brittle, so the improvements on the toughness after the cork particles can easily be perceived. Araldite 2020 is a two component adhesive (100/30 by weight), resin (component A) and hardener (component B). Component A is composed by diglycidyl ether of bisphenol A, (DGEBA) and diglycidyl ether of 1, 4 butanediol (DGEBOH). On the other hand, the component B is composed by isophorone diamine (IPDA).

Figure 2 represents the fracture toughness (G_{Ic}) of a three-point bending single-edge notched (SENB) specimens of epoxy reinforced with micro cork particles (different particle size) and neat resin. Fracture tests can be performed on bulk adhesive and/or adhesive joints. However, when bulk specimens are used a more precise determination of the adhesive properties is achieved. It can easily be observed that, when compared with neat resin, the incorporation of micro particles of cork leads to an increase in fracture toughness and that this effect is more significant when smaller particles are incorporated.

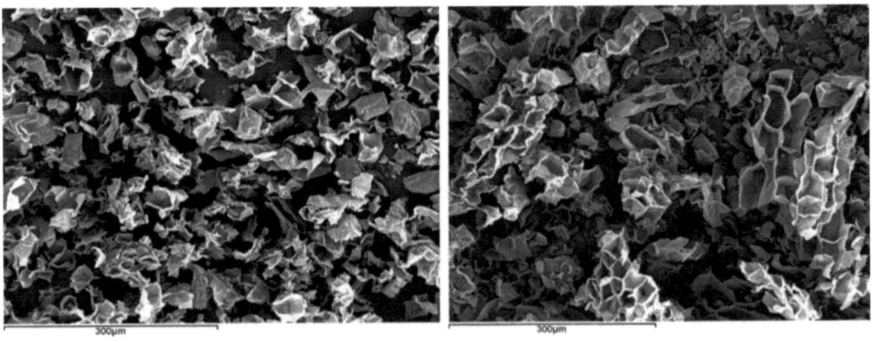

Fig. 1 Particle size distribution: *left* 38–53 μm; *right* 125–250 μm

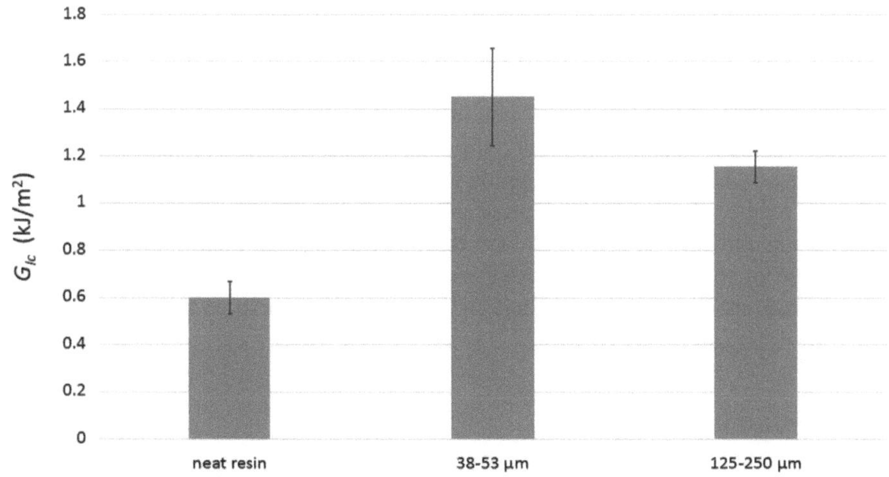

Fig. 2 Fracture toughness (G_{Ic}) of SENB specimens of epoxy reinforced with micro cork particles (different particle size) and neat resin

Some studies have demonstrated that instead of a precise particle size value, it might be more beneficial to have a well-defined particle size range. For instance, for bimodal particles, the composite presents a craze or mixed (craze/shear) deformation mode. The smaller particles (below a few microns of diameter) initiate the damage and the larger particles (above a few microns) stop and control the crack propagation [41, 42]. Figure 3 shows in detail the crack propagation process through two cork particles with different sizes. In the smaller particles, it is difficult to recognize the cork cellular structure, but it is clear that a particle is responsible for stopping the crack propagation. Larger particles can also stop crack propagation and also absorb impact, because cork cells are filled with gas (CO_2 + air).

The same behaviour was observed when tensile tests were performed, as for the same cork volume, particles with smaller dimension present higher strains. This fact leads to the conclusion that specimens with lower particle size are in fact more

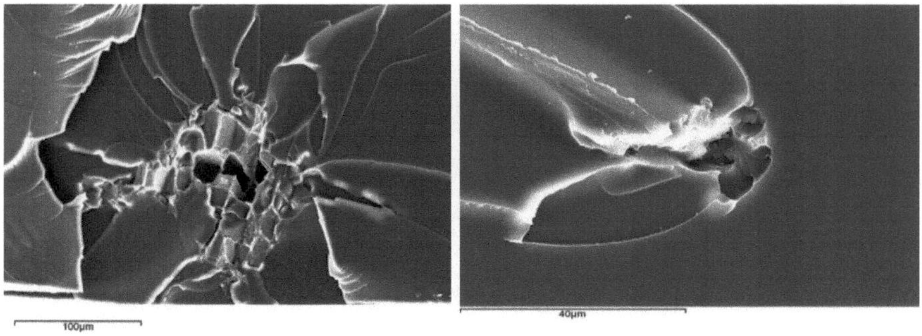

Fig. 3 Crack propagation through a cork particle. *left* 125–250 µm; *right* 38–53 µm

ductile than specimens with larger dimensions, which was seen previously in fracture toughness tests.

The mixing conditions are also a significant factor to take into account during composite production. In optimal conditions, the particles must be randomly distributed in the matrix, without any agglomeration. However, particles of micro-meter sizes are frequently predisposed to cluster during the composite fabrication, leading to mechanical properties far below the theoretical values. The particle size ratio of the matrix and the reinforcement is the main factor controlling the degree of micro-structural homogeneity. Thus, use of advanced mixing techniques, such as the use of centrifuge, is recommended to solve this issue.

3.2 Volume Fraction

The volume fraction or the amount of particles dispersed in a structural adhesive matrix is a significant parameter in controlling the resulting properties of the toughening adhesive [43–45].

The mechanics involved in the phenomena of reinforcing adhesives toughness are not only related to particle size, but also to the volume of material used. Due to all the variables involved, it is not simple to establish a direct relationship between the volume of material used and the toughness of the adhesive. Some authors show that toughness increases with increasing volume of particles, eventually reaching a maximum value. After this maximum value of toughness is reached, a decrease in toughness with an increase in the volume of particles occurs [15, 34, 35]. Barbosa et al. [10, 18] showed that similar behaviour is observed when using cork as reinforcement material. Figure 4 shows the fracture toughness of SENB specimens

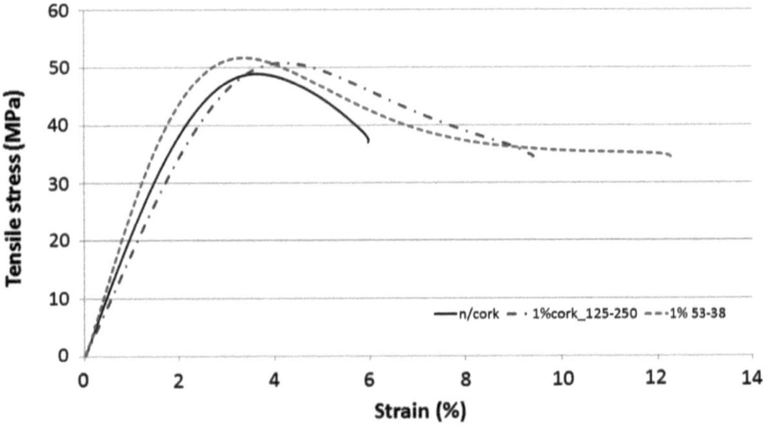

Fig. 4 Tensile stress-strain curves for specimens of neat resin and different particle size, for 1% volume

of epoxy reinforced with micro cork particles, for different amounts to neat resin, with 125–250 μm of particle size. Resin reinforced with micro particles of cork present a higher value of G_{Ic} compared to neat resin, increasing until reaching a maximum (at 1% volume); after this point, G_{Ic} starts to decrease. With increasingly higher volumes, there is a drastic reduction in the distance between particles, and it is believed that this fact is responsible for a reduction of G_{Ic}.

This behaviour is not only observed for fracture toughness, Barbosa et al. [22] studied the influence of amount of cork as reinforcement material in brittle resin and, as illustrated in Fig. 5, demonstrated that specimens with small amounts of cork show a higher maximum strain to failure than specimens without cork particles. Specimens with larger amounts of cork, between 2% and 5%, were found to exhibit lower values of strain. There seems to be an optimum volume of cork particles that leads to maximum adhesive ductility. Below 1% of cork, the particle concentration is insignificant and above 1%, the cork particles start to act as defects since the adhesive becomes more brittle. Between these two extremes, a middle point seems to exist where the cork particles act as a valuable reinforcement material. In the same study, the glass transition temperature (T_g) was measured, and it was observed that for 1% of cork its value was at a minimum, which is in accordance with the tensile test results (see Fig. 6).

Another relevant property that can be influenced by the cork amount is the ability of a material to absorb impacts. Cork has good impact absorbing behaviour due to its cell structure, which provides a damping effect. When cork cells are compressed, they absorb impact without fracturing. Figure 7 presents the energy absorbed during Charpy impact at the peak load and at rupture, for different cork volumes and Fig. 8 shows the displacement of the specimens during the test, also for the different volumes. It is clear that specimens of neat resin have different impact behaviour when compared with specimens with cork. Composites with 1%

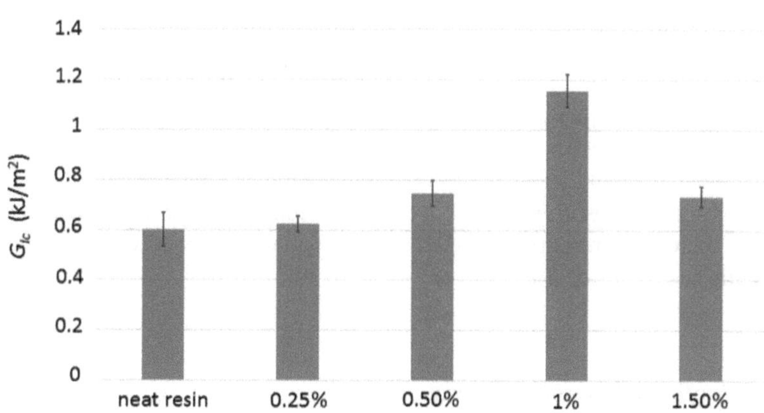

Fig. 5 Fracture toughness of SENB specimens of epoxy reinforced with micro cork particles (different amount for 125–250 μm particle size) and neat resin

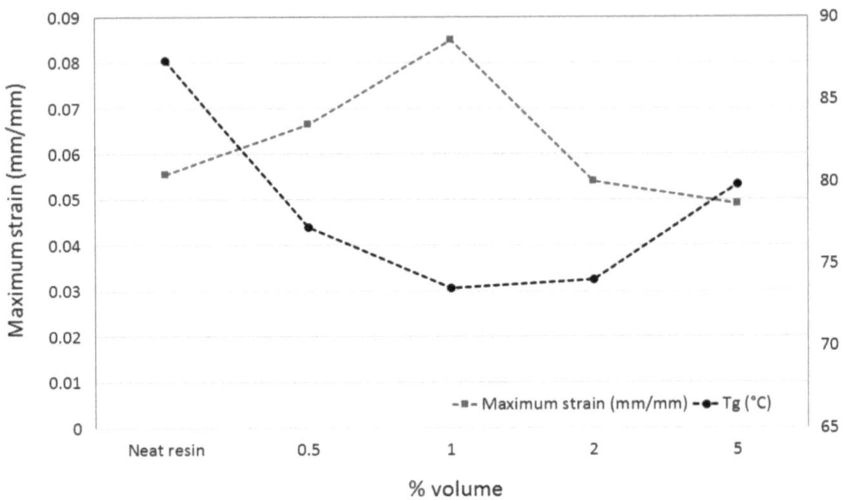

Fig. 6 Comparison between maximum strain and T_g for different volumes of cork particles

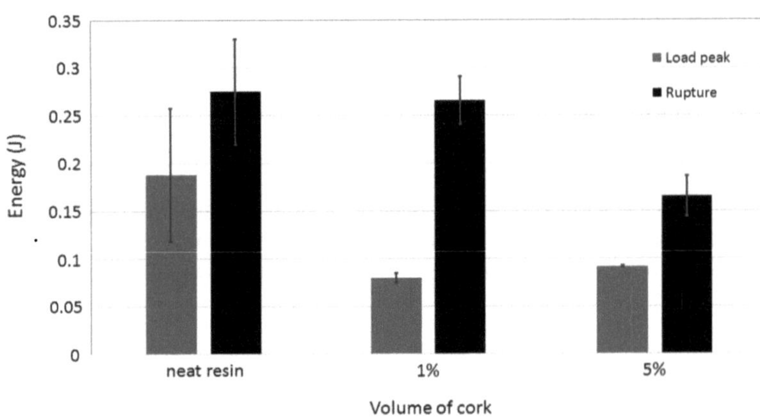

Fig. 7 Energy absorbed in Charpy impact tests, at the peak load and at rupture for specimens with 125–250 μm cork, for different cork volumes

of cork volume can absorb more energy at rupture and provide larger displacements at failure, therefore presenting a better impact behaviour when compared to specimens with neat resin or 5% cork.

3.3 Interface Particle/Matrix

The interface between the particles and the resin is a very important factor in the toughening process, as it is important to ensure that loads can be effectively

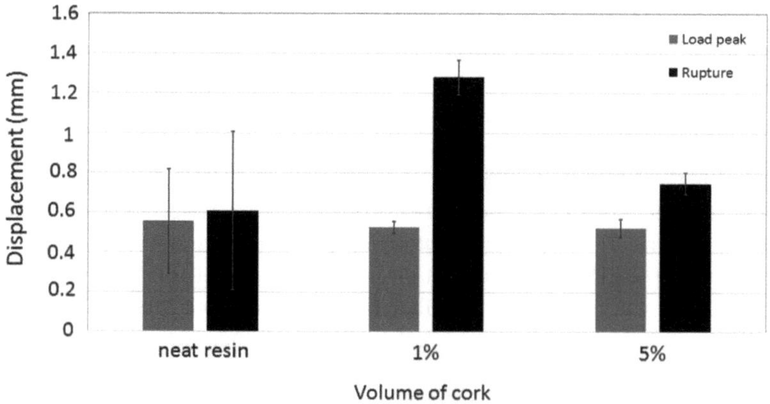

Fig. 8 Displacement in Charpy impact tests, at the peak load for specimens with 125–250 μm cork, for different cork volumes

transferred between the particles and the matrix. Therefore, it is crucial to have good wetting between resin and the particles, so that a strong bond can be guaranteed. When good adhesion is achieved, the reinforcing particles act as crack stoppers and not as defects in the resin matrix, with the applied stress being successfully transferred to the particles from the matrix [35]. It is known that weak bonds between particle and matrix result in lower fracture toughness when compared to strongly bonded particles. The bonding strength of the matrix/particle interface is, therefore, a central parameter determining which toughening mechanism is dominant in the filled system (since strengthening the particle/matrix adhesion increases the efficiency of pinning but suppresses crack tip blunting) [33].

Since particles chemistry determines both the rate of wetting and the strength of interaction with polymers, these features can be regarded as more important than the bulk properties. Hence, to guarantee proper interfacial interactions, their surface properties must sometimes be modified accordingly. Some degree of modification or treatment is frequently recommended for application to all surfaces prior to adhesive bonding. There are several techniques that can be used to modify the reinforcement's materials surface (particularly naturals) such as: graft copolymerization of monomers onto the surface, the use of maleic anhydride copolymers, alkyl succinic anhydride, stearic acid, etc. It is also known that the use of coupling agents such as silanes, titanes, zirconates, triazine compounds improves particle-matrix adhesion. Plasma treatments are also useful to treat polymeric surfaces [10, 46, 47].

When cork is used as a reinforcement material, it is important be aware that cork is hydrophobic due to the presence of suberin (one of its main components) and that this behaviour can decrease the adhesion between cork and resin [24, 47]. It is also important to consider that cork is a heterogeneous natural material that exhibits noticeable changes according to the cork board's origin (geographic, location of the board on the cork oak tree, etc.) and even the direction of cork growth.

The process of adhesion between the epoxy resin and cork particles may be dependent on physical (mechanical interlocking) and chemical bonds. Cork particles are inherently porous and have open cells which can be filled by the epoxy resin to achieve a mechanical interlocking effect, although the most durable bonds occur when the adhesive diffuses into cell walls to build chemical bonds with the cell's chemical components [18].

Some modifications can be performed on the cork surface in order to increase wettability, such as: atmospheric pressure plasma torch (APPT), low-pressure plasma chamber (LPPC) and silanization [47]. It is crucial to ensure that this type of treatments changes just the cork surface properties without altering its bulk properties. These treatments work by creating reactive chemical groups on the cork surface, which allow the resin to chemically bond with the cork cells.

A few studies were performed focusing on this process [18, 47]. The results show that that with atmospheric plasma treatment, the contact angle between cork surface and the water drop decreases. Plasma treatment is able to change the cork behaviour from hydrophobic to hydrophilic, independently of cell growth orientation (radial, tangential or axial). This ensures that the adhesion of the cork particles is uniform throughout the surface of the cork particles (see Fig. 9 and Table 1).

When analysing the surfaces that were treated with plasma by scanning electron microscopy (SEM), some differences were detected compared to the particles that were not treated. The thickness of the cell walls of the treated particles is smaller than that of the particles that did not have surface treatment, decreasing by approximately 3%. These results suggest that the plasma treatment is responsible for

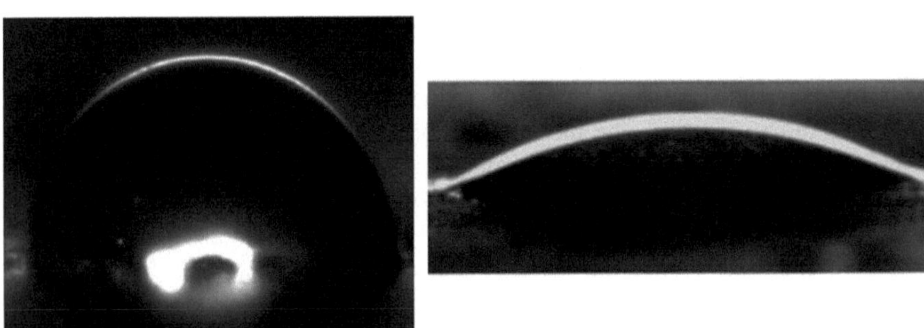

Fig. 9 Shape of the water drop on cork surface, (*left*) without treatment and (*right*) with atmospheric plasma treatment [18, 48]

Table 1 Contact angle (°) between water drop and cork surface

Section	Untreated specimens	Treated specimen
Radial	101	30
Tangential	99	33
Axial	103	37

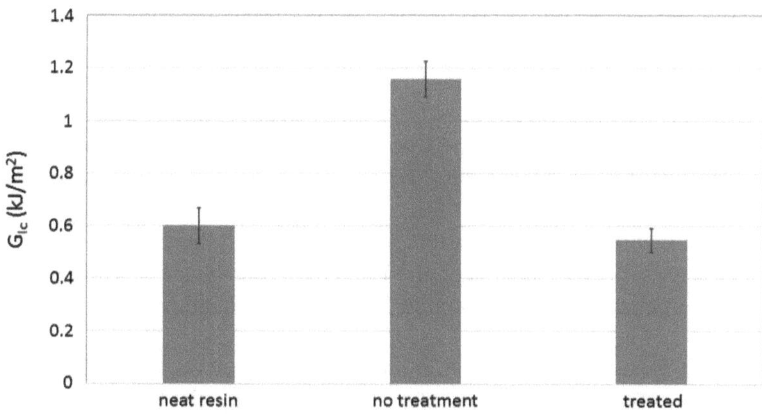

Fig. 10 Fracture toughness of SENB specimens of epoxy reinforced with micro cork particles (with and without surface plasma treatment for 125–250 μm particle size) and neat resin

an erosion of the cell walls. This might enable the introduction of resin inside the cork cells and thus compromising the mechanical properties of the composite. Figure 10 shows that the surface treatment can be responsible for a decrease in fracture toughness, fundamentally due this process of cell wall erosion. Specimens with cork which has a surface treatment present G_{Ic} values lower than those obtainable by the neat resin. In contrast, samples with cork which was not subjected to any surface treatment show higher values of G_{Ic} compared to those of the neat resin.

4 Kinetic and Chemical Analysis of Resin Reinforced with Micro Cork Particles

Resin composites, such as brittle resins with micro cork particles, are composed of reinforcing particles in an embedded brittle epoxy matrix. However, the ratio between epoxy/cork particles presents a great influence on the mechanical behaviour of the composite and on the cure conditions [48]. The influence of particles in the curing process of a resin will actually depend on the particle nature. While in some cases the particles do not change the degree of conversion, in other cases the particles can act as an accelerator or retarder of the reaction. The physical properties of the composite material after cure will then be dependent of the structure, time and cure temperature.

Taking into account this information, it is important to understand the influence of the particles in the cure process of these materials, in order to relate mechanical properties and the structure network, and therefore enable the creation of composites suitable for high performance applications [19].

The cure process of epoxy resins requires the transformation of monomers or pre-polymers with low molecular weight into 3D structure networks. In this process, the T_g of the material increases as a result of the enhanced cross-linking density. T_g is a very significant property of an adhesive, and is also very useful to use as a tool to establish a link between the mechanical properties and the cure process [19, 22]. Figure 11 presents the results of the conversion degree for neat resin and composites (0.5, 1, 2 and 5% cork) at 100 °C, for 30 min. At 100 °C, the adhesive is fully cured at 30 min, a temperature at which the material was cured for tensile tests and T_g. It is clear that the cork particle volume does not show any substantial influence on the conversion degree, neither delaying or activating the reaction.

In order to better understand the influence of the cork on the curing process of the neat resin, a Fourier transform infrared spectroscopy (FTIR) analysis was carried out. Epoxy resin (Araldite 2020) with micro cork particles was chemically evaluated before and after cure. Figure 12 shows the spectrum in the 4000–450 cm^{-1} wavenumber range for cork, resin (2020A) and hardener (2020B) before curing, and Araldite 2020 and composite with 0.5% cork after, curing. Cork is a bio-polymer composed mainly of suberin (40%), lignin (22%), polysaccharides (20%) and waxes and other extractives (15%), the typical FTIR spectrum is also shown in Fig. 12. In a first analysis, the spectrum before curing (Araldite 2020 A and B) shows several peaks resulting from either the resin or the hardener. Analysing the peaks of the spectra of the epoxy resin without and with 0.5% of cork, it is observed that there are no major differences. As previously noted, the volume of cork did not influence the degree of adhesive curing. The cork amount of 0.5% was selected merely for the sake of material savings. Analysing Fig. 12, it may be concluded that there is no formation of by-products after the addition of cork. In this composite resin/cork, the cork does not cause any biological or chemical reaction in the resin, neither is there any reaction between them. This is a key

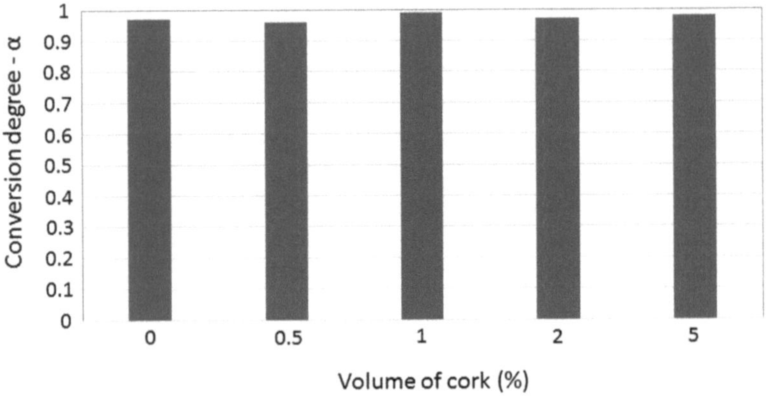

Fig. 11 Conversion degree for neat resin and composites (0.5, 1, 2 and 5% cork) at 100 °C for 30 min

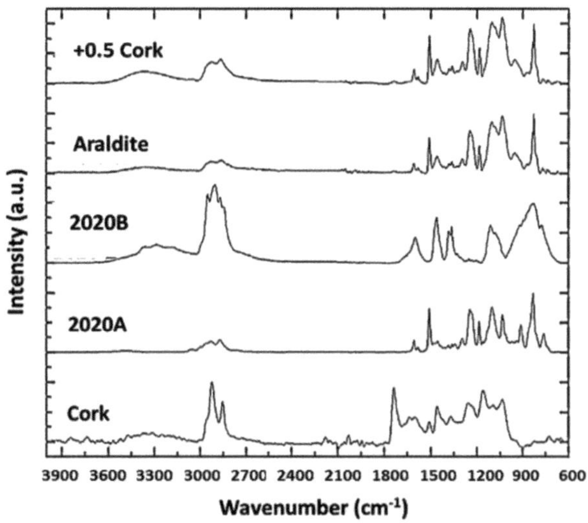

Fig. 12 FTIR spectrum, in the 4000–450 cm^{-1} wavenumber range for: Araldite 2020 (before and after cure), composite 0.5% cork/resin and cork particles [19]

information, since it attests that the effect of the cork particles on the mechanical properties of the resin is merely due to the mechanical interactions between the cork and the resin [18, 19, 22].

5 Hygrothermal Degradation

One of the major concerns of industrial users of structural adhesives (in particular epoxy resins) is the long term behaviour of the material when subjected to humid environments. The properties of epoxy resins are strongly affected by water absorption, which causes plasticization [49], decreases T_g and reduces the mechanical strength [50, 51]. Water can be present in epoxies in two distinct forms: as free water that fills the micro cavities of the network or as water bonded by strong interactions with polar segments (glycidyl epoxy rings and the nitrogen atoms of the amine compound). Plasticisation and swelling are two reversible moisture uptake effects, although there are irreversible processes such as micro-cracking and hydrolysis. The combination of all these effects leads to a noticeable degradation of most adhesive properties, not limited just to mechanical properties but also altering the thermal, physical and chemical properties [52].

In the last decade, the industrial use of materials reinforced with natural fibres or particles has been expanded significantly. These materials are of interest because they are not only strong but also cheap, light and originate from renewable sources. They can be used in a wide range of applications, from automotive and aerospace industry to sporting goods [21, 53]. Thereby it is imperative to understand how these new composite materials behave in the presence of an environment with high moisture content. As for the neat epoxy resins, there is also a degree of apprehension regarding the long term durability of these materials due to their susceptibility to moisture absorption and the effect on physical, chemical, mechanical and thermal properties [54]. As much of this degradation process occurs specifically in the resin-particle interface, the degradation of these composite materials is severely affected by the presence and nature of the reinforcement particles. Consequently, in order to estimate their long-term performance, it is vital to understand the behaviour of these composite materials under adverse conditions.

In studies focusing on the effect of moisture on reinforced composite materials (mainly in fibres), it was verified that the interface between matrix and natural material separates in the presence of humidity, leading to a decrease in the efficiency of the stress transfer between the matrix and the reinforcement. In some reinforced materials, the degree of degradation is directly proportional to moisture uptake, reducing the mechanical, physical and thermal properties of the material [21].

An analysis of the behaviour of the resin reinforced with micro particles of cork in the presence of moisture was also performed [21]. For bulk specimens of Araldite 2020 adhesive and Araldite 2020 with 1% of cork specimens immersed in deionized water at 50 °C, the average moisture uptake versus the square root of time over the thickness in deionized water at 50 °C, is presented in Fig. 13. In addition, a Fickian fit was applied to the experimental data points, which was found to be in good agreement with the experimental data. Regarding the absorption and desorption behaviour at 50 °C, both specimens were found to exhibit a similar behaviour, showing that the cork does not prevent water movement and does not promote increased water absorption.

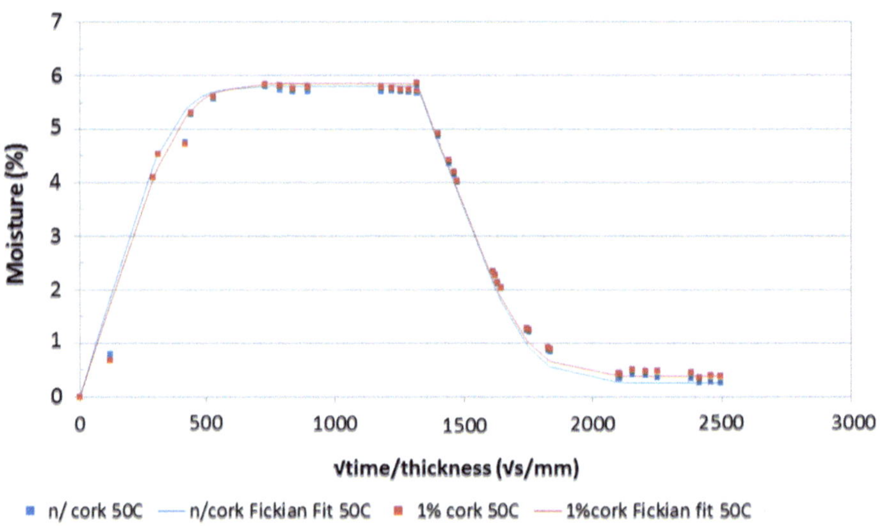

Fig. 13 Moisture absorption and desorption for 50 °C of neat resin and resin with 1%

In this research work [21], tensile tests were performed in moisture saturated and as cast specimens in order to scrutinise the effects of hygrothermal aging. Figure 14 shows typical tensile stress-strain curves obtained for these specimens. It was observed that the presence of cork favours the adhesive ductility in the as cast stage, presenting higher maximum strain to failure when compared to neat resin. When the effect of hygrothermal degradation on the specimens is analysed, it is observed that there are some significant changes in the mechanical behaviour. These specimens remained immersed in water until they reached saturation and were immediately tested after this point. It was observed that there is an increase in deformation with the incorporation of water molecules in the resin network. This fact is in agreement with several previous studies [55, 56], where it was described that hydrothermal aging affects mechanical properties, decreasing the tensile strength and Young's modulus of the epoxy resin. The mechanical properties of the adhesive are also influenced by the presence of micro cork particles: specimens with cork present a lower maximum strain and a lower Young's modulus, compared to specimens of neat resin. When water is absorbed, the composite has a lower deformation capacity (compared with the neat resin). On the other hand, after the desorption stage, the materials have a behaviour similar to the initial stage (as cast). However, in both conditions the resin reinforced with cork always presents a higher Young's modulus than that of the neat epoxy resin. Taking into account all these facts, it is possible to establish that the epoxy resin without cork is more susceptible to hydrothermal degradation, probably because water does not have as much capacity to broaden the molecular structure due to the impediment of cork particles.

In the case of materials reinforced with natural particles, in addition to the degradation of the polymeric material, there is also frequent degradation of the natural material. In some of these natural composites, the degradation of mechanical properties is significantly larger when compared to composites reinforced with

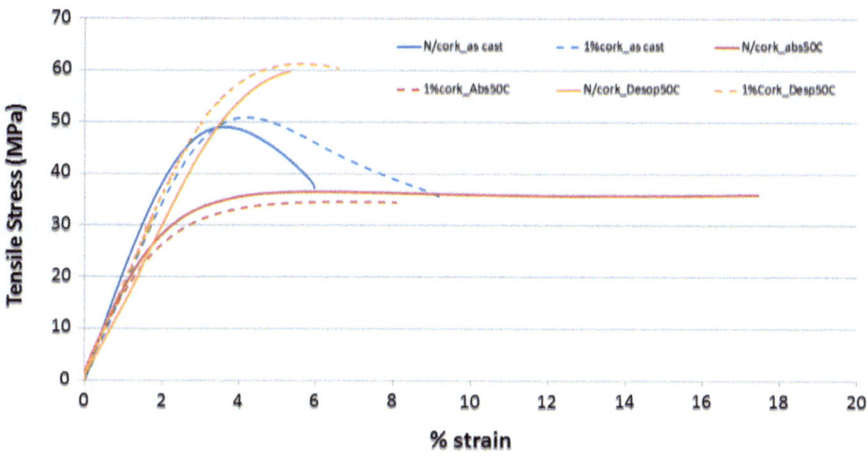

Fig. 14 Tensile stress–strain curves of specimens with different moisture stages (as cast and moisture saturation 50 °C) for different amounts of cork (neat resin and with 1% of cork)

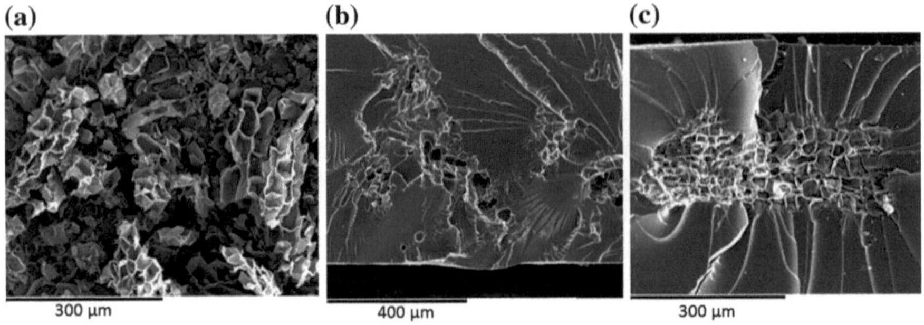

Fig. 15 Cork particles size 125–250 μm: **a** cork particles before inclusion in the resin; **b** Epoxy resin with cork particles as cast; **c** Epoxy resin with cork particles, after moisture saturation [21]

synthetic particles, mainly due to the higher moisture sorption behaviour and the organic nature of the natural material.

Figure 15 presents images of the cork particles in dissimilar states: before inclusion in the epoxy resin; after inclusion in the epoxy resin in the as cast condition and included in the epoxy resin after moisture saturation. It can be observed that, contrary to what typically occurs in natural reinforcements, the cork does not degrade in contact with moisture, maintaining its cellular structure intact. Good adhesion with the resin was identified as well.

6 Conclusions

This chapter gives an overview of recent works of the authors on the application of micro cork particles as reinforcement material for brittle epoxy. Some general conclusions can be drawn:

- Cork was found to change the mechanical properties of the epoxy resin. By varying the size and volume of the particles and modifying the interface between cork and resin, the mechanical properties can be optimized.
- Micro cork particles can increase the toughness of brittle epoxy adhesives. The particle volume and size present noticeable influence on the adhesive toughness. It was observed that specimens with 1% of cork present higher values of G_{Ic}, although with larger scatter. Since cork particles do not have a standardized geometry, its structure may diverge depending on biological and mechanical factors, which are particularly difficult to control. The fracture toughness of specimens reinforced with small particles (38–53 μm) was found to be higher than that of specimens reinforced with larger particles (125–250 μm).
- The surface treatments conventionally used to treat reinforcement particles were not effective in cork particles. Low-density plasma treatment decreases the

contact angle and increases the wettability of cork. There is an erosion of the surface which increases roughness, promoting adhesion between cork and adhesive, but is also responsible for an erosion of cell walls, leading to a decrease in cell wall thickness, giving lower mechanical performances.

- The behaviour during the curing of the neat resin and resin reinforced with cork is very similar. FTIR analysis demonstrates that curing of the adhesive leads to the formation of different functional groups of the resin, but the micro particles do not have any chemical reaction with the resin. Examining all the data from FTIR, it was understood that the cork particles have no direct influence on the curing of the adhesive, probably because the dimensions of incorporated particles remain at the micro level (125–250 mm), and do not affect the molecular structure of the adhesive.
- Considering hygrothermal degradation, neat resin and resin reinforced with cork were found to have very similar diffusion rates, since the presence of cork does not have significant influence on the absorption/desorption of moisture. With the moisture uptake, it is observed that there is a deterioration of the mechanical properties. The neat epoxy resin specimens show a greater variation in mechanical properties, compared with the samples containing 1% of cork. The cork behaviour is not compromised by the presence of water, because it is a rootless material, differing to what usually occurs in the natural fibres used to reinforce polymers.
- The use of micro cork particles as a reinforcing material can be an excellent application of a natural product, opening new and exciting development opportunities in this area and offer promising potential in the future.

Acknowledgements Financial support by Foundation for Science and Technology (PTDC/EME-TME/098752/2008 and SFRH/BD/88173/2012) are greatly acknowledged.

References

1. Packham, D.E.: Handbook of Adhesion. Wiley, England (2005)
2. Adams, R.D.: Adhesive Bonding—Science. Technology and Applications. Woodhead Publishing Limited, Cambridge (2000)
3. Bucknall, C.B.: Toughened Plastics. Springer Science-Business Media, London (1977)
4. Huang, Y., Hunston, D.L., Kinloch, A.J., Riew, C.K.: Mechanisms of toughening thermoset resins. In: Toughened Plastics I: Science and Engineering, pp. 1–35. American Chemical Society, Washington (1993)
5. Ramos, V.D., Costa, H.M., Soares, V.L.P., Nascimento, R.S.V.: Modification of epoxy resin: a comparison of different types of elastometer. Polym. Test **24**(3), 387–394 (2005)
6. Cardwell, B., Yee, A.F.: Toughening of epoxies through thermoplastic crack bridging. J. Mater. Sci. **33**(22), 5473–5484 (1998)
7. Tandon, G., Weng, G.: A theory of particle-reinforced plasticity. J. Appl. Mech. **55**(1), 126–135 (1988)
8. Martuscelli, E., Musto, P., Ragosta, G.: Advanced Routes for Polymer Toughening. Elsevier, Amsterdam (1996)

9. Gkikas, G., Barkoula, N.M., Paipetis, A.S.: Effect of dispersion conditions on the thermo-mechanical and toughness properties of multi walled carbon nanotubes-reinforced epoxy. Compos. B **43**(6), 2697–2705 (2012)
10. Barbosa, A.Q., da Silva, L.F.M., Banea, M.D., Öchsner, A.: Methods to increase the toughness of structural adhesives with micro particles: an overview with focus on cork particles. Materialwiss Werkst, **47**(4), 307–325 (2016)
11. Lange, F.: The interaction of a crack front with a second-phase dispersion. Philos. Mag. **22** (179), 0983–0992 (1970)
12. Withers, G.J., Yu, Y., Khabashesku, V.N., Cercone, L., Hadjiev, V.G., Souza, J.M., Davi, D.C.: Improved mechanical properties of an epoxy glass–fiber composite reinforced with surface organomodified nanoclays. Compos. B **72**, 175–182 (2015)
13. Petrie, E.M.: Handbook of adhesives and sealants. The McGraw-Hill Companies Inc, New York (2000)
14. Pethrick, R.A.: Design and ageing of adhesives for structural adhesive bonding–a review. In: Proceedings of the Institution of Mechanical Engineers, Part L: J. Mater. Des. Appl., 1464420714522981 (2014)
15. Johnsen, B.B., Kinloch, A.J., Mohammed, R.D., Taylor, A.C.: Sprenger, S, Toughening mechanisms of nanoparticle-modified epoxy polymers. Polymer **48**(2), 530–541 (2007)
16. Oksman, K., Skrifvars, M., Selin, J.-F.: Natural fibres as reinforcement in polylactic acid (PLA) composites. Compos. Sci. Technol. **63**(9), 1317–1324 (2003)
17. Hamza, T.A., Rosenstiel, S.F., Elhosary, M., Ibraheem, R.M.: The effect of fiber reinforcement on the fracture toughness and flexural strength of provisional restorative resins. J. Prosthet. Dentist. **91**(3), 258–264 (2004)
18. Barbosa, A.Q., da Silva, L.F.M., Öchsner, A., Abenojar, J., del Real, J.C.: Influence of the size and amount of cork particles on the impact toughness of a structural adhesive. J. Adhesion **88**(4–6), 452–470 (2012)
19. Barbosa, A.Q., da Silva, L.F.M., Abenojar, J., del Real, J.C., Paiva, R.M.M., Öchsner, A.: Kinetic analysis and characterization of an epoxy/cork adhesive. Thermochima Acta **604**, 52–60 (2015)
20. Barbosa, A.Q., da Silva, L.F.M., Öchsner, A., Abenojar, J., del Real, J.C.: Utilização de micro partículas de cortiça como material de reforço em adesivos estruturais frágeis. Ciência & Tecnologia dos Mater. **25**(1), 42–49 (2013)
21. Barbosa, A.Q., da Silva, L.F.M., Öchsner, A.: Hygrothermal aging of an adhesive reinforced with microparticles of cork. J. Adhes. Sci. Technol. **29**(16), 1714–1732 (2015)
22. Barbosa, A.Q., da Silva, L.F.M., Oechsner, A.: Effect of the amount of cork particles on the strength and glass transition temperature of a structural adhesive. Proc. Inst. Mech. Eng. L J. Mater. Des. Appl. **228**(4), 323–333 (2013)
23. Fortes, M.A., Pereira, H.: A Cortiça. IST Press, Lisboa (2004)
24. Silva, S.P., Sabino, M.A., Fernandes, E.M., Correlo, V.M., Boesel, L.F., Reis, R.L.: Cork: properties, capabilities and applications. Int. Mater. Rev. **50**(6), 345–365 (2005)
25. Pereira, H.: Chemical composition and variability of cork from Quercus suber L. Wood Sci. Technol. **22**(3), 211–218 (1988)
26. Úbeda, X., Pereira, P., Outeiro, L., Martin, D.A.: Effects of fire temperature on the physical and chemical characteristics of the ash from two plots of cork oak (Quercus suber). Land Degrad. Dev. **20**(6), 589–608 (2009)
27. Mano, J.F.: The viscoelastic properties of cork. J. Mater. Sci. **37**(2), 257–263 (2002)
28. Gil, L.: Cortiça: produção, tecnologia e aplicação. INETI, Lisboa (1998)
29. Gil, L.: New cork-based materials and applications. Materials **8**(2), 625–637 (2015)
30. Gil, L.: Cork composites: a review. Materials **2**(3), 776–789 (2009)
31. Abdallah, F.B., Cheikh, R.B., Baklouti, M., Denchev, Z., Cunha, A.M.: Effect of surface treatment in cork reinforced composites. J. Polym. Res. **17**(4), 519–528 (2010)
32. Singh, R., Zhang, M., Chan, D.: Toughening of a brittle thermosetting polymer: effects of reinforcement particle size and volume fraction. J. Mater. Sci. **37**(4), 781–788 (2002)

33. Kitey, R., Tippur, H.: Role of particle size and filler–matrix adhesion on dynamic fracture of glass-filled epoxyI. Macromeasurements. Acta Mater. **53**(4), 1153–1165 (2005)
34. Lauke, B.: On the effect of particle size on fracture toughness of polymer composites. Compos. Sci. Technol. **68**(15), 3365–3372 (2008)
35. Fu, S.-Y.: Effects of particle size, particle/matrix interface adhesion and particle loading on mechanical properties of particulate–polymer composites. Compos. B Eng. **39**(6), 933–961 (2008)
36. Nakamura, Y., Yamaguchi, M., Okubo, M., Matsumoto, T.: Effects of particle size on mechanical and impact properties of epoxy resin filled with spherical silica. J. Appl. Polym. Sci. **45**(7), 1281–1289 (1992)
37. Kim, B.C., Park, S.W.: Fracture toughness of the nano-particle reinforced epoxy composite. Compos. Struct. **86**(1), 69–77 (2008)
38. Azimi, H., Pearson, R., Hertzberg, R.: Fatigue of rubber-modified epoxies: effect of particle size and volume fraction. J. Mater. Sci. **31**(14), 3777–3789 (1996)
39. Rothon, R.: Particulate-Filled Polymer Composites. iSmithers Rapra Publishing (2003)
40. Nakamura, Y., Yamaguchi, M., Kitayama, A., Okubo, M., Matsumoto, T.: Effect of particle size on fracture toughness of epoxy resin filled with angular-shaped silica. Polymer **32**(12), 2221–2229 (1991)
41. Bagheri, R., Marouf, B., Pearson, R.: Rubber-toughened epoxies: a critical review. J. Macromol. Sci., Part C: Polymer Rev. **49**(3), 201–225 (2009)
42. Wrotecki, C., Heim, P., Gaillard, P.: Rubber toughening of poly (methyl methacrylate). Part II: effect of a twin population of particle size. Polym. Eng. Sci. **31**(4), 218–222 (1991)
43. Minfeng, Z., Xudong, S., Huiquan, X., Genzhong, J., Xuewen, J., Baoyi, W., Chenze, Q.: Investigation of free volume and the interfacial, and toughening behavior for epoxy resin/rubber composites by positron annihilation. Radiat. Phys. Chem. **77**(3), 245–251 (2008)
44. Huang, Y., Kinloch, A.: The toughness of epoxy polymers containing microvoids. Polymer **33**(6), 1330–1332 (1992)
45. Kinloch, A., Hunston, D.: Effect of volume fraction of dispersed rubbery phase on the toughness of rubber-toughened epoxy polymers. J. Mater. Sci. Lett. **6**(2), 137–139 (1987)
46. Herrera-Franco, P., Valadez-Gonzalez, A.: Mechanical properties of continuous natural fibre-reinforced polymer composites. Compos. A Appl. Sci. Manuf. **35**(3), 339–345 (2004)
47. Abenojar, J., Torregrosa-Coque, R., Martínez, M.A., Martín-Martínez, J.M.: Surface modifications of polycarbonate (PC) and acrylonitrile butadiene styrene (ABS) copolymer by treatment with atmospheric plasma. Surf. Coat. Technol. **203**(16), 2173–2180 (2009)
48. Abenojar, J., Martinez, M.A., Velasco, F., Pascual-Sanchez, V., Martin-Martinez, J.M.: Effect of boron carbide filler on the curing and mechanical properties of an epoxy resin. J. Adhes. **85** (4–5), 216–238 (2009)
49. Zhang, Y., Adams, R., da Silva, L.F.M.: Absorption and glass transition temperature of adhesives exposed to water and toluene. Int. J. Adhes. Adhes. **50**, 85–92 (2014)
50. Moy, P., Karasz, F.: The interactions of water with epoxy resins. Polym. Eng. Sci. **20**(4), 315–319 (1980)
51. Zhang, Y., Adams, R., da Silva, L.F.M.: Effects of curing cycle and thermal history on the glass transition temperature of adhesives. J. Adhes. **90**(4), 327–345 (2014)
52. Pavlidou, S., Papaspyrides, C.: The effect of hygrothermal history on water sorption and interlaminar shear strength of glass/polyester composites with different interfacial strength. Compos. A Appl. Sci. Manuf. **34**(11), 1117–1124 (2003)
53. Thwe, M.M., Liao, K.: Effects of environmental aging on the mechanical properties of bamboo–glass fiber reinforced polymer matrix hybrid composites. Compos. A Appl. Sci. Manuf. **33**(1), 43–52 (2002)
54. Fernández-García, M., Chiang, M.: Effect of hygrothermal aging history on sorption process, swelling, and glass transition temperature in a particle-filled epoxy-based adhesive. J. Appl. Polym. Sci. **84**(8), 1581–1591 (2002)

55. Ashcroft, I.A., Wahab, M.A., Crocombe, A.D., Hughes, D.J., Shaw, S.J.: The effect of environment on the fatigue of bonded composite joints. Part 1: testing and fractography. Compos. A Appl. Sci. Manuf. **32**(1), 45–58 (2001)
56. Wahab, M.A., Ashcroft, I.A., Crocombe, A.D., Hughes, D.J., Shaw, S.J.: The effect of environment on the fatigue of bonded composite joints. Part 2: fatigue threshold prediction. Compos. A Appl. Sci. Manuf. **32**(1), 59–69 (2001)

Magnetic Pulse Welding of Dissimilar Materials: Aluminum-Copper

I.V. Oliveira, A.J. Cavaleiro, G.A. Taber and A. Reis

1 Introduction

The joining of dissimilar materials is becoming increasingly important in several industrial applications. Pioneering structures are requiring high performance components, combining high strength-to-weight ratio, high electrical/thermal conductivity, improved corrosion resistance or high temperature strength, which can only be accomplished by the combination of dissimilar materials, with different key properties. Aluminum (Al) and copper (Cu) are two engineering materials widely used in the aerospace, transportation and electric power industries [1, 2]. Al/Cu bimetallic dissimilar joints are lighter and cheaper than the Cu/Cu assemblies and are especially suitable for electrical applications where the cross-section area of the conductor must be large to support high current. However, due to the difficulties in obtaining an electrically stable long-term bolted joint between these two dissimilar metals, on the last decade, efforts are being employed in enabling metallurgical welding between them [3].

I.V. Oliveira (✉) · A.J. Cavaleiro
Institute for Mechanical Engineering and Industrial Management (INEGI), Campus da FEUP, Rua Dr. Roberto Frias, 400, 4200-465 Porto, Portugal
e-mail: ioliveira@inegi.up.pt

A.J. Cavaleiro
e-mail: acavaleiro@inegi.up.pt

I.V. Oliveira · G.A. Taber
Materials Science and Engineering, The Ohio State University, 177 Watts Hall, 2041 College Rd, Columbus, OH 43210, USA
e-mail: taber.17@osu.edu

A. Reis
Faculty of Engineering of the University of Porto (FEUP), Rua Dr. Roberto Frias, 4200-465 Porto, Portugal
email: areis@fe.up.pt

L.F.M. da Silva (ed.), *Materials Design and Applications*, Advanced Structured Materials 65, DOI 10.1007/978-3-319-50784-2_31

Several welding processes are currently available to join two or more parts of similar and dissimilar materials, creating an assembled component with mechanical properties equal to the base materials. Still, fusion welding of dissimilar metals is often limited by their physical properties mismatch or due to its affinity to create intermetallics. For example, fusion bonding techniques such as laser spot welding and brazing have been used for dissimilar Al/Cu connections without adequate results, due to the formation of brittle phases in the weld [4, 5]. In fact, it is now well known that at temperatures higher than 120 °C the interaction between Al and Cu will result in brittle, low strength and high electrically resistant intermetallic compounds [3]. As a result, solid-state bonding processes have been experiencing a worldwide scientific and industrial interest and are being explored for the production of Al/Cu joints.

Friction stir welding (FSW) is one of the most popular solid state processes. It uses a non-consumable tool or one of the joining partners to generate heat through friction, leading the materials to a plasticized-viscous state, which are then stirred and forged behind the rotating tool pressed against the joining area, where it consolidates and cools down to form the weld. Despite the popularity of FSW due to its several advantages, it has been demonstrated by some authors [6, 7] that in FSW, the intermetallic compounds are still a problem. These works [6, 7] mention the existence of a relation between the properties of joints and the brittle intermetallic compounds and the time available for the formation of the compounds. Therefore, the formation of brittle intermetallic compounds can be minimized by the reduction of the time available for the interaction between the materials and minimizing the heat from the welding process [6, 7].

High-speed welding technologies such as explosive welding (EXW), magnetic pulse welding (MPW) or Vaporizing Foil Actuator Welding (VFA) are, therefore, promising alternatives. In explosive welding process, a flyer workpiece is accelerated against a fixed workpiece by the energy resulting from a controlled explosion. When the moving workpiece collides with the fixed component at a proper angle, complex mechanical and metallurgical actions occur, and a weld may take place. The collision velocity and angle are therefore the main key parameters. For welding to occur, a jetting action is required at the collision interface. The jet is the product of two angled surfaces colliding at high speed and is responsible for cleaning the metal surfaces and allows to pure metallic surfaces to join under extremely high pressure (in the order of GPa).

Magnetic Pulse Welding is a solid state welding process similar to EXW. Typically, the MPW system employs an electrical pulse power supply, capable to store electrical energy (1–100 kJ). During the discharge operation, a high-current pulse is released by the capacitor bank flowing through the cables to the welding coil, placed in a well-insulated workstation. Inside the tool coil, the parts to be welded are placed in lap joint configuration. The transient current flowing into the coil produces a magnetic field which, in turn, induces Eddy currents in the moving part, called flyer. The Eddy currents induced in the workpiece are responsible for limiting the penetration of the magnetic field into the workpiece and for creating its own induced magnetic field, opposite to the one generated in the coil. The interaction between the repulsive magnetic fields generates the so called Lorentz forces. These forces produce a magnetic pressure applied on the surface of the moving part. If this magnetic pressure is higher than the yield strength of the workpiece, the dynamic deformation of the flyer takes place.

Similar to EXW, under precisely controlled conditions, a high-pressure collision is created between the two surfaces of the metals to be bonded. The quality of the MPW joints is then dependent on the impact angle and velocity during joining. The impact velocity is related with the kinetic energy that is transformed to the energy used for bonding of the two materials meaning that is directly associated to the pressure that is responsible to move and deform the flyer workpiece. Insufficient velocity may result in not enough pressure to bond the materials, on the other hand, when the velocity is too high excessive heat can be generated. This excessive heat may lead to partial fusion of the materials, leading to thick intermetallic layers, with decreased properties. The impact velocity is mainly determined by the discharge energy, while the impact angle is function of geometrical properties such as gap or overlap distance. The weld point velocity is function of both flyer velocity and initial angle.

Bonding occurs when two clean metal surfaces, stripped of their oxide layers, are forced together into close contact. According to Stern et al. [8], the ability to generate atomic bonds depends on the compatibility at several levels such as, matching crystal structures, lattice parameters, and/or crystallographic orientations at the interface, electronic properties of the two metals, interface alloying and interface segregation of alloying elements or other diffusion effects [8]. If the compatibility between metals is guaranteed, the decrease in the system energy caused by the elimination of the two free surface energies is normally enough to cause bonding between the metals [8].

Various mechanisms have been introduced to describe the MPW bonding process. Authors claim that impact welding processes, such as EXW or MPW, can be classified as pressure weld operations which cause solid-state mechanical mixing between the materials, as a result of the severe plastic deformations at the interface. According to Stern et al. [8] and Ben-Artzy et al. [9], this theory is sustained by observed deformation microstructure features, such as structure refinement, equiaxed or elongated dislocation cells, slip bands and local recrystallization. Another reference [10] holds that, in MPW, the bonds occurs as a consequence of the local melting and solidification on a submicron scale. Ben-Artzy et al. [10] suggests that, during the MPW process, the kinetic energy of the flyer can be converted into heat by two mechanisms: the massive plastic deformation of the metals due to the impact and interface wave creation, and the severe deformation of the metal surfaces during the creation of the jet. This may cause local melting of the metal in the interface. In some cases, the melted layer is restricted to local pockets along the interface, while in others, a continuous layer of melted material and consequently intermetallic compound can be found. Frequently, both mechanisms can be identified in a single joint [8–11].

Frequently, a wavy interface morphology is observed in impact welding processes. Different theories have been suggested to explain the interface wave creation in both EXW and MPW joints: Jet Indentation Mechanism, flow instability mechanism, vortex shedding mechanism and stress wave mechanism [9–12].

Jet Indentation Mechanism [13] relies on the formation of waves by the re-reentrance of the jet on the steady workpiece, which forms a protuberance ahead of the stagnation point. As a result of the instability generated at this region together with the abrupt change in the velocity of the metals, the stagnation point "jumps" over the protuberance, causing the formation of a new protuberance. According to this theory, as this occurs periodically, a wavy interface is generated [9–12]. Flow instability mechanism also known as Kelvin–Helmholtz instability, states that when two fluids with different velocities interact, instabilities will occur at the interface [8–10]. According to some authors [14, 15], this principle can be applied to EXW or MPW as a result of the velocity discontinuity between the steady plate and the jet. Vortex shedding mechanism explains that the waves on the weld interface are formed by the existence of vortices in the material flow developed at a point where the fluid velocity is zero [9–12]. On the other hand, the Stress Wave Mechanism claims that the waves result from the successive wave reflections in the flying and stationary plates. Despite the attempts to clarify the wave formation mechanism for MPW, no consensus has been found.

Several researchers have examined metallurgical and geometrical aspects of Al/Cu EXW and MP welds [16–18]. The wave and intermetallic phase formation, pockets and films of molten and re-solidified material and inclusions of oxides are the most studied features. The formation of cracks, voids and pores, incomplete welding zones, strong plastic deformation as well as recrystallization zones, are also discussed is the literature.

Despite the existing knowledge, the lack of consensus regarding magnetic pulse welding (MPW) of Al/Cu dissimilar joints leave room to further investigations. The present contribution is an attempt to clarify the influence of process parameters on the joint strength and interface morphology for Al-Cu pairs.

2 Experimental Procedure

2.1 Electromagnetic Welding Setup, Specimens, and Welding Conditions

In this study, tubes were electromagnetically propelled against inner cylindrical rods in order to obtain a welded interface. The material of the flyer tube is the AA6063-T5 aluminum alloy while the rod is made of EN CW004A R290 copper alloy.

Welding experiments were carried out using a 25/25 Magnetic Pulse System (i.e. maximum stored energy of 25 kJ, and maximum working voltage of 25 V). The bank of capacitors with a total capacitance of 80 μF, total inductance of 0.1 μH and internal resistivity is 19 mK provides a current up to 400 kA.

The tool consists of a steel (40 CrMnNiMo 7) single turn coil which is suitable for welding cylindrical parts with an outer diameter of 20 mm. The tubular parts to

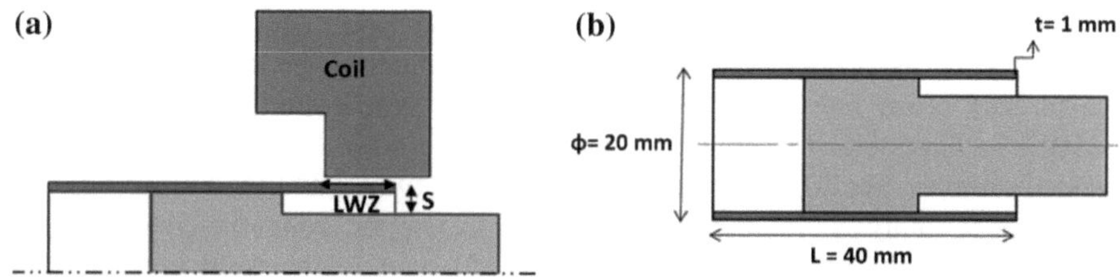

Fig. 1 **a** Dimensions of the flyer tube; **b** illustration of the assembly before welding

be welded are positioned in an overlap configuration as illustrated in Fig. 1a. The length of the tube inserted inside the workzone of the coil (LWZ) will be changed as well as the gap distance between the tube and the inner copper part. As shown in Fig. 1b, the tube has an external diameter of 20 mm, 1 mm thickness and 40 mm length. The diameter of the copper rod is adjusted in order to test different air gaps.

The weldability window was investigated regarding the effect of three main monitorable parameters, namely the charging voltage, the air gap width (S) and overlap distance (LWZ). In order to promote and test different air gaps, four internal rods were considered, with distinctive diameters: 13, 14, 15 and 16 mm. For each rod diameter, five discharge energies were applied: 5.76, 7.84, 10.24, 12.96 and 16 kJ corresponding to a charging voltage of 12, 14, 16, 18 and 20 kV, respectively. Furthermore, aiming to vary the initial impact angle by using the relation between the rod diameter and the overlap distance (LWZ), three distinctive overlaps were considered for each discharge energy value: 8, 10 and 11 mm.

A full factorial experimental matrix, taking into account all the possible combinations between parameters, was designed, containing a total of 60 assemblies. For statistical purposes, all the tests were conducted twice, meaning that a total of 120 experiments were performed.

2.2 *Weld Characterization*

The welding quality was evaluated by compression test and by interface microstructure examination.

The quasi-static compression experiments were conducted at room temperature and 2 mm/min using an Instron model 4507 servo-electric universal testing machine.

Scanning electron microscopy (SEM) was used for the characterization of the welding interface. The analyses were carried out at a 15 kV acceleration voltage in secondary and backscattered electron modes using a FEI QUANTA 400F high-resolution SEM equipped with a field emission gun.

The chemical composition of the interface was assessed by energy dispersive X-ray spectroscopy (EDS).

The phases present at the intermetallic layer were studied trough X-ray Diffraction by using synchrotron radiation at the Materials Research station of the Rossendorf beam line (ROBL) at the European Synchrotron Radiation Facility (ESRF) in Grenoble. For this analysis, slices of 1 mm thick were cut from the welded samples. Six different zones of the interface (with 100 μm × 100 μm) were subjected to analysis. Zone one corresponds to copper, zone six corresponds to aluminum and the remaining zones (2, 3, 4, 5) refer to the weld interface. The incident X-ray beam was monochromatized to 11.5 keV (λ = 0.1078 nm).

3 Results and Discussion

3.1 Mechanical Characterization

The static response of AA6063-T5 and EN CW004A R290 copper dissimilar joints welded by electromagnetic pulses was acquired for the 120 welded samples. Figure 2 shows the two behaviors identified during the compression tests. In Fig. 2a it is possible to observe that, after the compression test, the copper rod slipped out of the aluminum tube, i.e., the inner rod is physically separated from the outer tube. In Fig. 2b, on the other hand, is possible to observe the failure of the tube, meaning that the weld withstood an axial force larger than the buckling resistance of the aluminum. These results are in line with the respective force-displacement curves. Figure 2a shows an example of force-displacement graph measured during the compression test on samples in which the inner rod slipped out of the tube. The graph clearly shows a linear relationship between the force and displacement, indicating the development of elastic shear stresses. At a force of approximately 9 kN, the weld sheared and the copper part is physically disjointed from the aluminum tube as seen in Fig. 2b presents an example of a curve recorded during the compression test on samples in which the copper and the aluminum remain attached. As can be seen, the compression force increased

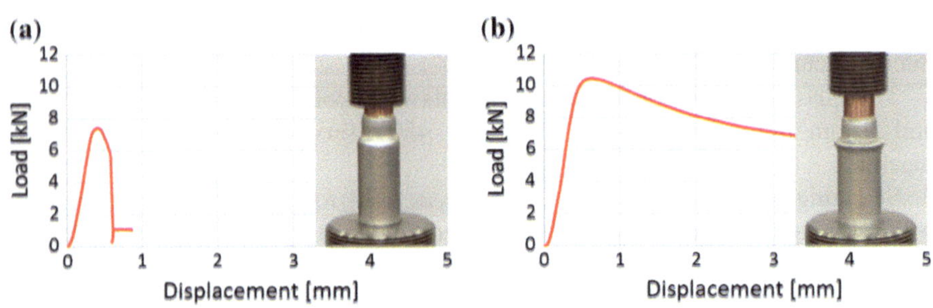

Fig. 2 Example of force-displacement curves recorded during the compression tests: **a** Samples in which the inner rod separated from the outer tube; **b** Samples in which the welds withstood an axial force larger than the buckling resistance of the tube

linearly with displacement (elastic deformation). At a force of 10.5 kN the aluminum tube started to buckle. Besides, it is noteworthy to mention that there is no sudden force decrease after the maximum force is reached. This sloped progression of the force-displacement curve can be explained by the fact that the weld does not fail and a significant amount of force is still required to further deform the outer tube. Summing up, the fact that the aluminum part fails before the weld when subjected to an axial force indicates that weld strength exceeds the buckling resistance of tube. The joint has a mechanical resistance higher than the base material.

The samples that remained joined after the compression test, in other words, the ones where the bulking of the aluminum tube was observed, were considered as "welded". 17 of the 60 samples presented a behavior indicative of an effective weld.

The results of the compression tests were used to define the weldability window, which give us an indication of the welding range, based on the values of discharge energy, gap and overlap distance—Fig. 3. The experimental conditions that lead to successful welds are contained by the two surfaces. It can be seen from the figure that the combination of different gap and overlap distances can lead to successful welds, if the selected energy leads to an appropriate impact velocity.

Still, it should be noted that this welding window only indicates the presence or absence of weld under a given set of input parameters; it does not quantify the strength of the weld. By the set of compression tests is possible to affirm that the samples present a resistance higher than the aluminum tube, but the real strength of the samples is not quantified. Different mechanical testing, such as peel tests, should be conducted in future analysis in order to it give a real indication regarding the strength of the welds.

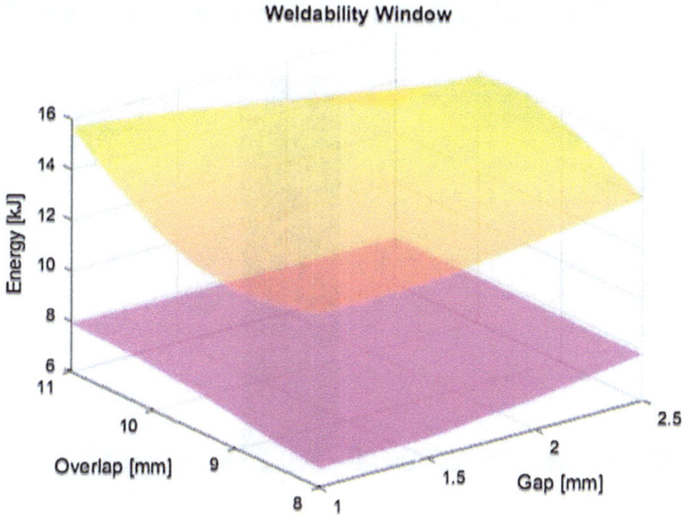

Fig. 3 Illustration of the weldability window of AA6063-T5/EN CW004A R290 for MPW with selected parameters

3.2 *Interfacial Features of Al/Cu Welds*

The interfaces of four samples, welded by using different values of energy were examined by SEM analysis. Figure 4 shows an overview of the weld interface and Fig. 5 a detailed view of the same zone. The Al/Cu interfaces reveal various morphologies when increasing the discharge energy. The formation of intermetallic phases rises with increasing pulse energies, leading to a less wavy and more irregular interface between aluminum and copper. The thickness of the nearly continuous intermetallic layer can be above 80 μm for the higher level of discharge energy—Fig. 4a. This probably means that, for the higher electromagnetic pulses, there is enough energy at the interface zone to melt the materials, creating a continuous intermetallic layer by the rapid solidification of the molten material. In fact,

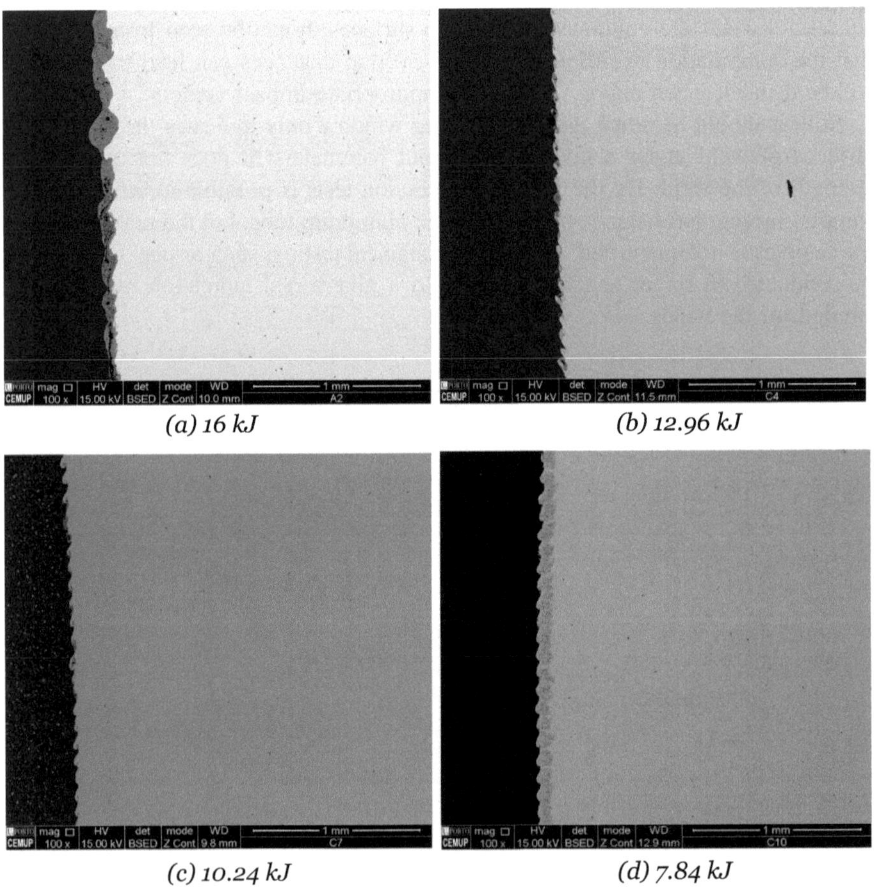

(a) 16 kJ (b) 12.96 kJ (c) 10.24 kJ (d) 7.84 kJ

Fig. 4 SEM micrographs of AA6063-T5/EN CW004A R290 MP welded interfaces for different values of discharge energy (overlap = 8 mm; Gap = 1 mm). Magnification 100×

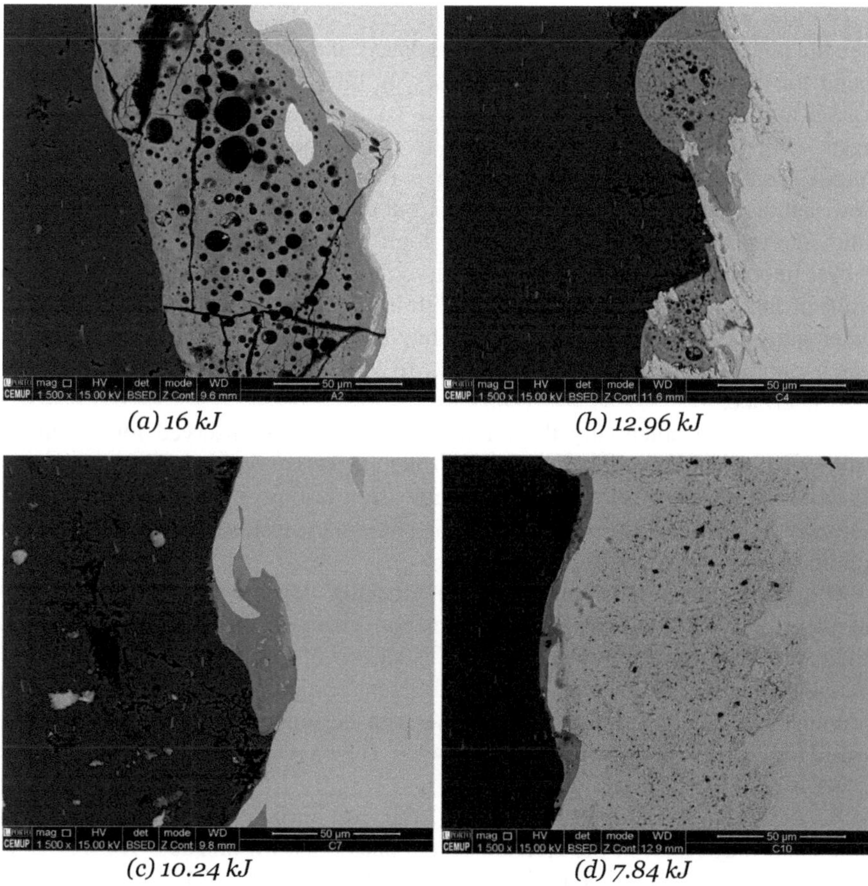

Fig. 5 SEM micrographs of AA6063-T5/EN CW004A R290 MP welded interfaces for different values of discharge energy (overlap = 8 mm; Gap = 1 mm). Magnification 1500×

some authors [16–21] state that most of the kinetic energy from the flyer is converted into Joule heat during the collision. The spatial and temporal temperature distribution during the impact weld determines if melting will occur, and therefore, if intermetallic compounds will form.

According to Ben-Artzy et al. [8] as a result of the collision and shock waves that shear and stir the two metals to be welded, the temperature at the interface can go up to 400°. In addition, the jetting action which is responsible for removing the surface oxide form the surfaces to be welded, could increase the interfacial temperature near the melting temperature of the metals, due to severe shearing deformation. Furthermore, if the impact angle is too small for the input discharge energy, the jet may be unable to escape, trapping the hot ejecta inside the interface. The combination of these phenomena will possibly generate enough temperature increase to melt a layer of both materials at the interface zone.

For low pulse energies, a periodic wavy interface with a relatively thin intermetallic phase was found. A maximum thickness of 6 μm of intermetallic layer was found for the lower electromagnetic pulses. In these cases, the intermetallic compounds are mainly concentrate in so called "melt pockets", typically located at the crests of the interface waves. The pockets of intermetallics are likely formed by intense local heating, melting, mixing and rapid solidification in wave vortices. Note that discontinuous pockets are found, but the most of the welded region is free from intermetallic phases. Furthermore, high volume of pores can be found in thicker intermetallic layers. The spherical voids at the interface can possible be the result of local melting and solidification shrinkage cavities at the Al/Cu interface. Other study [19] mentions that the voids might be the result of air in the initial gap which cannot escape in time during the MPW process, remaining trapped at the weld interface.

The pores are potentially the initiators of the cracks observed for the higher values of discharge energy. The cause of the cracks is not clear. They can be the result of the shear stress during the compression testing (note that these samples were subjected to mechanical testing) or caused by residual stress during rapid solidification of the molten material.

The chemical composition of the intermetallic layer was assessed for three samples by EDS. The chemical composition changes cross the interface were studied by EDS from different locations as shown in Fig. 6. Quantitative analysis pinpoints that the intermetallic layer has different portions of aluminum and copper, meaning that different intermetallic phases can be found. For 16 kJ for example, zone 2 has 40 wt% of Al and 60 wt% Cu while zone 3 revealed 65 wt% of Al and 35 wt% Cu.

According to Al–Cu phase diagram, the equilibrium solidification of these compound may result in the creation of the phases shown in Table 1.

To further confirm the phase identification at the interface, XRD was employed. The in situ study using synchrotron radiation was effective in resolving phase formation. Figure 7 shows the XRD data obtained. As can be seen, two intermetallic phases Al2Cu3 (ε) and Al2Cu (Θ), coexist at Al/Cu interface. Several studies [20, 21] show that the presence of such phases cause a significant increase in hardness at the Al/Cu interfaces, embrittling the joint.

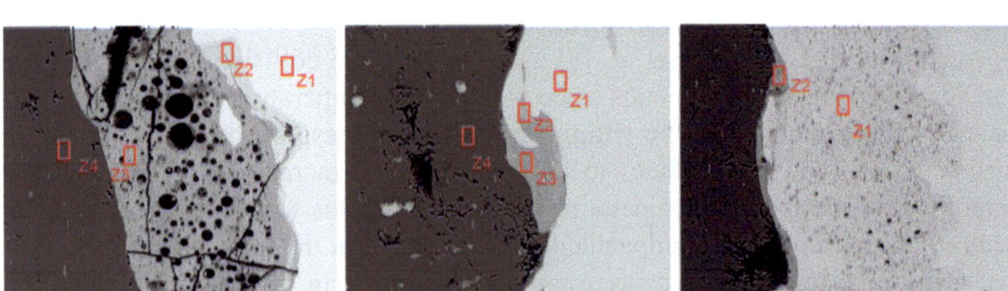

Fig. 6 EDS study on MPW welded Al-Cu interface-zones

Table 1 EDS study on MPW welded Al-Cu interface

	16 kJ		10.24 kJ		7.84 kJ	
Zone	At (%)	Probable phase	At (%)	Probable phase	At (%)	Probable phase
Z1	Cu	–	Cu	–	Cu	–
Z2	40% Al–60% Cu	ε2 or σ	50% Al–50% Cu	η1, η2 or π	65% Al–35% Cu	θ–Al2Cu
Z3	65% Al–35% Cu	θ–Al2Cu	65% Al–35% Cu	θ – Al2Cu	At (%)	Probable phase
Z4	Al	–	Al	–		

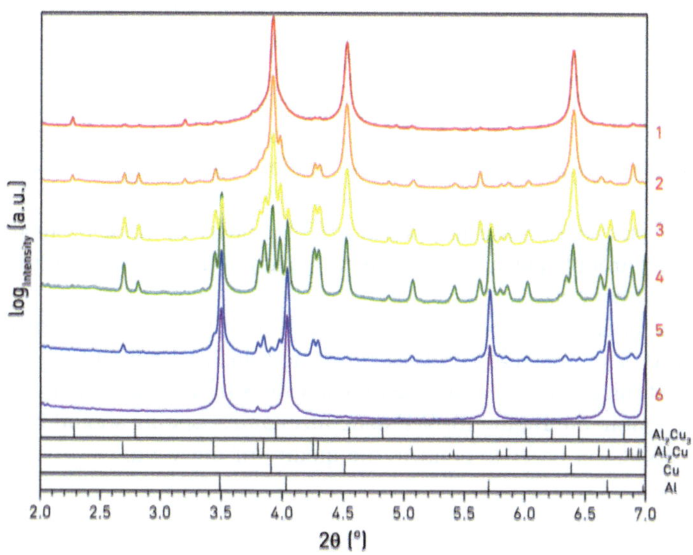

Fig. 7 XRD patterns of the Al/Cu interface

4 Conclusions

The weld of Al/Cu the dissimilar pair was investigated. Successful welds can be achieved by MPW with the combination of different process parameters. The strength of the Al/Cu welds was investigated by compression testing. The results from this test allow us to distinct the samples welded form the not welded ones, allowing the definition of the welding window. The zone contained by the limiting surfaces comprises the parameter combination that successful welds.

The morphology and composition of the AL/Cu interface was examined using SEM, EDS and XRD for samples with increasing energy. Evidences of melting at the interface zone was found, especially for the samples processed by higher values of discharge energy.

The melting and rapid solidification of interface material is the possible cause of the formation of intermetallic compounds and voids as well. The thickness, the composition and the geometry of the intermetallic layer is influenced by the pulse energy. Higher values of discharge energy seem to lead to a less wavy interface, thicker intermetallic layer and higher porosity level. For low pulse energies a periodic wavy interface with relatively thin intermetallic phase film seems be formed. In samples with a highly wavy interface, the intermetallics appear mainly concentrate in so called “melt pockets”. For less wavy interfaces, the intermetallics can be found as continuous layer with, a thickness dependent on process parameters.

The relationship between energy increase and melting is possibly related to multiple sources of heating. Higher values of pulse energy are associated with higher induced currents in the part, meaning that more Joule heating will be generated.

In addition, more energy means higher flyer velocity, the air existing in the gap between the parts to be welded is expelled, this aerodynamic drag (fluid friction) is proportional to the square of the velocity, further contributing to the generation of heat.

As the kinetic energy also increases with the square of velocity, the dissipation of this energy trough plastic work and jet generation also will contributes to an increase in temperature.

To reduce intermetallic phases, porosity and melt pockets, pulse energy should be minimized. The remaining process parameters such as gap and overlap distance should be optimized allowing the use of low values of discharge energy. It is worth noting here that in the future, further static mechanical testing should be carried out on welded joints, in order to confirm the quality of the weld.

References

1. Bhattacharya, T.K., Das, H., Jana, S.S., Pal, T.K.: Numerical and experimental investigation of thermal history, material flow and mechanical properties of friction stir welded aluminum alloy to dhp copper dissimilar joint. Int. J. Adv. Manuf. Technol., 1–15 (2016)
2. Kush, P.M., Vishvesh, J.B.: A review on dissimilar friction stir welding of copper to aluminum: Process, properties, and variants. Mater. Manuf. Process. **31**(3), 233–254 (2016)
3. Lee, W., Bang, K., Jung, S.: Effects of intermetallic compound on the electrical and mechanical properties of friction welded cu/al bimetallic joints during annealing. J. Alloy. Compd. **390**(12), 212–219 (2005)
4. Chuang, T.H., Tsao, L.C., Tsai, T.C., Yeh, M.S., Wu, C.S.: Development of a low-melting-point filler metal for brazing aluminum alloys. Metallurg. Mater. Trans. A **31** (9), 2239–2245 (2000)
5. Lee, S.J., Nakamura, H., Kawahito, Y., Katayama, S.: Effect of welding speed on microstructural and mechanical properties of laser lap weld joints in dissimilar Al and Cu sheets. Sci. Technol. Weld. Join. **19**(2), 111–118 (2014)

6. Ouyang, J., Yarrapareddy, E., Kovacevic, R.: Microstructural evolution in the friction stir welded 6061 aluminum alloy (t6-temper condition) to copper. J. Mater. Process. Technol. **172**(1), 110–122 (2006)
7. Yilbas, B.S., Sahina, A.Z., Kahramanb, N., Al-Garnia, A.Z.: Friction welding of st-al and al-cu materials. J. Mater. Process. Technol. **49**(3), 431–443 (1995)
8. Stern, A., Shribman, V., Ben-Artzy, A., Aizenshtein, M.: Interface phenomena and bonding mechanism in magnetic pulse welding. J. Mater. Eng. Perform. **23**(10), 3449–3458 (2014)
9. Ben-Artzy, A., Stern, A., Frage, N., Shribman, V., Sadot, O.: Wave formation mechanism in magnetic pulse welding. Int. J. Impact Eng. **37**(4), 397–404 (2010)
10. Ben-Artzy, A., Stern, A., Frage, N., Shribman, V.: Interface phenomena in aluminum: magnesium magnetic pulse welding. Sci. Technol. Weld. Join. **13**(4), 402–408 (2008)
11. Beyer, E., Brenner, B., Gobel, G., Herrmannsdrfer, T., Kaspar, J.: Insights into intermetallic phases on pulse welded dissimilar metal joints. In: 4th International Conference on High Speed Forming-2010
12. Hoseini, A., Behzad, T.: Weldability window and the effect of interface morphology on the properties of al/cu/al laminated composites fabricated by explosive welding. Mater. Des. **86**, 516–525 (2015)
13. Cowan, G.R., Bergmann, O.R., Holtzman, A.H.: Mechanism of bond zone wave formation in explosion-clad metals. Metallurg. Mater. Trans. B **2**(11), 3145–3155 (1971)
14. Hunt, J.N.: Wave formation in explosive welding. Philos. Mag. **17**(148), 669–680 (1968)
15. Robinson, J.L.: The mechanics of wave formation in impact welding. Philos. Mag. **31**(3), 587–597 (1975)
16. Gobel, G., Kaspar, J., Herrmannsdorfer, T., Brenner, B., Beyer, E.: Insights into intermetallic phases on pulse welded dissimilar metal joints. In: 4th International Conference on High Speed Forming-2010
17. Raoelison, R., Sapanathan, T., Buiron, N., Rachik, M.: Magnetic pulse welding of al/al and al/cu metal pairs: Consequences of the dissimilar combination on the interfacial behavior during the welding process. J. Manuf. Process. **20**, 112–127 (2015)
18. Wu, X., Shang, J.: An investigation of magnetic pulse welding of al/cu and interface characterization. J. Manuf. Sci. Eng. **136**(5), 051002 (2014)
19. Yu, H., Xu, Z., Fan, Z., Zhao, Z., Li, C.: Mechanical property and microstructure of aluminum alloy-steel tubes joint by magnetic pulse welding. Mater. Sci. Eng. A **561**, 259–265 (2013)
20. Tan, C.W., Jiang, Z.G., Li, L.Q., Chen, Y.B., Chen, X.Y.: Microstructural evolution and mechanical properties of dissimilar Al–Cu joints produced by friction stir welding. Mater. Des. **51**, 466–473 (2013)
21. Zhang, Y.: Investigation of magnetic pulse welding on lap joint of similar and dissimilar materials. PhD thesis, The Ohio State University, 2010

MIX
Papier aus verantwortungsvollen Quellen
Paper from responsible sources
FSC® C105338

If you have any concerns about our products,
you can contact us on
ProductSafety@springernature.com

In case Publisher is established outside the EU,
the EU authorized representative is:
Springer Nature Customer Service Center GmbH
Europaplatz 3, 69115 Heidelberg, Germany

Printed by Libri Plureos GmbH
in Hamburg, Germany